Multi-Objective Design of Antennas Using Surrogate Models

Multi-Objective Design of Antennas Using Surrogate Models

Slawomir Koziel
Adrian Bekasiewicz
Reykjavik University, Iceland

World Scientific
NEW JERSEY • LONDON • SINGAPORE • BEIJING • SHANGHAI • HONG KONG • TAIPEI • CHENNAI • TOKYO

Published by

World Scientific Publishing Europe Ltd.

57 Shelton Street, Covent Garden, London WC2H 9HE

Head office: 5 Toh Tuck Link, Singapore 596224

USA office: 27 Warren Street, Suite 401-402, Hackensack, NJ 07601

Library of Congress Cataloging-in-Publication Data
Names: Koziel, Slawomir, author. | Bekasiewicz, Adrian author.
Title: Multi-objective design of antennas using surrogate models / Slawomir Koziel (Reykjavik University, Iceland) & Adrian Bekasiewicz (Reykjavik University, Iceland).
Description: [Hackensack] New Jersey : World Scientific, [2016] | Includes bibliographical references.
Identifiers: LCCN 2016024714 | ISBN 9781786341471 (hc : alk. paper)
Subjects: LCSH: Antennas (Electronics)--Simulation methods. | Antennas (Electronics)--Mathematical models. | Antennas (Electronics)--Design and construction. | Multiple criteria decision making.
Classification: LCC TK7871.6 .K69 2016 | DDC 621.382/4--dc23
LC record available at https://lccn.loc.gov/2016024714

British Library Cataloguing-in-Publication Data
A catalogue record for this book is available from the British Library.

Desk Editors: Kalpana Bharanikumar/Mary Simpson

Typeset by Stallion Press
Email: enquiries@stallionpress.com

Printed in Singapore

To our families: Dorota, Anna Halina, Aleksandra, and Igor.

Preface

Full-wave electromagnetic (EM)-simulation models have become fundamental tools in the design of contemporary antenna structures. High-fidelity EM analysis ensures evaluation accuracy but it might be computationally expensive. Therefore, direct use of EM-simulations in the design process, especially for parametric optimization, is normally associated with high computational cost, often prohibitive. This makes simulation-driven antenna design a challenging task. An additional difficulty is the necessity of simultaneous handling of several, and often conflicting design objectives. If comprehensive information about the best possible trade-offs between these objectives is to be sought, genuine multi-objective optimization needs to be carried out. The most popular solution approaches, population-based metaheuristics, are too expensive in computational terms. Expedited multi-objective design can be realized using surrogate-assisted techniques. This book presents several ways of utilizing surrogate modeling methods (both local and global), variable-fidelity EM-simulations and design space reduction techniques for accelerated antenna design with respect to several criteria. The procedures formulated here are demonstrated using numerous examples of real-world antenna design problems. Furthermore, the book contains introductory material concerning numerical optimization, both conventional (gradient-based and derivative-free, including metaheuristics) and surrogate-based, as well as fundamentals of multi-objective optimization. Recommendations concerning practical aspects of surrogate-assisted multi-objective antenna optimization are also given.

Slawomir Koziel and Adrian Bekasiewicz
March 2016, Reykjavik, Iceland

About the Authors

Adrian Bekasiewicz received his M.Sc. degree in Electronic Engineering from Gdansk University of Technology, Gdansk, Poland, in 2011. He is currently pursuing his Ph.D. in Wireless Communication Engineering at Gdansk University of Technology, Gdansk, Poland. He is also a Research Associate with the School of Science and Engineering, Reykjavik University, Iceland. He has authored or co-authored over 150 peer-reviewed papers. His research interests include multi-objective optimization, metaheuristic algorithms, design of compact microwave antennas, and miniaturization of microwave/RF components.

Slawomir Koziel received his M.Sc. and Ph.D. degrees in Electronic Engineering from Gdansk University of Technology, Poland, in 1995 and 2000, respectively. He also received the M.Sc. degrees in Theoretical Physics and in Mathematics, in 2000 and 2002, respectively, as well as a Ph.D. in Mathematics in 2003, from the University of Gdansk, Poland. He is currently a Professor with the School of Science and Engineering, Reykjavik University, Iceland. His research interests include CAD and modeling of microwave and antenna structures, simulation-driven design, surrogate-based optimization, space mapping, circuit theory, analog signal processing, evolutionary computation and numerical analysis.

Acknowledgments

We would like to acknowledge the efforts of all those students, researchers, and colleagues who have helped us during the research work presented in this book. In particular, we would like to thank Dr. Stanislav Ogurtsov for inspiring discussions and help with EM modeling of antenna structures. We would also like to thank Computer Simulation Technology AG, Darmstadt, Germany, for making CST Microwave Studio available for our research purposes.

Contents

Chapter 1

Introduction

Antennas are essential components of wireless communication systems such as radio and television broadcasting, radar, mobile phones, satellite communications, airborne navigation, Bluetooth devices, RFID tags, wireless computer networks, to name just a few (Kraus and Marhefka, 2002). Design of modern antennas is a challenging task where an important step is adjustment of geometry and material parameters of the structure so as to satisfy given performance requirements concerning return loss, radiation pattern, gain, etc. (Gautam *et al.*, 2013; Nguyen *et al.*, 2013; Kuwahara, 2005). In many cases, geometrical constraints have to be taken into account because — for some applications (e.g., handheld and wearable devices; Guo *et al.*, 2012; Chahat *et al.*, 2011) — achieving compact designs might be of primary concern. For the sake of reliability, the process of dimension adjustment (also referred to as design closure; Koziel and Ogurtsov, 2014a) has to be based on accurate evaluation of antenna performance. The latter can only be ensured by full-wave electromagnetic (EM)-simulation. This is particularly important for compact antennas as well as other cases where EM couplings between the antenna itself and its environment (connectors, housing, feeding circuitry) may affect the device's operation. However, EM analysis of realistic and finely discretized antenna models might be computationally expensive. Typically, this is not a problem for design validation, but it may be prohibitive from the point of view of parametric optimization which requires multiple simulations of the antenna structure at hand.

Perhaps, the most widely used approach to simulation-driven antenna design is parameter sweeps guided by engineering experience. The choice of the parameters, their order, and the range of variability are decided using the designer's insight and knowledge. Owing to these factors, this interactive approach allows — in many cases — for yielding satisfactory designs in reasonable timeframe; in particular, to obtain better results than with brute-force optimization utilizing built-in optimization capabilities of commercial simulation tools. Nevertheless, parameter-sweep-based design is laborious and it does not guarantee optimum results (Chu *et al.*, 2013). One of the reasons is difficulty in exploring correlations between the antenna parameters. Another issue is that many antenna structures exhibit complex topologies with a large number of parameters that are highly interdependent (in how they affect antenna performance) and, therefore, have to be simultaneously adjusted. Furthermore, the optimum values of designable parameters are often counter-intuitive (Kurgan and Bekasiewicz, 2014; Koziel and Bekasiewicz, 2016a).

Automation of antenna design can be realized by means of numerical optimization of suitably defined objective function concerning performance figures of interest such as return loss, gain, etc., (and computed using an EM solver) with respect to the antenna geometry/material parameters (Koulouridis *et al.*, 2007; Bekasiewicz and Koziel, 2015a). The practical problem is a large number of objective function evaluations required by majority of conventional optimization techniques. This applies to gradient-based methods (Nocedal and Wright, 2006), derivative-free techniques such as pattern search methods or Nelder–Mead algorithm (Kolda *et al.*, 2003), and even more to population-based metaheuristics (genetic algorithms (GAs), Kerkhoff and Ling, 2007, particle swarm optimizers, Lizzi *et al.*, 2007, differential evolution, Guney and Basbug, 2013, etc.). Depending on the problem complexity (i.e., the design space dimensionality and nonlinearity of the antenna responses), hundreds, thousands, or even tens of thousands of objective function evaluations might be required to yield the optimized design, which is impractical unless the simulation cost of the structure is not of concern. For example, the most successful application of metaheuristic algorithms in antenna design was in array optimization that require global search capabilities (Talbi, 2009) and — in some cases — may be executed as the level of fast analytical array factor models (Ares-Pena *et al.*, 1999; Petko and Werner, 2007; Bevelacqua and Balanis, 2007; Selleri, 2008; Li *et al.*, 2010; Roy *et al.*, 2011; Bayraktar *et al.*, 2013). Apart from the high cost of EM analysis, other issues may arise such as numerical noise which is partially a result of adaptive meshing techniques used in contemporary EM solvers, where small changes of the design variables may result in a change of the mesh topology, and, consequently, discontinuity of the EM-simulated antenna responses as a function of designable parameters. The noise is particularly a problem for gradient-based methods that require smoothness of the objective function.

Reduction of the computational cost of EM-simulation-driven antenna design can be achieved in several ways. One of them is utilization of adjoint sensitivities (Director and Rohrer, 1969; Georgieva *et al.*, 2002) which allow evaluation of both the antenna response and its derivatives with respect to its designable parameters at little extra computational expense. Availability of cheap derivatives leads to substantial speedup of gradient-based optimization of EM-simulation models (Bandler and Seviora, 1972; Chung *et al.*, 2001). This was also demonstrated for antenna design (Toivanen *et al.*, 2009; Ghassemi *et al.*, 2010; Jacobsson and Rylander, 2010; Zhang *et al.*, 2012; Koziel and Bekasiewicz, 2015a). Nevertheless, it should be mentioned that adjoint sensitivities are not yet widespread in commercial EM solvers. So far, this technology has been implemented in Ansys HFSS (HFSS, 2012) and CST Microwave Studio (CST, 2013).

Surrogate-based optimization (SBO) (Simpson *et al.*, 2001; Bandler *et al.*, 2004a; Queipo *et al.*, 2005; Forrester and Keane, 2009; Yelten *et al.*, 2012; Koziel and Ogurtsov, 2014a) is another and probably the most promising approach to computationally efficient antenna optimization. In SBO, direct handling of the expensive high-fidelity model is replaced by iterative construction and re-optimization of its faster representation, referred to as a surrogate model. The surrogate is used as a prediction tool guiding the optimization process towards a better design. There are two primary ways of surrogate model construction. Function approximation models are obtained by approximating sampled EM-simulation data using techniques such as polynomial regression (Queipo *et al.*, 2005; Conn *et al.*, 2009), radial basis function interpolation (Gutmann, 2001; Forrester and Keane, 2009), artificial neural networks (ANN) (Haykin, 1998; Kim *et al.*, 2007), Kriging interpolation (Matheron, 1963; Forrester *et al.*, 2008), or support vector regression (SVR) (Angiulli *et al.*, 2000; Jacobs, 2012). Physics-based surrogates, on the other hand, are constructed by suitable correction of an underlying low-fidelity model (Bandler *et al.*, 2004a, 2004b; Cheng *et al.*, 2010; Bekasiewicz and Koziel, 2015b), which in case of antenna structures normally comes from coarse-discretization EM-simulations (Koziel and Ogurtsov, 2012a; Koziel *et al.*, 2015a).

The advantage of approximation models is their speed; a serious disadvantage is the initial setup cost, which — particularly for highly-dimensional design spaces — may be quite substantial (hundreds and thousands of training points and, consequently, EM-simulations). In many surrogate-assisted algorithms, the surrogate model is iteratively updated using additional training points allocated according to suitably defined infill criteria (Forrester and Keane, 2009; Koziel *et al.*, 2011a). These may be oriented towards design space exploration (aiming at improvement of the global accuracy of the surrogate; Forrester and Keane, 2009; Jones *et al.*, 1998) or its exploitation (aiming at improvement of the design in the vicinity of the current promising region of the space; Forrester and Keane, 2009; Jones *et al.*, 1998). The former models are becoming popular for global optimization when combined with population-based metaheuristics (Queipo *et al.*, 2005; Gazda *et al.*, 2012; Koziel and Ogurtsov, 2013a; Koziel and Bekasiewicz, 2015b). Surrogate-assisted search methods can also utilize variable-fidelity simulation models to speed up the design process (Koziel and Ogurtsov, 2011a, 2013b; Koziel and Bekasiewicz, 2016b; Koziel *et al.*, 2016a).

Physics-based surrogates typically exhibit much better generalization capability than the approximation ones, which translates into lower computational cost of the design process. On the other hand, physics-based SBO algorithms are normally used for local search because EM-simulation required by the underlying low-fidelity model (even if a coarse-mesh one) does not allow for global optimization of the surrogate (Koziel and Ogurtsov, 2012b; Koziel *et al.*, 2016a). Some of the popular SBO techniques in microwave and antenna engineering include space mapping (SM) (Bandler *et al.*, 2004a; Koziel *et al.*, 2008), adaptive response correction (Koziel *et al.*, 2009; Koziel and Ogurtsov, 2014b), manifold mapping (MM) (Echeverría and Hemker, 2008; Koziel *et al.*, 2013a), shape preserving response prediction (Koziel, 2010a, 2012), feature-based optimization (Koziel and Bandler, 2015; Koziel and Bekasiewicz, 2015d), etc. Hybridization of both types of surrogate models has also been reported recently in the context of antenna design (Koziel and Ogurtsov, 2013a; Koziel and Bekasiewicz, 2015b, 2015c).

An important aspect of contemporary antenna design is the necessity of simultaneous handling of several objectives. Although a typical design requirement for many antenna structures is to minimize the return loss over a frequency band of interest, other objectives, e.g., gain, radiation pattern, or the antenna size, may also be of interest (Aljibouri *et al.*, 2000; Chamaani *et al.*, 2010; Kuwahara, 2005). Having more than one objective considerably increases complexity of the design process because various criteria usually conflict with each other so that trade-off designs have to be sought. In many cases, especially if the designer's priorities are clearly defined or just to simplify the problem, a primary objective may be selected for optimization while the remaining ones are handled through constraints or suitably defined penalty function (Bekasiewicz and Koziel, 2015a; Koziel and Bekasiewicz, 2015a; Koziel *et al.*, 2016a). On the other hand, if *a priori* preference articulation is not possible, genuine multi-objective optimization becomes a necessity (Kuwahara, 2005; Jin and Rahmat-Samii, 2007). The goal of multi-objective optimization is to identify a Pareto front representing the best possible trade-offs between conflicting objectives, which gives comprehensive information about the capabilities of an antenna structure under design.

Perhaps, the most popular approaches for solving multi-objective antenna design problems are population-based metaheuristics such as GAs or particle swarm optimizers (PSO) (van Coevorden *et al.*, 2005; Ramos *et al.*, 2003; Goudos *et al.*, 2009; Lizzi *et al.*, 2007). Their attractiveness originates from a capability of finding the entire representation of the Pareto front in a single algorithm run. However, as mentioned before, population-based algorithms are very expensive in terms of the number of objective function evaluations necessary for the algorithm convergence. Consequently, their applicability is limited to situations when evaluation cost of the antenna structure at hand is not of primary concern. If high-fidelity discrete EM-simulations are involved in the design process, metaheuristic optimization is computationally prohibitive (John and Ammann, 2009; Afshinmanesh *et al.*, 2008; Chamaani *et al.*, 2010; Koziel and Bekasiewicz, 2016c).

Recently, techniques for cost-efficient multi-objective antenna design have been proposed that utilize an SBO paradigm (Koziel and Ogurtsov, 2013; Koziel *et al.*, 2014b; Koziel and Bekasiewicz, 2016c). It has been demonstrated that combination of methods such as space reduction, utilization of variable-fidelity EM-simulation models and response surface approximation surrogates, allow solving multi-objective antenna problems even in highly-dimensional design spaces (Koziel *et al.*, 2014c; Bekasiewicz *et al.*, 2014a; Koziel and Bekasiewicz, 2016d).

The topic of this book is multi-objective design of antenna structures using SBO. We begin, in Chapter 2, by formulating the antenna design as optimization problem, discussing computational models with the emphasis on distinction between EM-simulation models of various fidelities and their setup, as well as highlighting typical design challenges of contemporary antennas. In Chapter 3, we give a brief introduction to numerical optimization. In particular, we discuss gradient-based algorithms, derivative-free optimization methods, as well as outline the methods utilized for constrained optimization. Chapter 4 is focused on global optimization using population-based metaheuristics. Introduction to surrogate-based modeling and optimization is provided in Chapter 5, where the fundamentals of SBO are discussed along with introductory information about approximation- and physics-based surrogate modeling. In Chapter 6, we introduce the concept of Pareto optimality and Pareto front as well as discuss some popular solution approaches for multi-objective problems. Chapter 7 is an exposition of selected methods for surrogate-assisted design of antenna structures. Chapter 8 outlines techniques for design space reduction, which are important for handling multi-objective antenna design in highly-dimensional design spaces. Application case studies are provided in Chapter 9, where a number of real-world antenna design problems are discussed in detail. Various types of antennas are considered, including ultra-wideband planar structures, dielectric resonator antennas, as well as compact antennas. Selected topics and practical issues related to multi-objective design algorithms considered in the book are discussed in Chapter 10. Applications of the design techniques presented in this book for solving problems in other fields are outlined in Chapter 11. Chapter 12 contains a few numerical case studies concerning applications of multi-objective optimization. The book is concluded in Chapter 13, where recommendations for the readers interested in applying the presented algorithms and techniques in their antenna design are formulated, and possible future developments concerning, among others, automation of simulation-driven antenna design are discussed.

Chapter 2

Simulation-Driven Antenna Design

In this chapter, formulation and challenges of antenna design are discussed from the perspective of simulation-driven optimization. The emphasis is put on explaining the role of computer simulations in modern antenna engineering, formulating the design task as an optimization problem, as well as introducing the notation used throughout the book. Furthermore, we outline typical antenna design objectives. We also provide a brief discussion of computational models utilized in antenna optimization with the emphasis on a distinction between high- and low-fidelity electromagnetic (EM)-simulations, the latter utilized as auxiliary models to speed up the EM-driven design process. The chapter is concluded with a discussion of the challenges related to simulation-based design of antenna structures.

2.1 Simulation-Driven Design of Antenna Structures

Full-wave EM analysis is nowadays one of the most important tools of antenna engineering. It allows for accurate evaluation of antenna performance as well as accounting for its environment (e.g., connectors, feeds, housing, installation fixtures, etc.) and its effect on antenna characteristics. Because of the high cost of high-fidelity EM-simulation, it was initially only used for design verification. Today, EM-driven design and especially design closure (in particular, adjustment of geometry parameters of the antenna structure) is an academic and industry standard. It is possible to conduct the entire design process within the simulation environment. A large number of simulation software packages are available, including the general purpose EM solvers such as HFSS (HFSS, 2012), FEKO (FEKO, 2012), CST (CST, 2013), XFDTD (XFDTD, 2014), Momentum (Agilent,

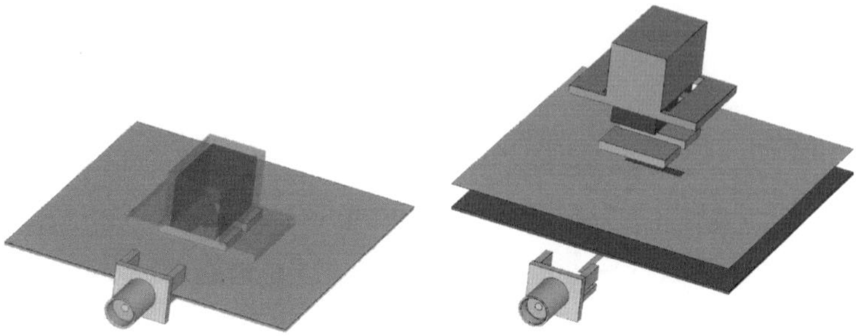

(a) (b)

Fig. 2.1. Geometry of a 5.5 GHz DRA (Petosa, 2007): (a) visualization of full-wave EM-simulation model of the structure; (b) highlight on decomposed antenna components, from the top: housing, dielectric resonator, ground plane with slot, substrate, driven element (microstrip line) and connector.

2011), or Sonnet ***em*** (Sonnet, 2013), and also antenna-specific design tools (e.g., Antenna Magus, 2015).

Here, we consider a few examples of contemporary antenna structures that will be used to explain typical characteristics and design objectives to be handled in the antenna design process. Figure 2.1 shows a 5.5 GHz dielectric resonator antenna (DRA) consisting of a dielectric resonator fed by a microstrip line through the ground plane slot (Petosa, 2007). The DRA is covered by polycarbonate housing. The antenna geometry is described by seven adjustable parameters. Figure 2.2 shows an example of an ultra-wideband planar (UWB) antenna designed to operate within a 4–10 GHz frequency range (Wu *et al.*, 2014).

It is a quasi-Yagi structure which consists of a dipole excited by a coplanar stripline (CPS) and two directors. The former is fed by a microstrip-to-CPS transition. Wideband operation of the structure is ensured by a microstrip impedance transformer. Two symmetrical stubs allocated on the edge of the ground plane allow obtaining more compact geometry. The structure is parameterized using 24 design variables. Finally, Fig. 2.3 shows a 100-element microstrip antenna array (Koziel and Ogurtsov, 2015). The lateral extensions of the structure are equal to the size of the microstrip patch. The antenna is covered using a plastic housing.

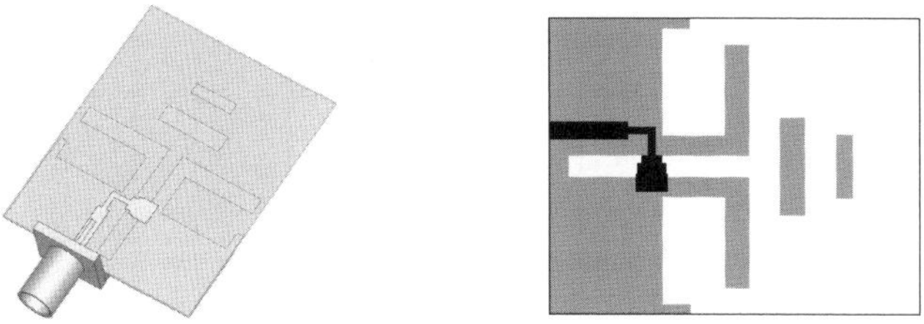

(a) (b)

Fig. 2.2. Realistic design of a quasi-Yagi antenna (Wu *et al.*, 2014) designed to work in a frequency band from 4 GHz to 10 GHz: (a) visualization of full-wave EM model of the structure; (b) highlight on geometrical details of the structure; the geometry etched on the top and bottom layers of the structure are marked with black and gray, respectively.

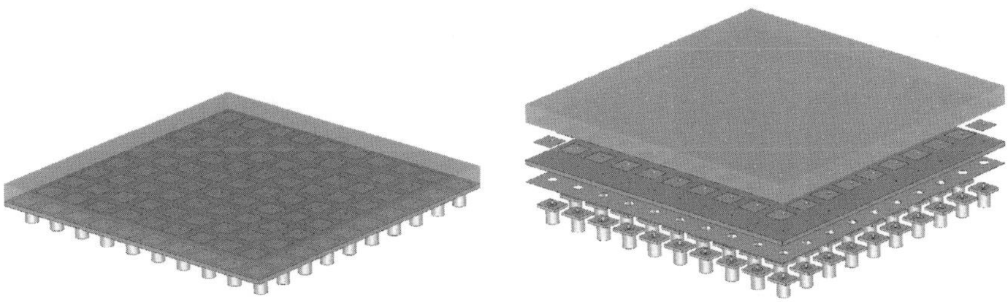

(a) (b)

Fig. 2.3. Realistic model of a 10×10 planar antenna array designed at 10 GHz: (a) visualization of a full-wave EM model; (b) highlight on antenna components, from the top: housing, microstrip patches, substrate with holes, ground plane with holes, SMA connectors.

Throughout this book, we will use a symbol $\boldsymbol{f}$ to denote a high-fidelity full-wave EM-simulation-model of the antenna of interest. Typically, $\boldsymbol{f}$ represents evaluation of antenna characteristics, such as reflection versus frequency (Koziel *et al.*, 2015a), voltage standing wave ratio (VSWR; Nguyen, Lee and Park, 2012), or complex impedance of the structure (Huynh and Lee, 1995). Often, field-related antenna characteristics are also of interest that include gain (Ding *et al.*, 2008), directivity (El-Khouly, Ghali and Khamis, 2007), radiation pattern (Chahat *et al.*, 2011), side-lobe level

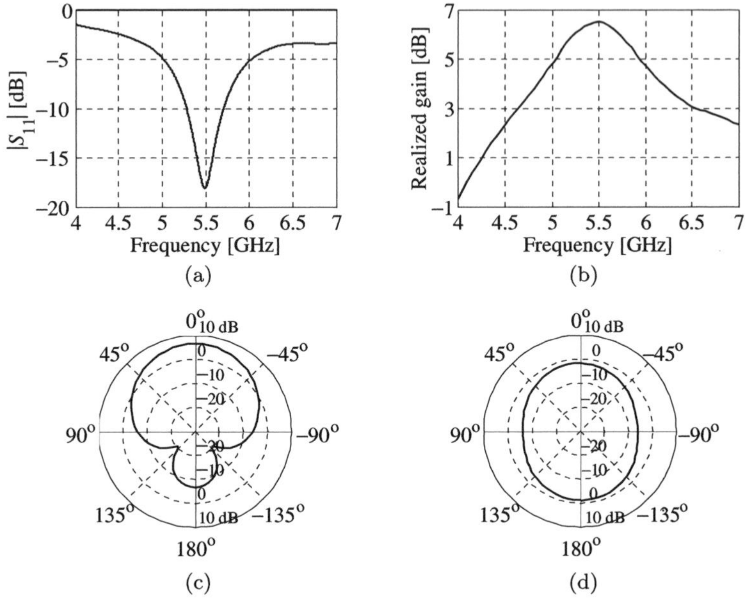

Fig. 2.4. Typical characteristics of the 5.5 GHz DRA evaluated at a particular design: (a) reflection; (b) realized gain for the zero zenith angle versus frequency; (c) realized gain at 5.5 GHz in the E-plane; (d) realized gain at 5.5 GHz in the H-plane.

(Koziel *et al.*, 2014a), etc. A vector $\boldsymbol{x} = [x_1\ x_2 \cdots x_n]^T$ will be used to denote antenna parameters (geometry and/or material ones) that are to be adjusted in the design process. Thus, $\boldsymbol{f}(\boldsymbol{x})$ will be used to denote the responses of the high-fidelity EM antenna model at a design $\boldsymbol{x}$. Typical responses of the antenna structures of Figs. 2.1–2.3 are shown in Figs. 2.4–2.6.

Antennas are key components of wireless communication systems. In order to be suitable for a specific application, the antenna structure has to satisfy performance requirements that are imposed by standardization institutions, depend on the properties of other components in the system, or come from geometrical limitations (e.g., maximum footprint).

The typical design requirements include achieving sufficient level of antenna reflection (typically, to ensure $|S_{11}|$ below -10 dB or VSWR ≤ 2) within a frequency band of interest (Nguyen *et al.*, 2012;

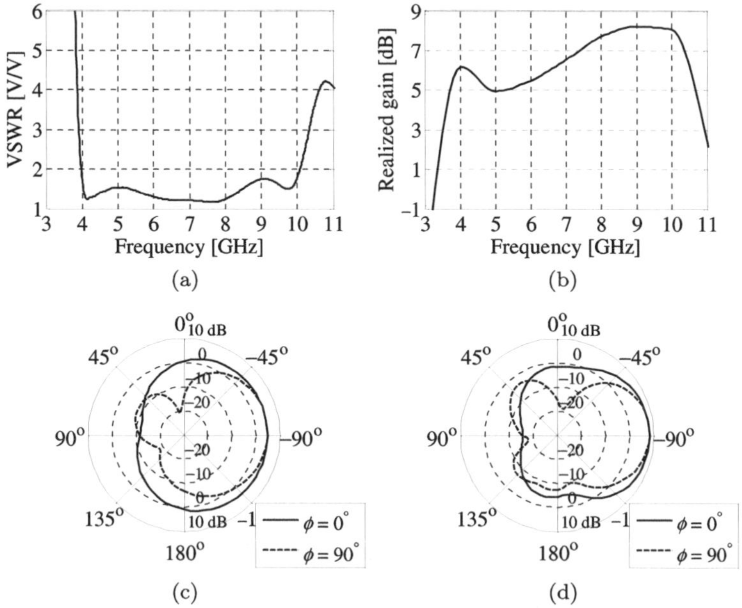

Fig. 2.5. Typical characteristics of quasi-Yagi antenna of Fig. 2.2 at a particular design: (a) VSWR; (b) realized gain for the zero zenith angle versus frequency; (c) realized gain at 5 GHz in the selected E-planes; (d) realized gain at 9 GHz in the selected E-planes.

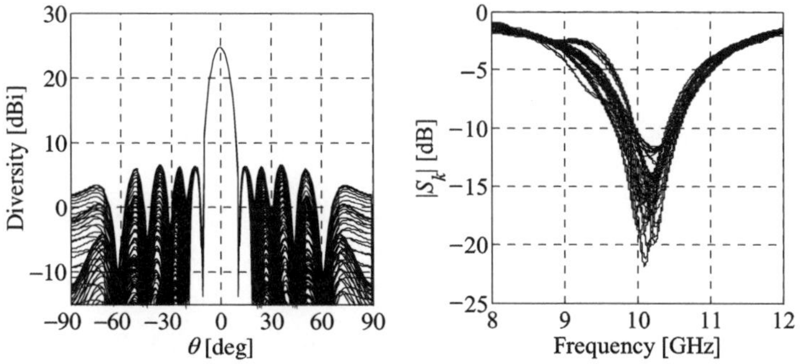

Fig. 2.6. Responses of the microstrip array antenna of Fig. 2.3 (here, at a design optimized for minimum side-lobe level): (a) directivity pattern shown for every 5° cuts; (b) active reflection coefficients.

Koziel *et al.*, 2015a). Field-related objectives include, among others, maximization of the in-band gain (Bekasiewicz *et al.*, 2014) or minimization of the axial ratio (Aljibouri *et al.*, 2000). The requirements can be also defined with respect to radiation pattern in order to ensure omnidirectional characteristics (Koziel and Bekasiewicz, 2015b) or reduce the side-lobe level (Koziel *et al.*, 2014a). The objectives can also be related to antenna geometry, e.g., minimization of lateral dimensions, the overall occupied area (Bekasiewicz *et al.*, 2014) or volume (Koziel and Bekasiewicz, 2015b), as well as obtaining a specific shape of the design.

The antenna design task can be formulated as the following nonlinear minimization problem (Koziel and Ogurtsov, 2011a):

$$\boldsymbol{x}^* = \arg\min_{\boldsymbol{x}} U(\boldsymbol{f}(\boldsymbol{x})), \tag{2.1}$$

where $\boldsymbol{x}^*$ is the optimum design to be found whereas U is the scalar merit function which encodes given design specifications. The objective function is implemented so that design improvement results in reducing the value of $U(\boldsymbol{f}(\boldsymbol{x}))$. In general, $U(\boldsymbol{f}(\boldsymbol{x}))$ is a vector of the form

$$U(\boldsymbol{f}(\boldsymbol{x})) = \left[U_1(\boldsymbol{f}(\boldsymbol{x})) \cdots U_{N_{\text{obj}}}(\boldsymbol{f}(\boldsymbol{x}))\right]^T, \tag{2.2}$$

where $U_k(\boldsymbol{f}(\boldsymbol{x})), k = 1, \ldots, N_{\text{obj}}$, is a kth design objective.

In practical design situations, the design task is formulated so that only one design requirement is explicitly handled, i.e., $N_{\text{obj}} = 1$. However, in this book, we are mostly interested in multi-objective design so that the problems with $N_{\text{obj}} > 1$ will typically be considered.

A specific definition of the merit function is problem dependent. In antenna design, U is often implemented as a minimax function with upper (and/or lower) specifications. Illustration of the minimax specifications for the reflection response is shown in Fig. 2.7(a), where the goal is to minimize the antenna response within a frequency range from 4 GHz to 10 GHz. The value of $U(\boldsymbol{f}(\boldsymbol{x}))$ (also referred to as minimax specification error) corresponds to the maximum violation of the design specifications within the frequency band of interest.

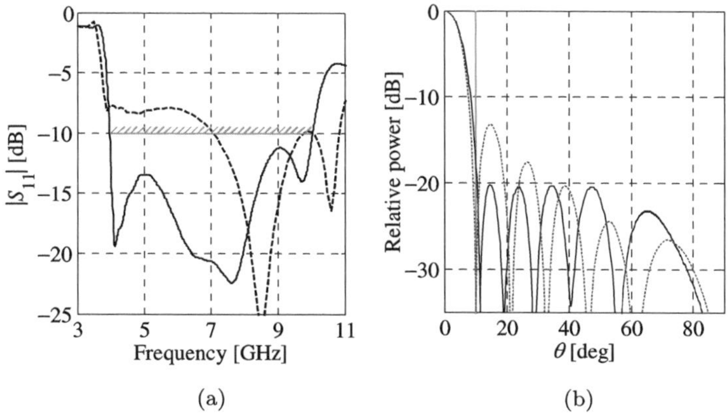

Fig. 2.7. Illustration of minimax specifications: (a) reflection minimization of the antenna of Fig. 2.2: the initial response (– –) violating the requirements of $|S_{11}| \leq -10\,\text{dB}$ for 4–10 GHz band, and the optimized response (—) that satisfies this requirement; (b) side-lobe level minimization of antenna of Fig. 2.3 (vertical line denoted the main beam width) at the initial design (- - -) and for the optimized array (—) indicating improvement of the side-lobe level by about 7 dB.

Figure 2.7(b) shows minimax specifications implemented for side-lobe level minimization of the array antenna.

Realistic antenna design problems are always constrained. Constraints are usually introduced to ensure physical consistency of the antenna structure under design (e.g., components of the structure do not overlap, etc.). Furthermore, constraints can be introduced to ensure that certain performance characteristics (reflection, radiation pattern, etc.) or physical dimensions (area, etc.) do not exceed the assumed values. In particular, constraints may be introduced to simplify the design problem which is inherently multi-objective so that only one (primary) objective is optimized directly whereas other objectives are controlled through appropriately defined constraints.

The following types of constraints can be considered:

- Lower and upper bounds for design variables, i.e., $l_k \leq x_k \leq u_k, k = 1, \ldots, n$; an alternative notation is $\boldsymbol{l} \leq \boldsymbol{x} \leq \boldsymbol{u}$, where $\boldsymbol{l} = [l_1 \ldots l_n]^T$, $\boldsymbol{u} = [u_1 \ldots u_n]^T$;

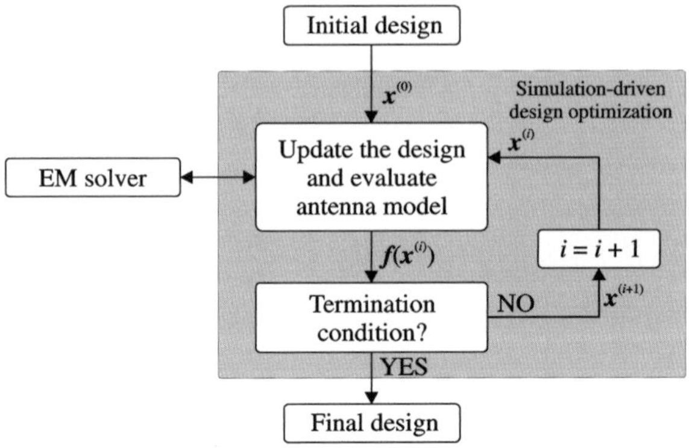

Fig. 2.8. Generic flowchart of simulation-driven design optimization of antenna structures.

- Inequality constraints, i.e., $c_{\text{ineq.k}}(\boldsymbol{x}) \leq 0, k = 1, \ldots, N_{\text{ineq}}$, where N_{ineq} is the number of constraints;
- Equality constraints, i.e., $c_{\text{eq.k}}(\boldsymbol{x}) = 0, k = 1, \ldots, N_{\text{eq}}$, where N_{eq} is the number of constraints.

Simulation-driven antenna optimization is normally an iterative process. The candidate designs generated by the optimization algorithm (based on, e.g., gradients of $U(\boldsymbol{f}(\boldsymbol{x}))$ with respect to $\boldsymbol{x}$) are evaluated using EM analysis to verify their quality and to provide information for further analysis. An overview of conventional (both local and global) as well as surrogate-assisted optimization techniques is provided in Chapters 3–5. Figure 2.8 shows a generic flowchart of the simulation-driven antenna design process.

2.2 Computational Models

The most popular and accurate technique for evaluating antenna performance is numerical full-wave EM analysis. As a matter of fact, for certain classes of antenna structures, full-wave simulation is the only way of reliable modeling and performance evaluation. Computational EM solvers can simulate models characterized by high level of fidelity with respect to physical realization of the structure at hand. They can accurately represent materials utilized for construction of the antenna as well as account for its feeding, components allocated within its vicinity, etc. (Kempel, 2007). This is essential for many real-world design problems. Prominent examples include simulations of broadband antennas (Valderas *et al.*, 2011), DSAs (Petosa, 2007), as well as structures with strong element coupling (Leung and Ng, 2005).

The core of computational EM methods is solving of Maxwell equations which is a numerically demanding task. Therefore, their application was held down for decades by limitations of available hardware. The situation changed when computers started offering reasonable computational power at affordable prices. Nowadays, a number of full-wave techniques suitable for the simulation of even very complex antenna models are available (both commercially and in public domain). The most popular, and potentially very accurate ones, are based on the finite-difference time-domain technique (FDTD) (Taflove and Hagness, 2005), the method of moments (MoM) (Harrington, 1993), and the finite-element method (FEM) (Jin, 2002). Each of them represents different numerical approach to EM-simulation and, consequently, their usefulness and efficiency may depend on the problem at hand. It should be noted that mentioned techniques are generic in a sense that they are not limited to antenna structures (cf. Chapter 11). Here, we discuss these solvers only from the application standpoint. Detailed description of theory behind the methods as well as guidelines for their implementation can be found in the literature (Davidson, 2010; Jin, 2002; Harrington, 1993; Taflove and Hagness, 2005).

FDTD is often considered the most natural and intuitive technique for solving Maxwell equations (Davidson, 2010). The method is capable of simulating the entire frequency range in a single run which makes it particularly attractive for broadband structures. Also, FDTD benefits from straightforward handling of material inhomogeneity and availability of GPU-based parallelization (CST, 2013). The latter enhances its applicability even to very complex design problems. At the same time, the technique is extremely expensive when applied for solving strongly resonant and/or graded structures

(Kempel, 2007). FDTD is very useful for simulation of broadband antennas such as monopoles, Vivaldi structures (Chamaani *et al.*, 2010), volumetric antennas (Giannakis *et al.*, 2016), Yagi–Uda-based designs (Wu *et al.*, 2014), etc. Its explicit (Taflove and Hagness, 2005) and implicit (Namiki, 1999; Clemens and Weiland, 2001) formulations have been implemented in software packages including, but not limited to, Remcom XFDTD (XFDTD, 2014), or CST Microwave Studio (CST, 2013), OptiFDTD (OptiFDTD, 2014). A definite advantage of the FDTD compared with the other two outlined below is that the FDTD can handle active and nonlinear materials and devices.

MoM is a frequency domain technique in its commonly used formulation (i.e., the problem is solved separately for each frequency point). Perhaps, its most important property is very efficient handling of structures such as wire or planar antennas (Peterson *et al.*, 1999). On the other hand, MoM is inefficient when inhomogeneous materials are considered and its numerical cost grows very fast with increase of the upper corner frequency, especially for volumetric applications (Kempel, 2007). The method is useful for structures such as patch antennas (Weigand *et al.*, 2003), spiral antennas (Khamas, 2008), log-periodic designs (Karr *et al.*, 2004) and many others. MoM implementations are available in packages such as HyperLynx (HyperLynx, 2015), ADS Momentum (ADS, 2011), or Sonnet Suite (Sonnet, 2003), to name just a few.

Similarly as MoM, FEM is typically based on the frequency domain formulation (Jin, 2002). The technique benefits from straightforward handling of complex geometries and inhomogeneous or dispersive materials (Davidson, 2010). Moreover, it can be efficient and accurate for handling resonant structures (Kempel, 2007). At the same time, FEM provides slightly better frequency scalability than MoM (Davidson, 2010). The method is useful for simulation of, e.g., DSAs (Almpanis *et al.*, 2008) and cavity-backed structures (Bayderkhani *et al.*, 2015). FEM is implemented in many EM packages, e.g., Ansys HFSS (HFSS, 2012) or CST Studio (CST, 2013). Also hybrids of FEM and MoM simultaneously featuring the strongest properties of both have been successfully implemented in FEKO (FEKO, 2012).

There are several factors that affect the computational cost of the EM-simulation model. These include discretization density of the structure, its geometrical complexity, the number of ports, utilized solver, optional computation of response derivatives, amount of post-processing (e.g., gain pattern calculation), etc. As mentioned before, we will use the symbol $\boldsymbol{f}$ to denote the high-fidelity EM antenna model. The high-fidelity model is understood as a model that provides sufficient accuracy. In particular, the design optimization process is supposed to be conducted at the level of $\boldsymbol{f}$. Simulation times of the high-fidelity antenna models range from a few minutes to many hours. For the sake of illustration, the antennas considered in Sec. 2.1 have been implemented in CST Microwave Studio. The simulation times below (as well as in the rest of the book) are obtained for dual Intel Xeon E5540 machine with 64 GB RAM:

- Antenna of Fig. 2.1: the high-fidelity model contains about 190,000 tetrahedral mesh cells and simulated using time domain solver. The simulation time is 19 min.
- Antenna of Fig. 2.2: the high-fidelity model contains about 2,000,000 hexahedral mesh cells and it is simulated using frequency domain solver. Evaluation time of the model is 20 min.
- Antenna of Fig. 2.3: the high-fidelity model contains about 35,700,000 mesh cells and it is simulated using a time-domain solver. The simulation time is 1,344 min.

Most of the design techniques discussed in this book exploit surrogate-based optimization (SBO) paradigm with the surrogate model constructed using an underlying low-fidelity EM antenna model $\boldsymbol{c}$. The low-fidelity model is a faster version of $\boldsymbol{f}$, typically evaluated using the same EM solver as the one utilized for $\boldsymbol{f}$. The most straightforward way of creating a low-fidelity model is by reducing the discretization density. In some solvers, this can be done through manipulation of just one control parameter (e.g., Lines per Wavelength, LPW, in CST; CST 2013, or Cells per Wavelength, in

Momentum, Agilent, 2011). In others, changes of several settings in the simulation setup are required to determine conditions for terminating the discretization process (e.g., Ansys HFSS, HFSS, 2012).

In addition to utilizing a coarser mesh, computational cost of low-fidelity EM models can also be reduced by simplification of their underlying physics. Possible changes include the following:

(a) Utilization of perfectly conducting metals;
(b) Neglecting dielectric substrate losses and material dispersion if their influence on the simulated response is not significant;
(c) Neglecting metallization thickness for planar structures;
(d) Ignoring moderate anisotropy of substrates;
(e) Feeding the structure from discrete sources rather than waveguide ports (HFSS, 2012; CST, 2013; FEKO, 2012; Taflove and Hagness, 2006).

Furthermore, the cost of model evaluation can be decreased by means of the following computational simplifications:

(a) Reduction of the computational domain size and utilization of simple absorbing boundaries with the finite-volume methods (Taflove and Hagness, 2006; Lin, 2006);
(b) Utilization of low-order basis functions with the finite-element and moment method solvers (Jin, 2002; Harrington, 1993; Makarov, 2002);
(c) Using relaxed criteria for solution termination such as the S-parameter error for the frequency domain methods with adaptive meshing, e.g., (HFSS, 2012; CST, 2013), and residue energy for the time-domain solvers (CST, 2013);
(d) Utilization of appropriate symmetry planes within the structure geometry (HFSS, 2012; CST, 2013; Sonnet, 2013).

It should be noted that despite splitting the aforementioned simplifications into two separate classes, as a matter of fact, they are all closely related. For instance, neglecting material dispersion and dielectric losses in the model simulated using a time-domain finite-volume-method simplifies the problem by reducing the number of unknowns (Taflove and Hagness, 2006).

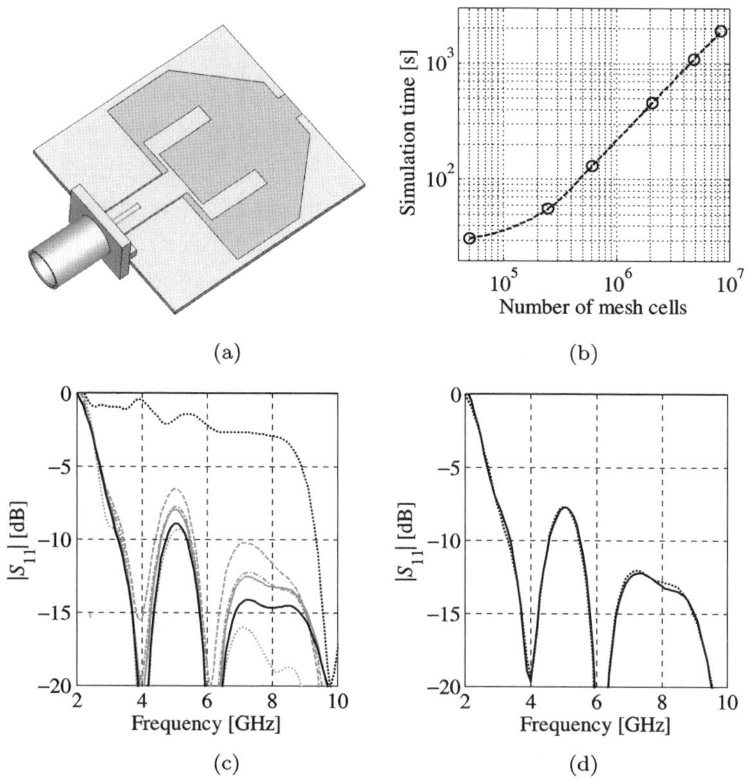

Fig. 2.9. Low-fidelity EM modeling illustrated using a uniplanar UWB antenna (Qing and Chen, 2009): (a) three-dimensional visualization of the antenna structure; (b) computational cost of the EM model as a function of the mesh density; (c) influence of the discretization level on accuracy of the reflection response; black lines represent the high-fidelity model (—) and the low-fidelity model that is too coarse (···); gray lines represent models containing 253,440 (···), 613,640 (– –), 2,097,186 (– ·) and 4,960,620 (—) cells; (d) reflection response for the low-fidelity model comprising 2,097,186 cells with (···) and without (—) additional simplifications (see the text for details).

The influence of mesh density and model simplifications on responses of the antenna structure is illustrated using a compact uniplanar antenna shown in Fig. 2.9(a) (Qing and Chen, 2009). A specific design of the structure model is implemented in CST Microwave Studio and simulated using its time domain solver (CST, 2013). The substrate material is a 0.762 mm thick Taconic RF-35. The high-fidelity model of the antenna contains 8,631,792 hexahedral mesh cells and its simulation time is 32 min. The relation between the number of mesh cells and CPU-time of the model is presented in Fig. 2.9(b), whereas the effect of the mesh density on the reflection characteristics is shown in Fig. 2.9(c). The results indicate that the two coarsely-discretized models with densest meshes (2,097,186 tetrahedrons and 4,960,620 tetrahedrons, respectively) are sufficiently close to the high-fidelity one. The antenna model composed of 1,124,200 cells can be considered as border one. The remaining models are poor, particularly the one with 51,612 cells which is unreliable and does not capture the critical features of the antenna response (e.g., resonances and their frequency allocation). In order to visualize the importance of physical and computational simplifications, the response of the model composed of 2,097,186 tetrahedrons (simulation time of 7.6 min) has been compared to the responses of its counterpart with neglected losses (both, for metallization and dielectric) and metallization thickness set to 0 mm. Also, the termination condition concerning the residual energy has been set to −25 dB (versus −35 dB for the reference model). The simulation results shown in Fig. 2.9(d) indicate that responses of both models are very similar. However, the evaluation cost of the simplified structure has been reduced by almost 70 percent (to only 2.4 min) compared to the model without simplifications. The simulation time of the model is only 7.5 percent of that of the high-fidelity one. The influence of simplifications on accuracy and numerical cost of the EM antenna models has also been discussed in Koziel and Ogurtsov (2014a).

As shown in the above example, appropriate control of the mesh density as well as utilization of other model simplifications allow for considerable reduction of the design evaluation cost. It should be noted that although the low-fidelity model can be made 10–50 times faster than the high-fidelity one, computational speedup is obtained at a cost of degraded accuracy (see Fig. 2.9). One of the consequences is that, due to inaccuracy, the low-fidelity model cannot directly replace the high-fidelity model in the design optimization process such as (2.1).

A possible workaround is an iterative approach where the solution to (2.1) is found through a series of approximations realized using surrogate-assisted methods (Koziel and Ogurtsov, 2014a; Queipo *et al.*, 2005). More details concerning these techniques can be found in Chapter 5. Here, we only mention that finding appropriate balance between the low-fidelity model speed and accuracy may be critical for the operation and performance of SBO algorithms. In particular, using the low-fidelity model that is too coarse increases the risk of failure of the algorithm (Koziel and Ogurtsov, 2012a). Also, although the model is faster, a larger number of algorithm iterations might be necessary (not only because of its inaccuracy and also because of other problems such as increased level of numerical noise; Koziel and Ogurtsov, 2012a).

On the other hand, more accurate (and more expensive) low-fidelity model improves the performance of the optimization process but at the expense of higher computational cost. Thus, finding the most appropriate trade-off between the speed and accuracy of the low-fidelity model is one of the critical aspects of EM-driven design involving variable-fidelity simulations. Nevertheless, rigorous and automated assessment and selection of the low-fidelity model (or models, in case of multi-fidelity methods, Leifsson and Koziel, 2015a) is still an open problem. One of the issues is to find a suitable measure of model discrepancies. It is not straightforward because some types of discrepancies (e.g., frequency shifts) can be easily corrected, yet they manifest themselves as large when using conventional measures (e.g., a norm). This means, among others, that a proper model selection is closely related to a specific optimization and surrogate model correction method (Duffy *et al.*, 2015).

The problem of automated low-fidelity model determination is still the subject of ongoing research. Until now, the most practical approach is visual inspection of the model discrepancies. Some initial results concerning automation of the process, based on correlation analysis, can be found in the literature (Duffy *et al.*, 2015; Koziel and Bekasiewicz, 2016e).

2.3 Challenges of Contemporary Antenna Design

Design of contemporary antenna structures is a challenging process that involves selection of appropriate evaluation tools and algorithmic techniques as well as requires considerable engineering experience. One of the most important reasons is that modern structures often feature complex geometries with a large number of parameters. These ensure sufficient flexibility of the design so that the imposed performance requirements can be fulfilled upon appropriate adjustment of the parameters. However, reliable evaluation of antenna structures can only be performed by means of full-wave EM-simulations. Consequently, simulation-driven design is nowadays a standard approach in antenna engineering.

As an industry (and also academic) standard, EM-driven antenna design is realized as a manual or semi-manual sweeping of selected geometrical dimensions guided by engineering experience (Gautam *et al.*, 2013; Wang *et al.*, 2014). Unfortunately, such methods are not only very laborious and also insufficient to handle modern structures where complex relationships between antenna dimensions (and/or material parameters) and its responses require simultaneous adjustment of several (or all) parameters (Bekasiewicz and Koziel, 2015a). Finding truly optimum designs by appropriate utilization of available degrees of freedom can only be realized using numerical optimization methods. Although desirable, direct solving of optimization tasks such as (2.1) might be impractical when using conventional algorithms that incur high, often prohibitive computational expense. As shown in Sec. 2.2, the CPU-time of single antenna simulation can be as long as many hours and thus optimization techniques, such as gradient-based search with numerical derivatives or derivative-free methods — both local (Nocedal and Wright, 2006; Conn *et al.*, 2009), and global (Horst *et al.*, 2000; Hendrix and Gazdag-Tóth, 2010) — require many antenna evaluations to converge. Depending on the problem dimensionality and other factors (e.g., nonlinearity of the antenna responses), the required number of simulations can be from dozens (a few variables for local methods), through hundreds

(10–20 variables, local methods) to many thousands (medium/large dimensionality, global optimization).

An additional challenge is that design of contemporary antennas typically involves handling several performance figures (Kuwahara, 2005; Chamaani *et al.*, 2010). Multi-objective optimization problems are fundamentally different from single-objective ones in the sense that solutions may not be comparable in the traditional sense. The optimal designs are typically sought in the Pareto sense and the optimization outcome is normally a set of solutions representing the best possible trade-offs between the objectives of interest. Needless to say, finding such designs requires considerably higher computational effort as the most popular solution approaches include population-based metaheuristics (Deb, 2001; Talbi, 2009).

The above discussion indicates that simulation-driven antenna design may not be practical when using conventional numerical optimization methods. In some cases, the speedup can be obtained by scaling up the computational resources and licenses for specialized EM packages (Jin and Rahmat-Samii, 2005; Hornby *et al.*, 2006). On the other hand, such a brute force approach is usually only available in large research facilities.

The focus of this book is on expedited design of antenna structures using appropriate combination of variable-fidelity EM models, surrogate modeling techniques, surrogate-assisted optimization, design space reduction methods, and — in some cases — population-based metaheuristics. The main topic is accelerated multi-objective optimization, although brief introduction to numerical optimization, including conventional gradient-based and derivative-free methods (Chapter 3) as well as population-based metaheuristics (Chapter 4) is also given. Furthermore, SBO techniques are discussed in Chapter 5 to the extent necessary to understand the remaining parts of the material. Multi-objective antenna design including methods and design cases is covered in Chapters 6–12.

Chapter 3

Introduction to Numerical Optimization

The primary topic of this book is surrogate-assisted of antenna structures; however, for the convenience of the reader, we provide a brief introduction to numerical optimization. In this chapter, we cover selected conventional single-objective optimization techniques, both gradient-based and derivative-free. Chapter 4 outlines global optimization using population-based metaheuristics. An important terminological distinction should be made here between conventional (or direct) and surrogate-based techniques. In direct methods, the expensive electromagnetic (EM)-simulation antenna model is handled directly in the optimization process, whereas in surrogate-assisted algorithms, majority of the operations are carried out using a fast replacement model (the surrogate) with only occasional reference to the expensive simulations. Figure 3.1 shows a general flow of the direct simulation-driven optimization process. It can be observed that each candidate solution (design) produced by the algorithm is evaluated using a computationally expensive high-fidelity simulation. Clearly, this might become a bottleneck if the number of such evaluations is large.

In this chapter, we briefly overview and discuss conventional numerical optimization techniques, including gradient-based and derivative-free methods. We also provide some information concerning constrained optimization. Population-based metaheuristic algorithms are covered in Chapter 4. The readers interested in further information on these techniques are referred to the rich literature on the subject (e.g., Nocedal and Wright, 2000; Yang, 2010a; Conn *et al.*, 2009).

3.1 Optimization Problem Formulation

Design optimization problem is formulated as

$$\boldsymbol{x}^* = \arg\min_{\boldsymbol{x}\in X} f(\boldsymbol{x}), \tag{3.1}$$

where $f(\boldsymbol{x})$ is a scalar objective function, whereas $\boldsymbol{x} \in X \subseteq R^n$; X is the objective function domain (also referred to as a design/search space). It should be emphasized that for practical antenna design problems, the objective function is a composition $U(\boldsymbol{f}(\boldsymbol{x}))$ of the merit function U and a vector-valued system response $\boldsymbol{f}(\boldsymbol{x})$, see also Chapter 2. The merit function translates the design specifications into a single number representing the performance of the system.

The problem (3.1) can also be formulated using explicit constraints, such as lower/upper bounds for design variables as well as other (both linear and nonlinear) inequality or equality constraints (cf. Chapter 2).

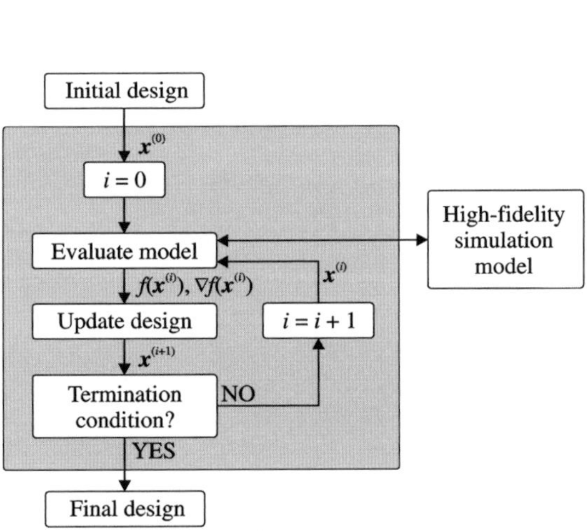

Fig. 3.1. Direct simulation-driven optimization flowchart. The candidate designs generated by the algorithm are evaluated through a high-fidelity simulation for verification purposes and to provide the optimizer with information to search for better designs. The search process may be guided by the model response only, or the response and its derivatives (if available).

3.2 Gradient-Based Optimization Techniques

Gradient-based methods belong to the most popular and widely used optimization techniques in engineering and science (Nocedal and Wright, 2006). The search process is based on the gradient $\nabla f(\boldsymbol{x})$ of the objective function $f(\boldsymbol{x})$, defined as

$$\nabla f(\boldsymbol{x}) = \left[\frac{\partial f}{\partial x_1}(\boldsymbol{x}) \;\; \frac{\partial f}{\partial x_2}(\boldsymbol{x}) \;\; \cdots \;\; \frac{\partial f}{\partial x_n}(\boldsymbol{x})\right]^T. \tag{3.2}$$

The importance of gradient is that it provides information about the local behavior of the function f in the vicinity of the point $\boldsymbol{x}$ (assuming that f is sufficiently smooth, i.e., at least continuously differentiable). More specifically, using first-order Taylor expansion of f around $\boldsymbol{x}$, one gets

$$f(\boldsymbol{x}+\boldsymbol{h}) \cong f(\boldsymbol{x}) + \nabla f(\boldsymbol{x})^T \cdot \boldsymbol{h} < f(\boldsymbol{x}), \tag{3.3}$$

for sufficiently small $\boldsymbol{h}$ as long as $\boldsymbol{h}$ is a descent direction, i.e., $\nabla f(\boldsymbol{x})^T \cdot \boldsymbol{h} < 0$. Note that the (locally) largest reduction of the function value is obtained for $\boldsymbol{h} = -\nabla f(\boldsymbol{x})$ which is referred to as the steepest descent direction.

The gradient can be utilized in the search process in two basic ways: (i) to move along a descent direction, or (ii) to exploit a local approximation model of the objective function constructed with ∇f. More information about the local behavior of the function can be obtained from the second-order derivatives, utilized by certain algorithms, specifically the Hessian $\boldsymbol{H}(\boldsymbol{x})$, which is defined as

$$\boldsymbol{H}(\boldsymbol{x}) = \begin{bmatrix} \dfrac{\partial^2 f}{\partial x_1 \partial x_1}(\boldsymbol{x}) & \cdots & \dfrac{\partial^2 f}{\partial x_1 \partial x_n}(\boldsymbol{x}) \\ \vdots & \ddots & \vdots \\ \dfrac{\partial^2 f}{\partial x_n \partial x_1}(\boldsymbol{x}) & \cdots & \dfrac{\partial^2 f}{\partial x_n \partial x_n}(\boldsymbol{x}) \end{bmatrix}. \tag{3.4}$$

Behavior of the gradient and Hessian can be used to identify the function's minimum. The points for which the gradient ∇f vanishes are called the *stationary points* of f. Also, vanishing of the gradient is a necessary condition for the minimum: if $\boldsymbol{x}^*$ is a local minimizer of f, then $\nabla f(\boldsymbol{x}^*) = 0$. However, the stationary point may or may not be a local minimizer. This depends on the second-order derivatives. In particular, if $\boldsymbol{x}^*$ is a stationary point and $\boldsymbol{H}(\boldsymbol{x}^*)$ is positive definite (Nocedal and Wright, 2006), then $\boldsymbol{x}^*$ is a local minimizer of f. On the other hand, if $\boldsymbol{x}^*$ is a local minimizer of f, then $\boldsymbol{H}(\boldsymbol{x}^*)$ is positive semidefinite.

3.2.1 *Gradient-Based Optimization Using Descent Methods*

In this section, we consider iterative algorithms, i.e., the methods that generate a sequence of vectors $\boldsymbol{x}^{(i)}$, $i = 0, 1, 2, \ldots$, being approximate solutions to the problem (3.1). The fundamental requirement is that $f(\boldsymbol{x}^{(i+1)}) < f(\boldsymbol{x}^{(i)})$, i.e., an improvement with respect to the cost function value is enforced between iterations. Let $\boldsymbol{x}^{(0)}$ be the initial solution. The generic structure of a descent method is the following (Nocedal and Wright, 2006):

1. Set $i = 0$;
2. Find a search direction $\boldsymbol{h}_d$;
3. Find a step length α;
4. Set $\boldsymbol{x}^{(i)} = \boldsymbol{x}^{(i)} + \alpha \boldsymbol{h}_d$;
5. Set $i = i + 1$;
6. If the termination condition is not satisfied, go to 2;
7. END

The search in the above algorithm is carried out along a descent direction $\boldsymbol{h}_d$ found in Step 2. As mentioned before, a descent direction is a vector $\boldsymbol{h}_d$ for which $\nabla f(\boldsymbol{x})^T \cdot \boldsymbol{h}_d < 0$ (cf. (3.3)). The length of the step, i.e., $\alpha \boldsymbol{h}_d$ is determined by a separate algorithm to ensure that $f(\boldsymbol{x}^{(i+1)}) < f(\boldsymbol{x}^{(i)})$ as explained later. The termination condition utilized in practical optimization algorithms may be based on convergence in argument, i.e., $\|\boldsymbol{x}^{(i+1)} - \boldsymbol{x}^{(i)}\| \leq \varepsilon_1$, vanishing of the gradient, i.e., $\|\nabla f(\boldsymbol{x}^{(i)})\| \leq \varepsilon_2$, convergence in the function value, i.e., $f(\boldsymbol{x}^{(i)}) - f(\boldsymbol{x}^{(i+1)}) \leq \varepsilon_3$, or a combination of the above.

Line search (Antoniou and Lu, 2007) is a procedure of finding the length of the step taken along the descent direction. More specifically, one considers a one-dimensional function of a length parameter α,

$$\varphi(\alpha) = f(\boldsymbol{x} + \alpha \boldsymbol{h}_d). \tag{3.5}$$

Two types of line search are utilized in practice. In *exact line search*, the step length α_e is found as $\alpha_e = \text{argmin}\{\alpha > 0 : f(\boldsymbol{x}+\alpha\boldsymbol{h}_d)\}$. *Soft line search* only requires satisfaction of certain (and usually rather loose) conditions such as the Wolfe conditions (Fletcher, 1987):

$$\varphi(\alpha_s) \leq \lambda(\alpha_s), \quad \text{where } \lambda(\alpha) = \varphi(0) + \rho\varphi'(0)\cdot\alpha \quad \text{with } 0 < \rho < 0.5 \tag{3.6}$$

and

$$\varphi'(\alpha_s) \geq \beta \cdot \varphi'(0) \quad \text{with } \rho < \beta < 1. \tag{3.7}$$

The first condition ensures sufficient decrease of the cost function f, whereas the second avoids making steps that are too short. Descent methods with line search satisfying (3.6), (3.7) are normally convergent to at least a first-order stationary point of f (i.e., where the function gradient vanishes). It is interesting to note that soft line search is generally more advantageous than exact line search in terms of the overall computational cost of the optimization process (the use of exact line search results in a smaller number but much more expensive iterations).

It should be emphasized that selecting the search direction to be the steepest descent direction, i.e., $\boldsymbol{h} = -\nabla f(\boldsymbol{x})$, in each iteration (so-called steepest descent method), leads to a poorly performing algorithm when close to the optimum. One of the famous examples is the Rosenbrock function $f(\boldsymbol{x}) = f([x_1\ x_2]^T) = (1 - x_1)^2 + 100(x_2 - x_1^2)^2$ (Rosenbrock, 1960). Figure 3.2(a) shows how the steepest descent method fails to find the optimum of this function because of a narrow and curved valley (the steps taken in subsequent iterations of the steepest-descent method with exact line search are perpendicular to each other). On the other hand, the method is useful in the initial stages of the optimization process, which leads to various hybrid methods.

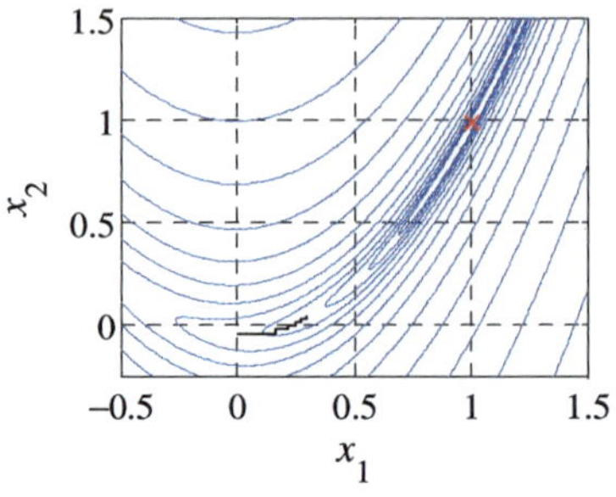

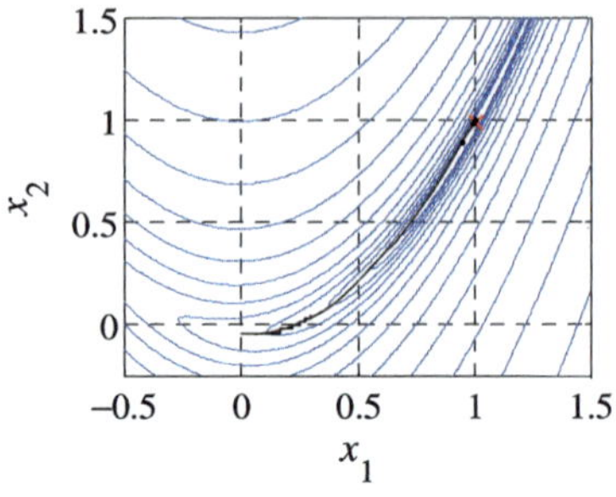

Fig. 3.2. Optimization of the Rosenbrock function (Rosenbrock, 1960): (a) steepest-descent methods with exact line search fails to find the optimum at $\boldsymbol{x}^* = [1\ 1]^T$ (marked $\times$): the narrow and curved valley containing the minimum slows down the algorithm convergence effectively preventing the method from reaching ($\boldsymbol{x}^*$); (b) conjugate-gradient algorithm with Fletcher–Reeves update method and soft line search: the optimum is found in less than 30 iterations.

In general, a better choice of the search direction is a combination of the previous direction $\boldsymbol{h}_{\text{prev}}$ and the current gradient, i.e.,

$$\boldsymbol{h} = -\nabla f(\boldsymbol{x}^{(i)}) + \gamma \boldsymbol{h}_{\text{prev}}, \tag{3.8}$$

which is exploited by the so-called conjugate-gradient methods (Yang, 2010a). Various conjugate-gradient methods differ in the way of updating the search direction. An important example is a Fletcher–Reeves (Fletcher and Reeves, 1964) method with

$$\gamma = \frac{\nabla f(\boldsymbol{x})^T \nabla f(\boldsymbol{x})}{\nabla f(\boldsymbol{x}_{\text{prev}})^T \nabla f(\boldsymbol{x}_{\text{prev}})}. \tag{3.9}$$

Another example is a Polak–Ribiére method (Polak and Ribiére, 1969) with

$$\gamma = \frac{(\nabla f(\boldsymbol{x}) - \nabla f(\boldsymbol{x}_{\text{prev}}))^T \nabla f(\boldsymbol{x})}{\nabla f(\boldsymbol{x}_{\text{prev}})^T \nabla f(\boldsymbol{x}_{\text{prev}})}. \tag{3.10}$$

In both cases, the new search direction $\boldsymbol{h}$ obtained from (3.8) is not guaranteed to be downhill (e.g., when the soft line search is used to find the iteration step) so that safeguards such as defaulting to $\boldsymbol{h} = -\nabla f(\boldsymbol{x})$ are necessary in case $\nabla f(\boldsymbol{x})\boldsymbol{h}^T \geq 0$. Figure 3.2(b) shows the Fletcher–Reeves conjugate-gradient method (with soft line search) optimizing the Rosenbrock function. Conjugate-gradient methods are characterized by linear convergence rate (Crowder and Wolfe, 1972).

Instead of using line search, descent methods can be realized with trust region framework (Conn *et al.*, 2000), where the iteration step is determined by means of the local model of f created using first- (or, optionally, higher-) order derivatives. The first-order Taylor expansion of f is defined as

$$f(\boldsymbol{x}+\boldsymbol{h}) \approx q(\boldsymbol{h}) \quad \text{with } q(\boldsymbol{h}) = f(\boldsymbol{x}) + \nabla f(\boldsymbol{x})\boldsymbol{h}^T. \tag{3.11}$$

The second-order Taylor expansion (Bischof *et al.*, 1993) is given by

$$f(\boldsymbol{x}+\boldsymbol{h}) \approx q(\boldsymbol{h}) \quad \text{with } q(\boldsymbol{h}) = f(\boldsymbol{x}) + \nabla f(\boldsymbol{x}) \cdot \boldsymbol{h}^T + \frac{1}{2}\boldsymbol{h}^T \boldsymbol{H}(\boldsymbol{x})\boldsymbol{h}. \tag{3.12}$$

Assuming that the function f is sufficiently smooth, the model $q(\boldsymbol{h})$ is a good approximation of $f(\boldsymbol{x}+\boldsymbol{h})$ for sufficiently small vectors $\boldsymbol{h}$. In the trust region framework, the candidate iteration step size $\boldsymbol{h}_{tr}$ is determined as

$$\boldsymbol{h}_{tr} = \arg \min_{\boldsymbol{h}, \|\boldsymbol{h}\| \leq \delta} q(\boldsymbol{h}), \tag{3.13}$$

where $\delta > 0$ is the trust region radius.

The step $\boldsymbol{h}_{tr}$ is accepted if it leads to the improvement of the cost function, i.e., $f(\boldsymbol{x}+\boldsymbol{h}_{tr}) < f(\boldsymbol{x})$. Otherwise it is rejected. The trust region radius δ is adjusted using a so-called *gain ratio* defined as

$$r = \frac{f(\boldsymbol{x}) - f(\boldsymbol{x}+\boldsymbol{h})}{q(0) - q(\boldsymbol{h})}. \tag{3.14}$$

A typical adjustment scheme would be $\delta \leftarrow 2\delta$ if $r > 0.75$ and $\delta \leftarrow \delta/3$ if $r < 0.25$. It should be noted that for both models (3.11) and (3.12), the condition $f(\boldsymbol{x}+\boldsymbol{h}_{tr}) < f(\boldsymbol{x})$ would be satisfied if δ is sufficiently small.

3.2.2 *Newton and Quasi-Newton Methods*

Given the cost function f is at least twice continuously differentiable, it can be locally represented using its second-order Taylor approximation (3.12). If the Hessian $\boldsymbol{H}(\boldsymbol{x})$ of f at $\boldsymbol{x}$ is positive definite, then the model $q(\boldsymbol{h})$ has a unique minimizer for $\boldsymbol{h}$ such that $\nabla q(\boldsymbol{h}) = 0$, i.e., where

$$\nabla f(\boldsymbol{x}) + \boldsymbol{H}(\boldsymbol{x})\boldsymbol{h} = 0. \tag{3.15}$$

This observation gives rise to the Newton's method (Antoniou and Lu, 2007), in which the next iteration point is obtained by solving (3.15) as follows:

$$\boldsymbol{x}^{(i+1)} = \boldsymbol{x}^{(i)} - [\boldsymbol{H}(\boldsymbol{x}^{(i)})]^{-1}\nabla f(\boldsymbol{x}^{(i)}). \tag{3.16}$$

The algorithm (3.16) is well defined assuming that $\boldsymbol{H}(\boldsymbol{x})$ is non-singular. Moreover, if the Hessian is positive definite for all iterations and the starting point is sufficiently close to the optimum, the method usually converges very quickly (quadratically) to a minimum.

Unfortunately, the basic Newton algorithm as formulated in (3.16) is not a practical method. Its fundamental disadvantages include lack of global convergence for many problems, possibility of converging to a maximum or a saddle point, and possibility of (3.16) to be ill-conditioned. Furthermore, the Newton method requires analytical second-order derivatives.

The major practical difficulty, i.e., the Hessian not being positive definite, can be alleviated by using so-called damped Newton method, where (3.16) is replaced by

$$\boldsymbol{x}^{(i+1)} = \boldsymbol{x}^{(i)} - [\boldsymbol{H}(\boldsymbol{x}^{(i)}) + \mu\boldsymbol{I}]^{-1}\nabla f(\boldsymbol{x}^{(i)}), \tag{3.17}$$

where $\boldsymbol{I}$ is the identity matrix. It can be observed that if $\mu > 0$ is sufficiently large, the matrix $\boldsymbol{H}(\boldsymbol{x})+\mu\boldsymbol{I}$ is positive definite because it is dominated by the term $\mu\boldsymbol{I}$. The step $\boldsymbol{h}_\mu$ found as a solution to the problem $[\boldsymbol{H}(\boldsymbol{x}) + \mu\boldsymbol{I}]\boldsymbol{h}_\mu = -\nabla f(\boldsymbol{x})$ is a minimizer of the following model:

$$q_\mu(\boldsymbol{h}) = q(\boldsymbol{h}) + \frac{1}{2}\mu\boldsymbol{h}^T\boldsymbol{h} = f(\boldsymbol{x}) + \nabla f(\boldsymbol{x}) \cdot \boldsymbol{h}^T + \frac{1}{2}\boldsymbol{h}^T[\boldsymbol{H}(\boldsymbol{x}) + \mu\mathbf{I}]\boldsymbol{h}. \tag{3.18}$$

Additionally, $\boldsymbol{h}_\mu$ is a descent direction for f at $\boldsymbol{x}$. Also, the large steps are penalized in (3.17) because of the term $\mu\boldsymbol{h}^T\boldsymbol{h}/2$. If μ is very large, the algorithm defaults to the steepest descent method as $\boldsymbol{h}_\mu \approx -\nabla f(\boldsymbol{x})/\mu$.

```
x = x^(0); μ = μ^(0); found = false; k = 0;
while ~found & k ≤ k_max
  while f''(x)+μI not positive definite
    μ = 2*μ;
  end
  Solve (f''(x)+μI)h_dn = -f'(x); Compute gain factor r;
  if r > δ
    x = x + h_dn; μ = μ*max{1/3,1-2(2r-1)^3};
  else
    μ = μ*2;
  end
  k = k+1; Update found;
end
```

Fig. 3.3. Pseudocode of the LM algorithm.

In practical implementations, large values of μ are utilized at the early stages of the iteration process (3.17), and μ is reduced as the algorithm converges, which allows for finding the minimizer of f. A popular method exploiting this approach is Levenberg–Marquardt (LM) algorithm (Bates and Watts, 1988), where the value of μ is increased at the beginning of each iteration until $\boldsymbol{H}(\boldsymbol{x})+\mu\boldsymbol{I}$ becomes positive definite, and then updated based on the value of the gain ratio r (3.14). The typical updating formula is: $\mu \leftarrow \mu \cdot \max\{1/3, 1-(2r-1)^3\}$ if $r > 0$, and $\mu = 2\mu$ otherwise (Nielsen, 1999). The flow of the LM algorithm is shown in Fig. 3.3.

Damped Newton methods resolve all of the issues of the basic Newton's algorithm except one: the potentially high cost of obtaining second-order derivatives of the objective function f. This issue is addressed by the so-called quasi-Newton methods, where the exact Hessian (or, even better, its inverse) is approximated using appropriate updating formulas. Perhaps, the best and the most popular updating formula is a Broyden–Fletcher–Goldfarb–Shanno (BFGS) one (Nocedal and Wright, 2006). BFGS preserves positive symmetry and definiteness of the approximation. It is defined as follows for Hessian approximation $\boldsymbol{B}$:

$$\boldsymbol{B}_{\text{new}} = \boldsymbol{B} + \frac{1}{\boldsymbol{h}^T\boldsymbol{y}}\boldsymbol{y}\boldsymbol{y}^T - \frac{1}{\boldsymbol{h}^T\boldsymbol{u}}\boldsymbol{u}\boldsymbol{u}^T, \tag{3.19}$$

where $\boldsymbol{h} = \boldsymbol{x}_{\text{new}} - \boldsymbol{x}, \boldsymbol{y} = \nabla f(\boldsymbol{x}_{\text{new}}) - \nabla f(\boldsymbol{x}), \quad \boldsymbol{u} = \boldsymbol{B}\boldsymbol{h}.$

Approximation of the inverse of Hessian $\boldsymbol{D}$ is defined as

$$\begin{gathered}\boldsymbol{D}_{\text{new}} = \boldsymbol{D} + \kappa_1 \boldsymbol{h}\boldsymbol{h}^T - \kappa_2(\boldsymbol{h}\boldsymbol{v}^T + \boldsymbol{v}\boldsymbol{h}^T), \\ \text{where } \boldsymbol{h} = \boldsymbol{x}_{\text{new}} - \boldsymbol{x}, \quad \boldsymbol{y} = \nabla f(\boldsymbol{x}_{\text{new}}) - \nabla f(\boldsymbol{x}), \quad \boldsymbol{v} = \boldsymbol{B}\boldsymbol{y} \\ \kappa_2 = \frac{1}{\boldsymbol{h}^T\boldsymbol{y}}, \quad \kappa_1 = \kappa_2(1 + \kappa_2(\boldsymbol{y}^T\boldsymbol{v})).\end{gathered} \tag{3.20}$$

The quasi-Newton algorithm with the BFGS formula uses soft line search with loose conditions (e.g., $\rho = 0.0001$ and $\beta = 0.9$, cf. (3.6), (3.7)).

Quasi-Newton algorithms belong to the most efficient methods for unconstrained optimization. On the other hand, conjugate gradient methods may be better when the number of design variables is large: Newton-like methods rely on matrix operations while conjugate gradient ones only use vector operations, which is cheaper; also Newton-type methods require more storage ($n \times n$ matrix versus a few vectors for conjugate gradients).

3.2.3 *Remarks on Constrained Optimization*

The methods and algorithms outlined so far in this chapter allow solving unconstrained optimization problems. However, majority of practical problems are constrained, with typical constraints being lower/upper bounds for design variables, as well as linear/nonlinear equality and inequality constraints. In this section, some remarks concerning constrained optimization methods are provided. The reader interested in more details is referred to one of numerical optimization textbooks (e.g., Bertsekas, 1982; Nocedal and Wright, 2006).

Formulation of a constrained optimization problem has been presented in Sec. 2.1. For the purpose of this section, we consider the problem (3.1) assuming N inequality and M equality constraints of the form,

$$g_k(\boldsymbol{x}) \le 0, \quad k = 1, \ldots, N, \tag{3.21}$$

$$h_k(\boldsymbol{x}) = 0, \quad k = 1, \ldots, M. \tag{3.22}$$

The set of points satisfying the above constraints is called a feasible region. Constrained optimization seeks for the minimum of the function f within its feasible domain.

The concepts of the optimum and optimality conditions are more complex than for the unconstrained case. The first-order necessary conditions (so-called Karush–Kuhn–Tucker or KKT conditions; Kuhn and Tucker, 1951) for an optimum of a constrained problem (3.1), (3.21), (3.22) state that — at the local optimum $\boldsymbol{x}^*$ of f — there exist constants $\mu_1, \ldots, \mu_N$, and $\lambda_1, \lambda_2, \ldots, \lambda_M$, such that

$$\nabla f(\boldsymbol{x}^*) + \sum_{k=1}^{N} \mu_k \nabla g_k(\boldsymbol{x}^*) + \sum_{k=1}^{M} \lambda_k \nabla h_k(\boldsymbol{x}^*) = 0 \tag{3.23}$$

and

$$g_k(\boldsymbol{x}^*) \leq 0, \quad \mu_k g_k(\boldsymbol{x}^*) = 0, \quad k = 1, \ldots, N, \tag{3.24}$$

where

$$\mu_k \geq 0, \quad k = 1, \ldots, N, \tag{3.25}$$

and provided that all the functions (both objective and constraints) are continuously differentiable. The function at the left-hand side of (3.25) is referred to as the Lagrangian function, whereas coefficients μ and λ are called Lagrange multipliers.

The two general approaches to handling constrained optimization can be distinguished. Assuming that the unconstrained minimum of f is not in the interior of the feasible region, the constrained minimum must be on its boundary. So-called active set methods search for the optimum along the boundary. Inequality constraints are handled by keeping track of the set of active constraints while moving downhill along the boundary. Another group of methods is based on approaching the optimal solution in an iterative way, either from within the feasible region (so-called interior point methods) or by possible use of infeasible points (however, not by moving along the feasible region boundary). In the latter approaches, the objective function is modified and corresponding unconstrained optimization problems are solved in each iteration.

One of the simplest ways of handling constraints is by means of penalty functions. In this approach, the original problem (3.1) with constraints (3.21), (3.22) is replaced by

$$\arg\min_{\boldsymbol{x}} \phi(\boldsymbol{x},\boldsymbol{\alpha},\boldsymbol{\beta}) = \arg\min_{\boldsymbol{x}} \left\{ f(\boldsymbol{x}) + \sum_{k=1}^{N} \alpha_k \bar{g}_k^2(\boldsymbol{x}) + \sum_{k=1}^{M} \beta_k h_k^2(\boldsymbol{x}) \right\}, \tag{3.26}$$

where $\boldsymbol{\alpha} = [\alpha_1 \ldots \alpha_N], \boldsymbol{\beta} = [\beta_1 \ldots \beta_M], \alpha_k, \beta_k \gg 1$, are penalty factors, and $\bar{g}_k(\boldsymbol{x}) = \max\{0, g_k(\boldsymbol{x})\}$ (inequality only contributes to ϕ if they are active). In practice, one solves a sequence of problems (3.26) for increasing values of penalty factors using the solution of the previous iteration as the starting points for the next one. Figure 3.4 shows an example of penalty-function-based optimization of the function $f(x, y) = x^2 + 4y^2$ with an equality constraint $h_1(x, y) = (x + 2)^2 + (y + 2)^2 - 2$. The unconstrained minimum is $[0\ 0]^T$, whereas the constrained minimum is $[1.389\ -0.725]^T$. The sequence of Figs. 3.4(a)–(f) show the contour plots of the penalized cost function (3.26) for the increasing value of the penalty factor β_1 (by one order of magnitude in each iteration).

The barrier method replaces the original constrained problem (3.1) by the following formulation (for simplicity, only inequality-type constraints are assumed here):

$$\arg\min_{\boldsymbol{x}} \phi(\boldsymbol{x},\mu) = \arg\min_{\boldsymbol{x}} \left\{ f(\boldsymbol{x}) - \mu \sum_{k=1}^{N} \ln(-g_k(\boldsymbol{x})) \right\}, \tag{3.27}$$

where $\mu > 0$ is a barrier parameter. The search starts from a feasible point. A sequence of unconstrained problems is solved with μ decreasing to zero so that the minimum of $\phi(\boldsymbol{x},\mu)$ converges to a solution of f.

Other approaches for constrained optimization are augmented Lagrangian methods (Tapia, 1978). They are a class of algorithms that are similar to penalty methods in the sense that the original constrained problem is solved as a sequence of suitably formulated unconstrained tasks. The unconstrained objective is the Lagrangian of the constrained problem with an additional penalty term (so-called augmentation). Assuming — for the sake of simplicity — equality-only constraints $h_k(\boldsymbol{x}), k = 1, \ldots, M$, the augmented Lagrangian

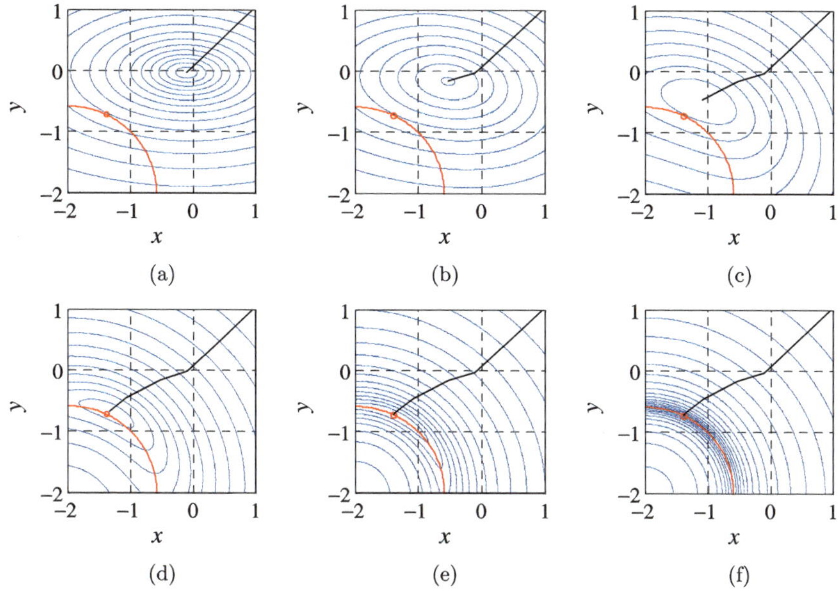

Fig. 3.4. Constrained optimization of the function $f(x, y) = x^2 + 4y^2$; the equality constraint is $h_1(x, y) = (x + 2)^2 + (y + 2)^2 - 2$. The unconstrained minimum is $[0\ 0]^T$, the constrained minimum is $[1.389 - 0.725]^T$. Plots (a)–(f) show the contour plots of the penalized cost function (3.26) for the increasing values of the penalty factor β_1. Constrained minimum marked using a circle.

algorithm generates a new approximation of the constrained solution to (3.1) as

$$\boldsymbol{x}^{(i+1)} = \arg\min_{\boldsymbol{x}} \phi^{(i)}(\boldsymbol{x})$$

$$= \arg\min_{\boldsymbol{x}} \left\{ f(\boldsymbol{x}) + \frac{\mu^{(i)}}{2} \sum_{k=1}^{M} h_k^2(\boldsymbol{x}) - \sum_{k=1}^{M} \lambda_k h_k(\boldsymbol{x}) \right\}. \quad (3.28)$$

The starting point is the previous approximation $\boldsymbol{x}^{(i)}$. In each iteration, the coefficients μ_k are updated as follows: $\lambda_k \leftarrow \lambda_k - \mu^{(i)} h_k(\boldsymbol{x}^{(i+1)})$. The coefficients λ_k are estimates of the Lagrange multipliers (cf. (3.23)), and their accuracy increases as the optimization process progresses. The value of μ is increased in each iteration, however, unlike in the penalty method, it is not necessary to ensure $\mu \rightarrow \infty$.

One of the most popular techniques for solving constrained optimization problems is nowadays sequential quadratic programming (SQP; Han, 1977; Bertsekas, 1982). At each iteration of SQP, the following quadratic programming subproblem is utilized to compute a search direction $\boldsymbol{h}^{(i)}$:

$$\boldsymbol{h}^{(i)} = \arg\min_{\boldsymbol{h}} \left\{ f(\boldsymbol{x}^{(i)}) - \nabla f(\boldsymbol{x}^{(i)})^T \boldsymbol{h} + \frac{1}{2}\boldsymbol{h}^T \boldsymbol{H}(\boldsymbol{x}^{(i)})\boldsymbol{h} \right\}, \quad (3.29)$$

so that

$$g_k(\boldsymbol{x}^{(i)}) + \nabla g_k(\boldsymbol{x}^{(i)})^T \boldsymbol{h} \leq 0, \quad k = 1, \ldots, N \quad (3.30)$$

and

$$h_k(\boldsymbol{x}^{(i)}) + \nabla h_k(\boldsymbol{x}^{(i)})^T \boldsymbol{h} = 0, \quad k = 1, \ldots, M. \quad (3.31)$$

Here, $\boldsymbol{x}^{(i)}$ is a current iteration point, whereas $\boldsymbol{H}$ is a symmetric, positive definite matrix (preferably an approximation of the Hessian of the Lagrangian of f, cf. (3.23)). The new approximation $\boldsymbol{x}^{(i+1)}$ is subsequently obtained using line search.

3.3 Derivative-Free Optimization

Despite strong theoretical foundations of gradient-based techniques and their good performance on smooth objective functions, they cannot be applied to a growing number of practical optimization problems. One of the reasons is that derivative information may not be available or expensive to compute (e.g., through finite differentiation of an expensive objective function). Another even more important issue is that often when gradient-based search does not perform well then the objective function is noisy, which is often the case for objective functions evaluated using computer simulation such as finite-element analysis. One of the sources of the numerical noise is utilization of adaptive meshing techniques in many commercial solvers, which may lead to considerable changes of mesh topology even for very slight changes of the geometry parameters of the structure under analysis. Another reason is termination of the simulation process upon achieving certain (required) resolution of the solution in terms of convergence (usually based on residual energy, etc.).

Yet another issue is that increased complexity of contemporary engineering systems, including antenna structures, often results in non-convex, multi-modal and rugged functional landscapes. For such cases, global optimization may be necessary to find a satisfactory design.

The aforementioned difficulties have significantly increased — in the recent years — popularity of derivative-free optimization techniques in which the search process does not rely on derivative data (Conn *et al.*, 2009). These techniques include local methods, i.e., mostly pattern search algorithms (Kolda *et al.*, 2003) and similar algorithms, global optimization methods such as population-based metaheuristics (Yang, 2010a), as well as surrogate-based optimization (SBO) algorithms (Queipo *et al.*, 2005; Koziel *et al.*, 2011a). Metaheuristics will be covered in Chapter 4, whereas SBO methods will be outlined in Chapter 5. Here, we briefly present the concept of the pattern search (Sec. 3.3.1) as well as the Nelder–Mead algorithm (Sec. 3.3.2).

3.3.1 *Pattern Search*

Pattern search is an important and popular class of optimization algorithms in which the search is restricted to a predefined grid. The objective function is evaluated on a stencil determined by a set of directions that are suitable from geometric or algebraic point of view. The initial grid is subsequently modified (in particular, refined) during the optimization run. In the simplest versions of the pattern search algorithm, the grid may be rectangular, although computationally more efficient versions usually rely on the grids generated using so-called positive spanning directions (Davis, 1954) that ensure local improvement of the objective function (assuming its continuity) for sufficiently refined grid while using only $n + 1$ neighboring points (versus $2n + 1$ points for rectangular grids with n being the design space dimension).

Figure 3.2 shows the conceptual illustration of the pattern search (Kolda *et al.*, 2003). For the sake of simplicity, the optimization process in Fig. 3.5 is restricted to the rectangular grid and thus explores

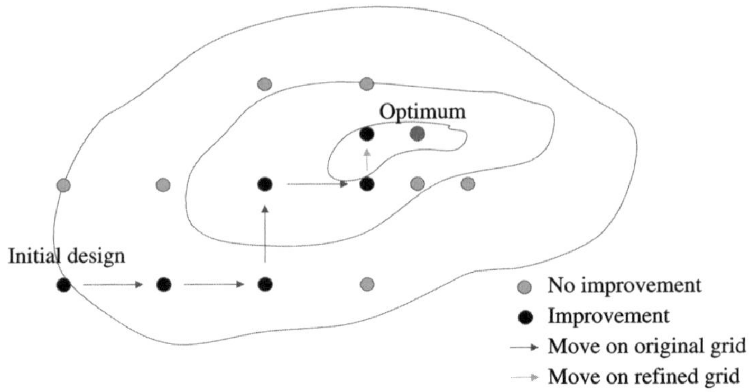

Fig. 3.5. Conceptual illustration of the pattern search process, here using a rectangular grid. The exploratory movements are restricted to the grid around the initial design. In the case of failure of making a successful move, the grid is refined to allow smaller steps (Koziel and Ogurtsov, 2014a). Practical implementations of pattern search routines typically use more sophisticated strategies (e.g., grid-restricted line search).

a grid-restricted vicinity of the current design. If the operations performed on the current grid fail to improve the objective function, the grid is refined to allow smaller steps. A typical termination criterion involves reaching a required resolution (e.g., a user-defined minimum stencil size). There are numerous variations of the pattern search algorithms (see, e.g., Kolda *et al.*, 2003, for a review).

While being relatively robust, pattern search and similar methods usually converge slowly when compared to gradient-based algorithms. On the other hand, they do not need derivative information and, even more importantly, they are relatively immune to numerical noise. Interest in these types of methods has recently been revived, not only due to the development of rigorous convergence theory (e.g., Conn *et al.*, 2009), but also because pattern search may easily benefit from availability of parallel computing resources.

3.3.2 *Nelder–Mead Algorithm*

The Nelder–Mead algorithm is a derivative-free downhill simplex procedure for unconstrained optimization (Nelder and Mead, 1965).

The search process is based on manipulating a finite set of candidate solutions forming a simplex. A simplex or n-simplex is a convex hull of a set of $n+1$ affinely independent points $\boldsymbol{x}^{(i)}$ in a vector space (normally, we consider n-simplex in an n-dimensional space). The set $X = \{\boldsymbol{x}^{(i)}\}_{i=1,\dots,k}$ is affinely independent if the vectors $\boldsymbol{v}^{(i)} = \boldsymbol{x}^{(i+1)} - \boldsymbol{x}^{(1)}, i = 1,\dots,k-1$, are linearly independent. A convex hull $H(X)$ of X is defined as a set of all convex combinations of the points from X, i.e., $H(X) = \{\sum_{i=1,\dots,k} a_i\boldsymbol{x}^{(i)}\colon \boldsymbol{x}^{(i)} \in X,\ a_i \geq 0,\ \sum_{i=1,\dots,k} a_i = 1\}$. 1-simplex, 2-simplex, and 3-simplex are, respectively, a line segment, a triangle, and a tetrahedron (see Fig. 3.6).

The Nelder–Mead algorithm follows the procedure that consists of processing and updating the simplex vertices based on the corresponding objective function values (see also Fig. 3.7):

- Order the vertices so that $f(\boldsymbol{x}^{(1)}) \leq f(\boldsymbol{x}^{(2)}) \leq \cdots f(\boldsymbol{x}^{(n+1)})$. An auxiliary point $\boldsymbol{x}^{(0)}$ is then defined to be center of gravity of all vertices but $\boldsymbol{x}^{(n+1)}$;
- Compute a *reflection* $\boldsymbol{x}^r = (1+\alpha)\boldsymbol{x}^{(0)} - \alpha\boldsymbol{x}^{(n+1)}$ (a typical value of α is 1); If $f(\boldsymbol{x}^{(1)}) \leq f(\boldsymbol{x}^r), < f(\boldsymbol{x}^{(n)})$ reject $\boldsymbol{x}^{(n+1)}$ and update the simplex using $\boldsymbol{x}^r$;
- If $f(\boldsymbol{x}^r) < f(\boldsymbol{x}^{(1)})$, compute an *expansion* $\boldsymbol{x}^e = \rho\boldsymbol{x}^r + (1-\rho)\boldsymbol{x}^{(0)}$ (a typical value of ρ is 2); If $f(\boldsymbol{x}^e), < f(\boldsymbol{x}^r)$ reject $\boldsymbol{x}^{(n+1)}$ and update the simplex using $\boldsymbol{x}^e$;
- If $f(\boldsymbol{x}^{(n)}) \leq f(\boldsymbol{x}^r) < f(\boldsymbol{x}^{(n+1)})$, compute a *contraction* $\boldsymbol{x}^c = (1+\gamma)\boldsymbol{x}^{(0)} - \gamma\boldsymbol{x}^{(n+1)}$ (a typical value of γ is 0.5); If $f(\boldsymbol{x}^c) \leq f(\boldsymbol{x}^r)$ reject $\boldsymbol{x}^{(n+1)}$ and update the simplex using $\boldsymbol{x}^c$;
- If $f(\boldsymbol{x}^c) > f(\boldsymbol{x}^r)$ or $f(\boldsymbol{x}^r) \geq f(\boldsymbol{x}^{(n+1)})$ shrink the simplex: $\boldsymbol{x}^{(i)} = \boldsymbol{x}^{(1)} + \sigma(\boldsymbol{x}^{(i)} - \boldsymbol{x}^{(1)})$, $i = 1,\dots,n+1$ (a typical value of σ is 0.5).

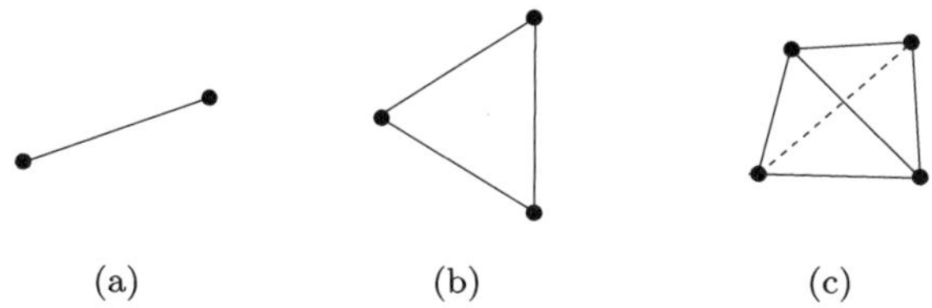

Fig. 3.6. The concept of simplex: (a) 1-simplex, (b) 2-simplex, (c) 3-simplex.

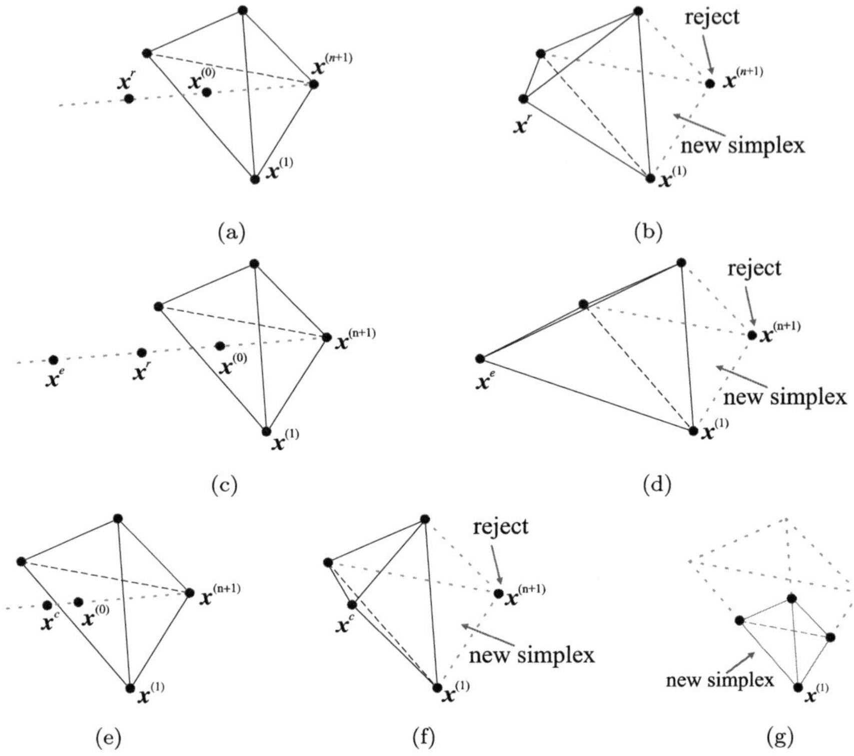

Fig. 3.7. Operations of the Nelder–Mead algorithm: (a) reflection: $\boldsymbol{x}^r = (1+\alpha)\boldsymbol{x}^{(0)} - \alpha\boldsymbol{x}^{(n+1)}$, (b) acceptance of the reflection point $\boldsymbol{x}^r$ if $f(\boldsymbol{x}^{(1)}) \leq f(\boldsymbol{x}^r) < f(\boldsymbol{x}^{(n)})$, (c) expansion: $\boldsymbol{x}^e = \rho\boldsymbol{x}^r + (1-\rho)\boldsymbol{x}^{(0)}$, (d) acceptance of the expansion point $\boldsymbol{x}^e$ if $f(\boldsymbol{x}^e) < f(\boldsymbol{x}^r)$, (e) contraction: $\boldsymbol{x}^c = (1+\gamma)\boldsymbol{x}^{(0)} - \gamma\boldsymbol{x}^{(n+1)}$, (f) acceptance of the contraction point $\boldsymbol{x}^c$ if $f(\boldsymbol{x}^c) \leq f(\boldsymbol{x}^r)$, (g) shrinking of the simplex if none of the above operations led to the objective function improvement.

The above procedure describes one iteration of the algorithm, which is continued until the termination condition is met (typically, reduction of the simplex size below a user-defined threshold). The Nelder–Mead algorithm is very popular. It is characterized by a relatively low computational cost (only a few objective function evaluations per iteration), however, its convergence is rather slow.

3.4 Summary

Optimization techniques described in this chapter and direct handling of EM-simulation model may be considered as practical solutions for solving certain engineering problems including those related to antenna design. Clearly, a choice of a particular algorithm is problem dependent. For example, gradient-based optimizers are more suitable for optimizing smooth objective functions, particularly if it is possible to obtain cheap sensitivity information (e.g., through adjoints; Georgieva *et al.*, 2002; Koziel and Bekasiewicz, 2015a).

Derivative-free methods are suitable for handling analytically intractable (noisy, discontinuous) functions. Note that the methods considered in this chapter are local ones so that they are only suitable for finding the local minimum, typically the closest to the initial design. Global optimization can be realized using, among others, metaheuristic algorithms, described in Chapter 4.

Despite their usefulness, direct methods have an important disadvantage related to the computational cost of the optimization process. Conventional optimization algorithms (e.g., gradient-based schemes with numerical derivatives) require tens, hundreds or even thousands of objective function calls per run, depending on the number of design variables. As mentioned in Chapter 2, high-fidelity EM antenna models tend to be very expensive, up to a few hours per simulation, depending on the structure complexity. Consequently, direct solving of the problem (3.1) by coupling the optimization algorithm with the EM solver as the objective function evaluator is usually impractical. Also, availability of massive computing resources is not always translated into computational speedup due to a growing demand for simulation accuracy, both because of utilizing more accurate models (e.g., with finer discretization of the structure under consideration) and growing complexity of the systems (e.g., including the antenna environment into the simulation process; Kempel, 2007). A separate issue is numerical noise, which is always present in

EM-based objective functions and might be a problem particularly for gradient-based methods.

Surrogate-assisted optimization is one of the most promising ways of handling computationally expensive EM-driven design problems. The key concept is to replace direct optimization of the high-fidelity EM model by optimization of its cheap and analytically tractable representation referred to as a surrogate model. Introduction to surrogate-based methods is provided in Chapter 5. Also, the multi-objective antenna design procedures — the main topic of this book — are heavily based on surrogate modeling.

Chapter 4

Global Optimization Using Population-Based Metaheuristics

Optimization methods outlined in Chapter 3 are local ones, i.e., they normally allow for finding an optimum that is located in the vicinity of the initial solution. Unfortunately, in many practical problems, objective functions with multiple optima have to be handled. Furthermore, the functional landscape of the problem at hand is often unknown in terms of the nonlinearity of the objective function, importance of particular variables, and also the number and the location of the optima. At the same time, estimating a reasonably good starting point is often very difficult. In all these cases, utilization of local methods usually leads to unsatisfactory results and global search may be necessary.

Global optimization is a broad subject with numerous methods and algorithms developed over the last decades (Goldberg, 1989; Horst *et al.*, 2000; Horst and Tuy, 1990; Hansen and Walster, 2003). The algorithms for global optimization can be roughly categorized into deterministic (Horst and Tuy, 1996; Zhigljavsky and Žilinskas, 2008) and stochastic ones (Goldberg, 1989; Talbi, 2009). The most successful deterministic methods include cutting plane methods, branch and bound methods, and interval methods (Luenberger, 2003; Jaulin *et al.*, 2001; Hansen and Walster, 2003). Stochastic methods include simulated annealing (van Laarhoven and Aarts, 1989), Monte Carlo methods (Kroese *et al.*, 2011), as well as population-based approaches (Goldberg, 1989; Talbi, 2009).

In this chapter, we focus on population-based metaheuristics as the most popular solution approaches to global optimization nowadays (Talbi, 2009; Yang, 2010). Their popularity is a result of several attractive features, including simple implementation, versatility, ability to handle "difficult" functions (discontinuous, noisy), as well as easy parallelization. From the point of view of this book, population-based methods are of interest because some of their multi-objective versions are utilized as components of surrogate-assisted algorithms presented in Chapter 7 and applied for antenna design.

In the following sections, we briefly introduce population-based metaheuristics and outline the most important classes of algorithms, including evolution strategies (ES), genetic and evolutionary algorithms (GAs and EAs), particle swarm optimizers (PSO), differential evolutions (DEs), as well as mention a few recent techniques such as firefly algorithm, cuckoo search, etc. The chapter is concluded with remarks concerning general characteristics of population-based algorithms and their applicability for design optimization of antenna structures.

4.1 Introduction to Population-Based Metaheuristics

Metaheuristics belong to a group of derivative-free global optimization methods. They have been developed to handle problems that are challenging for conventional algorithms (including gradient-based ones), and involve non-differentiable, discontinuous or noisy objective functions, as well as the presence of multiple local optima. Vast majority of metaheuristic algorithms are based on observation or mimicking of natural processes (e.g., biological or social systems). Most of these methods process sets (or populations) of potential solutions to the optimization problem at hand (often referred to as individuals or agents). In the course of the optimization process, the method-specific interactions between individuals as well as selection methods are applied in order to both identify and explore promising regions of the design space. Some older metaheuristic algorithms (e.g., simulated annealing; Kirkpatrick *et al.*, 1983; van Laarhoven and Aarts, 1989) process single solution rather than a set of individuals.

Perhaps, the most popular type of metaheuristics are GAs (Goldberg, 1989) and EAs (Back *et al.*, 2000). Other methods include ES (Beyer, 2001), PSO (Kennedy *et al.*, 2001), DE (Price *et al.*, 2005), and, more recently, firefly algorithm, cuckoo search, bat algorithm and others (Yang, 2010a, 2014).

We will use the symbol P to denote a set of individuals (population) processed by the algorithm. The typical flow of the

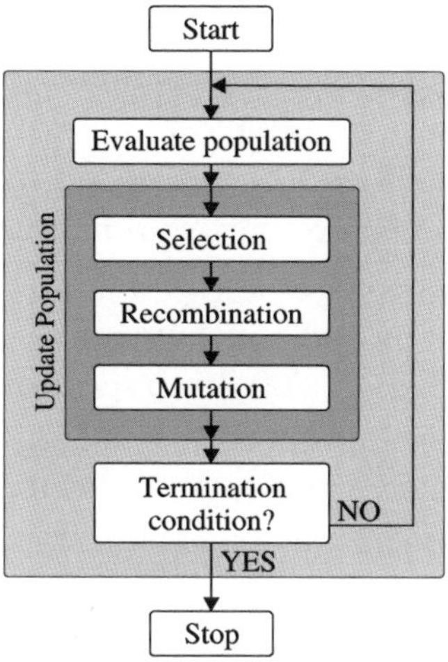

Fig. 4.1. Generic flow of a population-based metaheuristic algorithm.

metaheuristic algorithm is as follows (see also Fig. 4.1):

1. Initialize population P;
2. Evaluate population P;
3. Select parent individuals S from P;
4. Create a new population P from the parent individuals S by applying recombination operators;
5. Introduce local perturbations in individuals of P by applying mutation operators;
6. If termination condition is not satisfied go to 2;
7. END.

In most algorithms, the population is initialized randomly. In Step 2, the individuals in the population are evaluated, and the corresponding values of the objective function determine their "fitness". In many algorithms, in particular GAs and EAs, an important step is the selection of the subset of individuals to form a new population. Depending on the algorithm, the selection process can be deterministic (i.e., by picking up the best individuals, ES) or partially random (probability of being selected depends on the fitness value, but there is a chance to survive even for poor individuals, EAs). A typical example of a popular selection procedure is a tournament selection, where out of the set of candidates randomly picked from the current population, the one with the smallest value of the objective function (or the highest "fitness") will be picked up for the new population. This kind of procedure favors better individuals, yet gives a chance for worse ones to be selected.

Selection pressure (e.g., in tournament selection adjusted by the candidate set size; Tan *et al.*, 2005) is important because it allows one to control the convergence properties of the algorithm. It is essentially the amount of preference given to the best individuals to survive; if it is too low, the algorithm may converge slowly and the search may turn into a random search; if it is too high, the best individuals quickly take over the population and one may face a premature convergence which usually negatively affects the algorithm performance (the search concludes before a global optimum can be identified).

In certain algorithms, such as PSO or DE, no selection is used (i.e., individuals are modified from iteration to iteration but never die out).

Modification of individuals is performed by the two major types of operators: exploratory ones (e.g., crossover in EAa or ES) and exploitative ones (e.g., mutation in GAs). The purpose of exploratory operators is to combine information contained in the parent individuals to create an offspring. An example of such an operator used by EAs with natural (floating point) representation is an arithmetic crossover, where a new individual $\boldsymbol{c}$ can be created as a convex combination of the parents $\boldsymbol{p}_1$ and $\boldsymbol{p}_2$, i.e., $\boldsymbol{c} = \alpha \boldsymbol{p}_1 + (1 - \alpha)\boldsymbol{p}_2$, where $0 < \alpha < 1$ is a random number (Eiben and Smith, 2003). These types of operators allow making large "steps" in the design space and, therefore, allow the algorithm to explore the new and promising regions (Michalewicz, 1996).

Exploitative operators introduce small perturbations (e.g., $\boldsymbol{p} \leftarrow \boldsymbol{p} + \boldsymbol{r}$, where $\boldsymbol{r}$ is a random vector selected according to a normal probability distribution with zero mean and certain, problem-dependent variance). These operators allow exploitation of a given region of the design space improving local search properties of the algorithm (Deb, 2001; Coello Coello *et al.*, 2007). In some algorithms, e.g., PSO, the difference between both types of operators is not as clear: modification of the individual may be based on the best solution found so far by that given individual as well as the best solution found by the entire population (Clerc, 2006; Kennedy *et al.*, 2001). On the other hand, the operators used by the algorithms such as PSO should be interpreted as exchange of information between individuals that affect their behavior rather than the individuals themselves.

In Secs. 4.2–4.6, we briefly outline several classes of metaheuristic algorithms, specifically, ES, GAs, EAs, PSO and DE. Section 4.7 characterizes a few more recent algorithms, whereas Sec. 4.8 summarizes the chapter and provides a short discussion of population-based methods in the context of design optimization of antenna structures.

4.2 Evolution Strategies

ES are chronologically the first population-based search methods. They were developed in mid-1960s in Germany by I. Rechenberg and H.P. Schwefel and applied for solving continuous optimization problems, specifically, aerodynamic shape optimization (Rechenberg, 1965; Schwefel, 1968).

There are several variations of ES available (Hansen *et al.*, 2015). The initial versions were simple schemes using only mutation operators. Later on, recombination operators were introduced to allow exchange of information between individuals (Schwefel and Rudolph, 1995). More recent versions of ES utilize covariance matrix adaptation (CMA–ES) to shape the mutation distribution and adjust it to a particular landscape of the optimized cost function (Hansen and Kern, 2004). A review of various ES algorithms can be found in the work of (Hansen *et al.*, 2015).

ES process a population of μ individuals so that, in each iteration, λ offsprings are created from the set of parent individuals using problem-specific recombination operators. The new set of parent individuals is chosen either from offsprings (so-called (μ, λ)-ES) or from offsprings and parents (so-called $(\mu + \lambda)$-ES). Each individual

initialize population $P_\mu = \{(\boldsymbol{x}^{(1)},\boldsymbol{s}^{(1)}),\ldots,(\boldsymbol{x}^{(\mu)},\boldsymbol{s}^{(\mu)})\}$;
while ~termination_condition
  select a parent population $P_\lambda = \{(\boldsymbol{x}^{(1)},\boldsymbol{s}^{(1)}),\ldots,(\boldsymbol{x}^{(\lambda)},\boldsymbol{s}^{(\lambda)})\}$ from P_μ;
  for $j = 1:\lambda$
    mutate $\boldsymbol{s}^{(j)}$;
    mutate $\boldsymbol{x}^{(j)}$ using updated value of $\boldsymbol{s}^{(j)}$;
  end
  select a new population P_μ from P_λ ((μ,λ)-selection) or from
    $P_\lambda \cup P_\lambda$ ((μ+λ)-selection);
end

Fig. 4.2. Pseudocode of the basic version of the ES algorithm.

contains both the solution data (design parameters) and the set of strategy parameters that determine its further mutation process.

The basic flow of the ES algorithm is shown in Fig. 4.2. The parent population is selected randomly. The new population is chosen using deterministic truncation selection based on objective function values of individuals. The mutation operators for unconstrained real-valued search space are defined as

$$(\boldsymbol{x}, \boldsymbol{s}) \leftarrow \begin{cases} \boldsymbol{s} \leftarrow \boldsymbol{s} \cdot e^{\tau N(0,1)}, \\ \boldsymbol{x} \leftarrow \boldsymbol{x} + \boldsymbol{s} \cdot N(0, \boldsymbol{I}), \end{cases} \tag{4.1}$$

where $N(0,1)$ and $N(0,\boldsymbol{I})$ are (0,1) normally distributed random scalars and vectors ($\boldsymbol{I}$ stands for the identity matrix of the size equal to the number of design variables n), respectively, and τ is the rate of self-adaptation (learning parameters). Parameter $\boldsymbol{s}$ controls the strength of the design parameter mutation. It is suggested (Schwefel, 1988) that τ is proportional to $n^{-1/2}$. In more advanced implementations of ES, the identity matrix $\boldsymbol{I}$ in $N(0, \boldsymbol{I})$ may be replaced by a general covariance matrix with evolving entries (Hansen *et al.*, 2015), which allows for taking into account correlations between variables, thus leading to more efficient search for the optimum.

Later versions of ES introduced recombination (see Fig. 4.3) to allow exchange of information between individuals (Beyer, 2001).

```
initialize population P_μ = {(x^(1),s^(1)),...,(x^(μ),s^(μ))};
while ~termination_condition
  for j = 1:λ
    select a parent set P_ρ = {(y^(1),t^(1)),...,(y^(ρ),r^(ρ))} from P_μ;
    create (x^(j),s^(j)) by recombining (y^(1),t^(1)) to (y^(ρ),r^(ρ));
    mutate s^(j);
    mutate x^(j) using updated value of s^(j);
  end
  select a new population P_μ from P_λ ((μ,λ)-selection) or from
    P_λ ∪ P_λ ((μ+λ)-selection);
end
```

Fig. 4.3. Pseudocode of the ES algorithm with recombination.

The two most typical recombination operators used by ES are the arithmetical averaging,

$$\boldsymbol{x} = \frac{1}{\rho}\sum_{j=1}^{\rho} \boldsymbol{y}^{(j)}, \tag{4.2}$$

where $\boldsymbol{y}^{(j)}$ are parent individuals, and coordinate-wise random transfer

$$\boldsymbol{x} = [x_1 \dots x_n]^T = \left[y_1^{(r_1)} \dots y_n^{(r_n)}\right]^T, \tag{4.3}$$

where r_1 to r_n are randomly selected from the set $\{1, 2, \dots, \rho\}$.

Some recommendations and rules of thumb concerning practical usage of ES include (Beyer and Schwefel, 2002) the following:

- Typical μ/λ ratios in (μ, λ)-selection in continuous search spaces are from 1/7 to 1/2;
- Using $(\mu + \lambda)$-selection in conjunction with variation operators (allowing reaching any point of the search space) guarantee stochastic convergence to a global solution;
- Evolution is usually performed on a phenotypic level;
- Mutation should respect reachability condition (each point of the search space should be reachable in a finite number of steps).

Recombination is applied whenever possible and useful. The main goal of recombination is conservation of the common components of the parents.

4.3 Genetic Algorithms

GAs have been developed in late 1970s, mostly by Holland (1975), De Jong (1975) and Goldberg (1989). Initially, GAs processed populations of individuals represented as binary strings. Nowadays, integer or floating point representation is preferred. GAs traditionally emphasize the role of crossover operator. They are considered to be good heuristics for combinatorial problems, although they may be used for continuous optimization.

4.3.1 *Algorithm Structure and Representation*

An algorithmic flow of GAs is very similar to a generic structure of population-based metaheuristics shown in Fig. 4.1. Typically, GAs process individuals that are represented in certain internal format (referred to as *genotype*). Recombination and mutation operations are performed on a genotype; however, assessment of individual's fitness is performed on decoded individuals (referred to as *fenotype*). Traditional GAs utilize bit-string representation (Goldberg, 1989). Gray coding is often utilized instead of standard binary coding because small changes of the genotype imply small changes of the fenotype with the Gray coding (for standard coding, the fenotype changes depend on the position of the bit in the string). Currently, natural representation is preferred, e.g., floating point representation is used for continuous optimization problems, integer representation is used for combinatorial optimization (e.g., traveling salesman problem), etc. In many cases, representation is tailored to the problem in order to facilitate genetic operations or ensure feasibility of individuals (Michalewicz, 1996; Caorsi *et al.*, 2000; Lai and Jeng, 2006; Ding *et al.*, 2008).

4.3.2 *Crossover*

Crossover is a fundamental operator in GAs. It is designed to exchange information between individuals. Figure 4.4 shows several variations of two-parent crossover for bit-string representation. Crossover is explorative, i.e., it makes large steps into an area

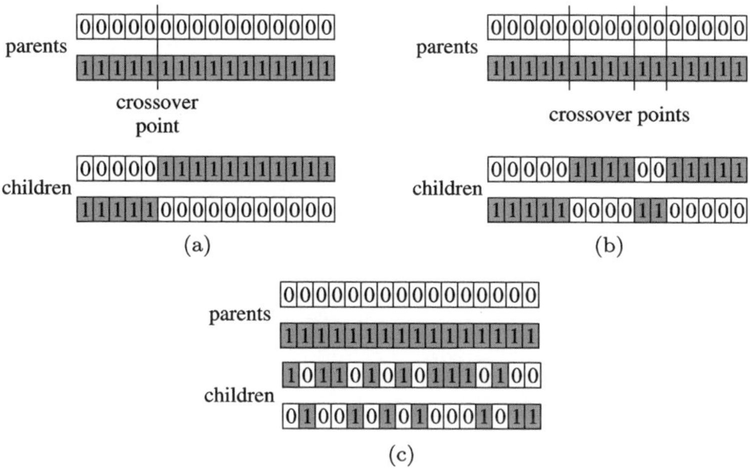

Fig. 4.4. Two-parent crossover for bit-string GAs: (a) one-point crossover, (b) multi-point crossover and (c) uniform crossover (bits are changed randomly).

somewhere in between the areas occupied by the parents. Consequently, it helps discovering the promising area in the search space. Typical crossover rate, p_c, is high (e.g., 0.8), so that most of the parents exchange their data. Crossover may be generalized to involve more parents; however, no significant advantages of this approach have been reported (Coello Coello *et al.*, 2007). An important remark is that crossover does not produce new genetic information and, therefore, cannot be used as the only operator in a GA.

Crossover operators for floating point representation are arranged differently. Let the parents be denoted as $\boldsymbol{x} = [x_1 \ldots x_n]^T$ and $\boldsymbol{y} = [y_1 \ldots y_n]^T$, whereas the child be denoted as $\boldsymbol{z} = [z_1 \ldots z_n]^T$. The following three basic types of crossover operators can be distinguished (Eiben and Smith, 2003):

- Discrete: $z_i = x_i$ or y_i (random selection);
- Intermediate: $z_i = ax_i + (1 - a)y_i$ with $0 \leq a \leq 1$ (a selected randomly);
- Arithmetic: $\boldsymbol{z} = a\boldsymbol{x} + (1 - a)\boldsymbol{y}$ with $0 \leq a \leq 1$ (a selected randomly).

Variations and mixtures of the above operators are also used in practical implementations of GAs.

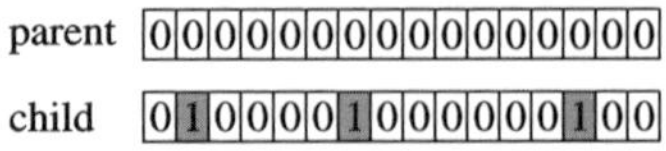

Fig. 4.5. Mutation operator for bit-string GAs. Each gene is flipped independently with a probability p_m.

4.3.3 *Mutation*

Mutation introduces small, random changes into the genotype. The mutation rate p_m is normally small, typically between $1/N$ and $1/L$ (where N and L are the population size and the chromosome length, respectively). Bit-flip mutation operator for traditional GAs is shown in Fig. 4.5. Mutation is an exploitative (local) operator, therefore it facilitates optimization within a promising area using available information encoded in the chromosome. Mutation produces new genetic information and therefore, in principle, may be used as the only operator in a GA. However, it is recommended to use both mutation and crossover in general (Goldberg, 1989).

The mutation operator for floating point representation (we assume that the individual is denoted as $\boldsymbol{x} = [x_1 \ldots x_n]^T$) can be written as $x_i \to x_i' = x_i + \Delta x_i$, where Δx_i is a random deviation that may be drawn with uniform or non-uniform (e.g., normal) distribution; in both cases, x_i' is eventually curtailed to the range if x_i (Eiben and Smith, 2003). Non-uniform distributions that promote smaller changes are typically preferred, e.g., (Michalewicz, 1996),

$$\Delta x_i = \begin{cases} (x_{i.\max} - x_i) \cdot (2(r - 0.5))^\beta & \text{if } r > 0.5, \\ (x_{i.\min} - x_i) \cdot (2(0.5 - r))^\beta & \text{otherwise}, \end{cases} \tag{4.4}$$

where $r \in [0, 1]$ is a random number; β is normally larger than 1 (e.g., 3) and may increase in later stages of the algorithm run.

4.3.4 *Selection*

An important component of a GA is a selection scheme. Selection is a process of choosing individuals that will become parents to create the new population. Selection is based on the fitness value of the individuals (normally, better individuals are given higher value of fitness; for minimization of the objective function f, the fitness of the individual $\boldsymbol{x}$ could be $-f(\boldsymbol{x})$). In GAs, selection is partially stochastic. It promotes better individuals, however, it should contain an element of randomness to avoid premature convergence of the algorithm. One of the first selection schemes used in GAs was a roulette wheel selection (Coello Coello *et al.*, 2007), where the probability of an individual to be selected was proportional to its fitness value. Obvious disadvantages of this scheme (e.g., a risk of premature convergence due to the population being taken over by a single highly fit individual, or reduction of the selection pressure when all fitness values are on a similar level) led to the development of better schemes.

One of the most popular selection schemes is a tournament selection which works as follows (Bäck, 1996):

1. Randomly choose the subset of k individuals from the current population (k is a tournament size).
2. Select an individual with the highest fitness value within the subset.
3. Repeat Steps 1 and 2 N times in order to create the parent set.

Tournament selection is free from problems of roulette wheel selection. The selection pressure can be adjusted by changing the tournament size (from pure random selection for $k = 1$ to fully deterministic selection for $k = N$; the latter would lead to instant convergence of the algorithm).

4.3.5 *Elitism*

Elitism is a technique frequently used in GAs and other types of metaheuristics (Zitzler and Thiele, 1999; Lim and Isa, 2014; Yang, 2014) to avoid losing the best individual(s) found so far: the set of best individuals (typically one) is stored and deterministically inserted into the new population (cf. Fig. 4.6). Various versions of this technique have been considered; for example, there are algorithms where individuals may survive within the population up to a certain number of iterations (Tan *et al.*, 2005).

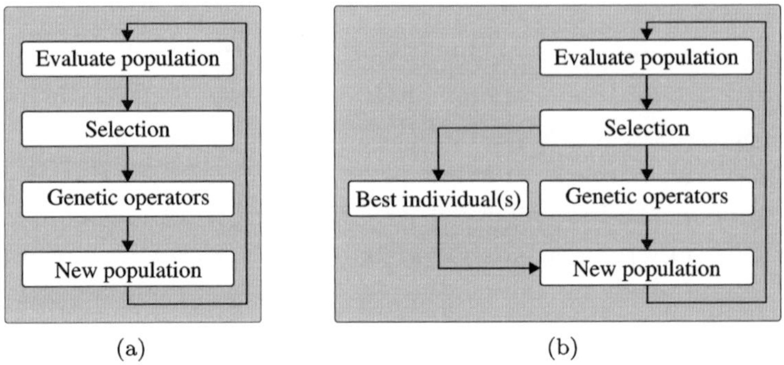

Fig. 4.6. Flowchart of a GA without (a) and with elitism (b).

4.3.6 *Selected Topics*

Here, we briefly discuss various population models utilized in GAs, as well as the issue of controlling selection pressure.

There are three basic approaches to replacing the old population by a new one (Coello Coello *et al.*, 2007):

1. Generational models: each individual survives for exactly one generation, and the entire set of parents is replaced by the offspring.
2. Steady-State models: one offspring is generated per iteration, and one member of the population is replaced.
3. Generation gap: the fraction of the population is replaced varying from 1.0 (as in generational models) to $1/N$ (as in steady-state models).

The most popular model is a generational one, however, other models may exhibit certain advantages. One of these is, lower computational cost of the algorithm iteration for the generation gap models and, obviously, the steady-state one.

As mentioned before, selection pressure is the amount of preference given to the better individuals to survive. The major controlling factor here is the selection scheme used in an algorithm. Selection pressure is critical for controlling the operation and performance of GAs. In particular, selection pressure that is too high may lead to premature convergence, i.e., the situation when all individuals become (almost) identical before the search space has been properly explored.

Production of a new genetic material is a mechanism opposite to selection pressure, and it is mainly dependent on the mutation rate. Normally, we want to keep a balance between the selection pressure and production of a new genetic material so that the premature convergence is avoided, but the evolution is not a random search either. Practical methods for controlling the balance between selection pressure and the production of a new genetic material are based on monitoring diversity of the population and adjusting mutation rate accordingly (self-adaptation) (Hesser and Männer, 1991). Diversity of the population may be measured using standard deviation of individuals (details depend on a particular representation used in the algorithm). Mutation rate can be reduced at the final stages of the algorithm run in order to improve exploitation at the expense of exploration and, therefore, obtain more precise localization of the optimal solution.

4.4 Evolutionary Algorithms

GAs are often considered to be a special case of a broader class of methods called EAs. As a matter of fact, GAs using floating-point representation are considered by some to be EAs but not GAs.

In general, the features that distinguish EAs from other types of optimization algorithms are the following:

- The use of population;
- The use of stochastic search operators;
- The use of stochastic selection.

There is a little theory explaining why EAs work. Early, semi-heuristic results for bit-string-based GAs (so-called schemata theorem by Holland, 1975) suggesting that GAs work by discovering, exploiting and combining "building blocks" (groups of closely interacting genes), have been later criticized (partially because of success of algorithms using natural representations; Eiben and Smith, 2003; Liu *et al.*, 2014a; Ding and Wang, 2013; Alander *et al.*, 2004; Yeung *et al.*, 2008). The results obtained using Markov Chain analysis (where GAs are treated as stochastic processes; Horn, 1993; David,

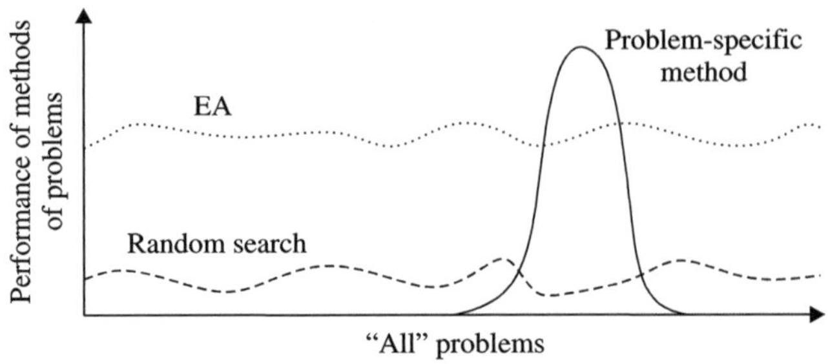

Fig. 4.7. Robustness of EAs versus pure random search and problem-specific methods.

1994; Suzuki, 1995) consider algorithm convergence to a global solution in a probabilistic sense; however, practical significance of such results is minor.

EAs are considered robust problem solvers in the following sense (cf. Fig. 4.7):

- They do not perform as good as problem-specific methods for problems for which the latter methods were designed;
- They perform better than a random search "on average";
- They perform better than problem-specific methods "on average".

These features are generally common for other types of population-based metaheuristics as well (cf. Secs. 4.5–4.7).

Engineering practice indicates that "standard" off-the-shelf implementation of an EA may not perform well. It is recommended that EAs should be individually designed for solving a given problem. In particular, adding a problem-specific knowledge to the algorithm is essential to obtain good performance (e.g., by using suitable representation and tailored operators).

4.5 Particle Swarm Optimization

PSO belongs to the most popular metaheuristic algorithms nowadays. PSO is based on mimicking the swarm behavior such as fish or bird schooling in nature (Kennedy, 1997; Kennedy *et al.*, 2001). It was first developed in mid-1990s and since then, many of its variations have been applied to virtually every area of optimization, computational intelligence, and engineering design (Lizzi *et al.*, 2008; Poli *et al.*, 2014; Pavlidis *et al.*, 2005; Rocca *et al.*, 2012; Manica *et al.*, 2009).

In a PSO algorithm, a set of individuals (traditionally referred to as swarm) are moving around the design space and interact with other individuals by means of social influence and social learning. The swarm is typically modeled by particles in a multi-dimensional space that are represented by their positions and velocities. Individual particles communicate good positions to each other and adjust their own velocity and position based on this data. The movement of a swarming particle consists of two major components: a stochastic component and a deterministic component. Each particle is attracted toward the position of the current global best $\boldsymbol{g}$ and its own best location $\boldsymbol{x}_i^*$ found during the optimization run. At the same time, it has a tendency to move randomly.

Let $\boldsymbol{x}_i$ and $\boldsymbol{v}_i$ denote the position of the velocity vectors of the ith particle. Both vectors are updated according to the following rules:

$$\boldsymbol{v}_i \leftarrow \chi[\boldsymbol{v}_i + c_1\boldsymbol{r}_1 \cdot (\boldsymbol{x}_i^* - \boldsymbol{x}_i) + c_2\boldsymbol{r}_2 \cdot (\boldsymbol{g} - \boldsymbol{x}_i)], \tag{4.5}$$

$$\boldsymbol{x}_i \leftarrow \boldsymbol{x}_i + \boldsymbol{v}_i, \tag{4.6}$$

where $\boldsymbol{r}_1$ and $\boldsymbol{r}_2$ are vectors with components being uniformly distributed random numbers between 0 and 1; $\cdot$ denotes component-wise multiplication. The control parameter χ is typically equal to 0.7298 (Kennedy *et al.*, 2001); c_1 and c_2 are acceleration constants determining how much the particle is directed towards good positions (they represent a "cognitive" and a "social" components, respectively); usually, $c_1, c_2 \approx 2$. The sum of $c_1 + c_2$ should not be smaller than 4.0; as the values of these parameters increase, χ gets smaller as does the damping effect.

The initial locations of all particles should distribute relatively uniformly so that they can sample over most regions, which is especially important for multi-modal problems. The initial velocity of a particle can be taken as zero.

There are many different variants and extensions of the standard PSO algorithm (Clerc and Kennedy, 2002; Kennedy *et al.*, 2001; Fong *et al.*, 2016). For example, various modifications concerning the parameter χ (also called the inertia weight) have been considered to better control the convergence rate of PSO (Bansal *et al.*, 2011; Nickabadi *et al.*, 2011). There were also multiple attempts to combine PSO with other metaheuristics such as GAs (Kao and Zahara, 2008), simulated annealing (Da and Xiurun, 2005), or ant colony optimization (Shelokar *et al.*, 2007) and also with local search methods (Noel, 2012).

4.6 Differential Evolution

DE was developed in late 1990s by Storn and Price (1997). It is a vector-based EA, and can be considered as a further development to GAs. It is a derivative-free stochastic search algorithm with self-organizing tendency. The design parameters are represented as n-dimensional vectors (also called agents). The new agents are allocated in the search space by combining the positions of other existing agents. More specifically, an intermediate agent is generated from two agents randomly chosen from the current population. This temporary agent is then mixed with a predetermined target agent. The new agent is accepted for the next generation if and only if it yields reduction in objective function.

The basic DE algorithm uses a random initialization. A new agent $\boldsymbol{y} = [y_1 \dots y_n]$ is created from the exisiting one $\boldsymbol{x} = [x_1 \dots x_n]$ as follows:

1. Three agents $\boldsymbol{a} = [a_1 \dots a_n]$, $\boldsymbol{b} = [b_1 \dots b_n]$ and $\boldsymbol{c} = [c_1 \dots c_n]$ are randomly extracted from the population ($\boldsymbol{a} \neq \boldsymbol{b} \neq \boldsymbol{c} \neq \boldsymbol{x}$);
2. A position index $p \in \{1, \dots, N\}$ is determined randomly (N is the population size);
3. The new agent $\boldsymbol{y}$ is determined as follows (separately for each of its components $i = 1, \dots, n$):

   a. select a random number $r_i \in (0, 1)$ with uniform probability distribution,

b. if $i = p$ or $r_i < CR$ let $y_i = a_i + F(b_i - c_i)$, otherwise let $y_i = x_i$; here, $0 \leq F \leq 2$ is the differential weight and $0 \leq CR \leq 1$ is the crossover probability, both defined by the user,
c. if $f(\boldsymbol{y}) < f(\boldsymbol{x})$, then replace $\boldsymbol{x}$ by $\boldsymbol{y}$; otherwise reject $\boldsymbol{y}$ and keep $\boldsymbol{x}$.

The above DE scheme is often referred to as *DE/rand/1/bin*, meaning that the donor vector $\boldsymbol{a}$ is selected randomly ("rand"), there is one pair of vectors selected to calculate the mutation differential ("1"), and that the recombination is binomial ("bin"). Some of DE variations, using the terminology explained above are as follows (Mezura-Montes *et al.*, 2006a, 2010):

- *DE/rand/p/bin* — explained above; p denotes the number of pairs of vectors used to calculate the mutation differential (in particular, p can be equal to 1);
- *DE/rand/p/exp* — as above but with exponential crossover (Noman and Iba, 2005);
- *DE/best/p/bin, DE/best/p/exp* — a donor vector is the best solution found so far in the population;
- *DE/current-to-rand/p/bin, DE/current-to-best/p/exp* — variants with arithmetic recombination;
- *DE/current-to-rand/p/bin* — a variant with combined arithmetic-discrete recombination.

It should be emphasized that although DE resembles some other stochastic optimization techniques, unlike traditional EAs, DE perturbs the solutions in the current generation vectors with scaled differences of two randomly selected agents. As a consequence, no separate probability distribution has to be used, and thus the scheme presents some degree of self-organization. Additionally, DE is simple to implement, uses very few control parameters, and has been observed to perform satisfactorily in multi-modal optimization problems (Chakraborty, 2008; Rocca *et al.*, 2011a).

Many research studies have been focused on the choice of the control parameters F and CR as well as the modification of the updating scheme of the step 3b. Interested reader is referred to the work of (Price *et al.*, 2005) for details.

4.7 Other Methods

Metaheuristic optimization has been constantly growing over the last three decades. With the major methods and algorithms developed before 1980s (e.g., ES, GAs, EAs) and 1990s (PSO, DE, Ant Colony optimization), the number of new algorithms as well as improvement of the existing ones proposed in the literature has been accelerating. Examples include bee algorithm (Yang, 2005), bacteria foraging algorithm (Gazi and Passino, 2004), harmony search (Geem *et al.*, 2001), firefly algorithm (Yang, 2008), bat algorithm (Yang, 2010a), gravitational search algorithm (Rashedi *et al.*, 2009). Most of these techniques are based on observations of specific natural or social phenomena, however, the general concepts are usually similar (cf. Sec. 4.1). A good exposition of some of the recent methods can be found in the work of (Yang, 2010a).

4.8 Summary

In this chapter, the fundamentals of global optimization using population-based metaheuristic algorithms have been presented. Several major techniques have also been described, including genetic and EAs, PSO, as well as DE. Furthermore, some practical aspects of population-based optimization have been discussed, such as balancing selection pressure to control the algorithm convergence. Population-based metaheuristics have numerous variants, yet they all share similar basic principles as mentioned earlier in the chapter. Majority of the methods described here are for unconstrained optimization (apart from box constraints that can be easily implemented into the algorithms). Constrained metaheuristic optimization is a research area in its own (Mezura-Montes *et al.*, 2010; Koziel and Michalewicz, 1998) and exceeds the scope of this chapter.

As mentioned at the beginning, population-based methods have several important advantages that make them attractive tools for solving a variety of real-world problems. These include global search properties, capability of handling "difficult" functions (non-differentiable, discontinuous, noisy, multi-modal, etc.), as well as easy implementation and parallelization. Also, most major metaheuristics can (and have been) extended to allow multi-objective optimization (cf. Chapter 6).

From the point of view of antenna design, the fundamental disadvantage of population-based methods is their considerable computational cost. In some cases, when reasonably good analytical models are available (e.g., antenna array design with analytical array factor models, Aljibouri *et al.*, 2008; Jin and Rahmat-Samii, 2005; Rocca *et al.*, 2011b), utilization of metaheuristics is well justified. However, for majority of problems when the antenna structure is evaluated using full-wave EM analysis, computational cost of metaheuristic optimization is usually prohibitive. The problem can be addressed by means of surrogate-assisted methods (Bandler *et al.*, 2004a; Forrester and Keane, 2009; Queipo *et al.*, 2005) including both data-driven models (Koziel and Ogurtsov, 2013c; Koziel *et al.*, 2013b) and variable-fidelity EM-simulations (Jacobs and Koziel, 2014; Koziel, 2009). Exposition of such methods in the context of multi-objective antenna design can be found in Chapters 7 and 8.

Chapter 5

Surrogate-Based Modeling and Optimization

In the context of electromagnetic (EM)-simulation-driven design of antenna structures, the major disadvantage of conventional numerical optimization techniques outlined in Chapters 3 and 4 is their high computational cost, typically measured in hundreds (for local methods) or thousands and tens of thousands of objective function evaluations (for population-based metaheuristics). The exception is gradient-based search with adjoint sensitivities (Ghassemi *et al.*, 2013; Koziel and Ogurtsov, 2012c; Koziel *et al.*, 2013c, 2014d; Koziel and Bekasiewicz, 2015a, 2016f), where the optimization process can be conducted in reasonable time even for relatively large number of designable parameters. Nevertheless, it seems that the most promising approaches in terms of expedited simulation-driven design are those exploiting surrogate models. In this chapter, we provide a brief introduction to surrogate-based optimization (SBO). In particular, we outline the SBO concept, discuss various surrogate modeling techniques, as well as overview surrogate-assisted optimization methods both approximation- and physics-based. The algorithms for cost-efficient multi-objective optimization of antenna structures considered later in this book are largely based on SBO paradigm and exploit specific SBO methods as the building blocks of the antenna design frameworks.

5.1 Surrogate-Based Optimization: Brief Introduction

Conventional optimization techniques have been developed to handle specific classes of objective functions. For example, gradient-based algorithms require that the function is sufficiently smooth (e.g., at least continuously differentiable up to a certain order, Nocedal and

Wright, 2006). Population-based metaheuristics work well with multimodal and even discontinuous functions, however, due to the large number of objective functions required for convergence, individual evaluation cost of the function is supposed to be low. Nowadays, majority of engineering design is conducted using computer simulation models. Despite of clear advantages such as design automation and a possibility of reducing the role of costly prototyping, their downside — from the point of view of simulation-driven design — is potentially high cost. In case of antennas, the typical simulation times are from a few minutes for simple structures to many hours for complex antenna systems at fine discretization of the respective structures (Afshinmanesh *et al.*, 2008; Bekasiewicz and Koziel, 2015a; Chamaani *et al.*, 2010; Hanninen, 2012; Koziel *et al.*, 2015a). In many cases, it makes the use of conventional optimization routines virtually prohibitive.

Practical necessity led to development of alternative design methods. Among these, SBO techniques (Forrester and Keane, 2009; Leifsson and Koziel, 2015a; Koziel *et al.*, 2011a; Koziel and Leifsson, 2013; Koziel and Ogurtsov, 2014a; Queipo *et al.*, 2005; Yelten *et al.*, 2012) seem to be the most promising for solving design optimization problems involving expensive simulation models. The core of SBO is to replace direct optimization of the original (high-fidelity) model by an iterative process yielding a sequence $\boldsymbol{x}^{(i)}$, $i = 0, 1, \ldots,$ of approximate solutions to the original optimization problem (3.1) that, hopefully, converge to it. The sequence is generated using a fast yet reasonably accurate representation of the high-fidelity model, referred to as a surrogate. In each iteration, the surrogate model is updated using the high-fidelity model evaluation at one (e.g., the most recent) design or several designs selected using an appropriate infill strategy (Forrester *et al.*, 2008; Koziel *et al.*, 2011a; Koziel and Ogurtsov, 2014a). A popular way of generating the new design is optimization of the current surrogate model. In this case, the SBO process can be written as (Koziel *et al.*, 2011a)

$$\boldsymbol{x}^{(i+1)} = \arg\min_{\boldsymbol{x}} U(s^{(i)}(\boldsymbol{x})), \tag{5.1}$$

where $s^{(i)}$ is the surrogate model at the ith iteration (see Fig. 5.1). The initial design $\boldsymbol{x}^{(0)}$ is typically obtained using engineering insight or by optimizing any available lower-fidelity model.

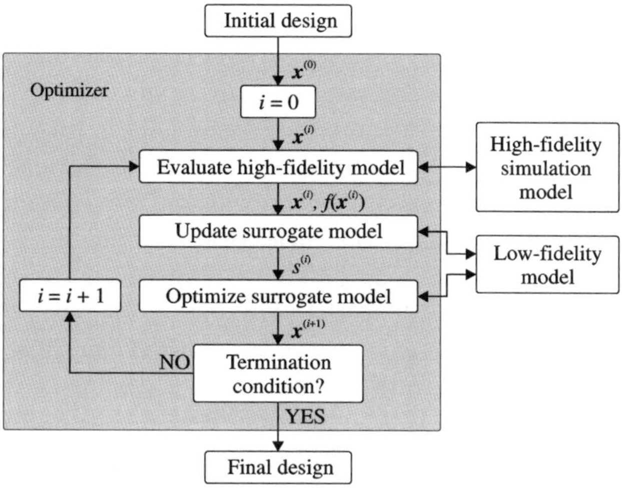

Fig. 5.1. Conceptual illustration and flow of the SBO process. The surrogate model s is iteratively updated and optimized to yield a sequence $\boldsymbol{x}^{(i)}, i = 0, 1, \ldots,$ of approximate solutions to the original problem $\boldsymbol{x}^* = \arg\min\{\boldsymbol{x}{:}U(f(\boldsymbol{x}))\}$. The high-fidelity model f is evaluated for verification purposes and to provide data that can be used to update the surrogate. The number of iterations needed in SBO is often substantially smaller than for conventional optimization techniques.

Assuming that the surrogate model is very fast, the cost of solving (5.1) can be neglected so that the overall optimization cost is merely determined by the number of high-fidelity model evaluations. On the other hand, the number of iterations required by a typical SBO algorithm is substantially smaller than for most of the conventional methods when solving (3.1) (Koziel *et al.*, 2006a). Consequently, the total optimization cost of the SBO procedures tends to be low. Fast surrogates can be obtained from analytical models, simplified physics models (e.g., equivalent circuit representations in microwave engineering, Bandler *et al.*, 2004a), or as data-driven models (also referred to as response surface approximation, RSA, models; e.g., Liu *et al.*, 2014b). RSA models can be constructed beforehand (by extensive sampling of the design space, Koziel and Ogurtsov, 2011b) or iteratively (by adaptive sampling techniques, e.g., Forrester and Keane, 2009).

In the case of antennas, the surrogate model is usually constructed using coarse-discretization EM-simulations (Koziel and Ogurtsov,

2012a; Koziel *et al.*, 2016b) because reliable equivalent circuit models, leaving alone analytical models, are hardly available. Some comments concerning low-fidelity model setup for antenna structures are provided in Chapter 2. An important issue here is finding a proper balance between the model accuracy and speed (Koziel and Ogurtsov, 2012a; Leifsson and Koziel, 2015b). In practical antenna design cases, the achievable time evaluation ratio between the high- and low-fidelity models is between 5 and 50 or so. One of the consequences is that the computational cost of the surrogate model evaluation when solving (5.1) cannot be neglected. As a matter of fact, it may become a significant contributor to the overall optimization cost, which leads to additional challenges concerning the development and implementation of the SBO algorithms (Forrester and Keane, 2009; Koziel *et al.*, 2011a; Koziel and Ogurtsov, 2014a).

In the context of local optimization, it is important that the surrogate is consistent with the high-fidelity model at the current design $\boldsymbol{x}^{(i)}$, i.e., $s^{(i)}(\boldsymbol{x}^{(i)}) = f(\boldsymbol{x}^{(i)})$ (zero-order consistency, Alexandrov and Lewis, 2001), and $J[s^{(i)}(\boldsymbol{x}^{(i)})] = J[f(\boldsymbol{x}^{(i)})]$ (first-order consistency), where $J[\cdot]$ stands for the model Jacobian. If this is the case, by embedding the surrogate-based algorithm in a trust region framework (Conn *et al.*, 2000),

$$\boldsymbol{x}^{(i+1)} = \arg \min_{\boldsymbol{x},\ ||\boldsymbol{x}-\boldsymbol{x}^{(i)}|| \le \delta^{(i)}} U(s^{(i)}(\boldsymbol{x})), \tag{5.2}$$

the convergence of the algorithm (5.1) to a local optimum of f is guaranteed (Alexandrov *et al.*, 1998) assuming that the models involved are sufficiently smooth. In (5.2), $\delta^{(i)}$ is the trust-region radius at iteration i, and it is updated in each iteration based on a gain ratio $\rho = [U(f(\boldsymbol{x}^{(i+1)})) - U(f(\boldsymbol{x}^{(i)}))]/[U(s(\boldsymbol{x}^{(i+1)})) - U(s(\boldsymbol{x}^{(i)}))]$, i.e., the actual versus surrogate-predicted objective function improvement (Conn *et al.*, 2000); $\delta^{(i)}$ is increased if the improvement is consistent with the prediction and decreased otherwise. Given the first-order consistency, high-fidelity model objective improvement is guaranteed for sufficiently small values of δ.

Satisfaction of the first-order consistency condition requires access to high-fidelity model gradients. Furthermore, additional assumptions concerning the smoothness of the functions involved are also necessary for convergence (Echeverría and Hemker, 2008). Convergence of various SBO algorithms can also be ensured under various other scenarios, see e.g., space mapping (Koziel *et al.*, 2006a, 2008b), manifold mapping (MM) (Echeverría and Hemker, 2008) or surrogate management framework (Booker *et al.*, 1999).

The SBO process can be implemented either as a local or as a global search algorithm. In the case of local optimization, trust-region-like convergence safeguards are often used (cf. (5.2)) and at least zero-order consistency conditions are ensured by appropriate construction of the surrogate model (Koziel *et al.*, 2010a). For global search, the surrogate is usually constructed in the larger portion of the design space. The new solutions are typically obtained using global methods such as evolutionary algorithms. The surrogate model itself is updated using appropriate statistical infill criteria based on the expected improvement of the objective function or minimization of the (global) modeling error (Forrester and Keane, 2009; Liu *et al.*, 2014b).

5.2 Surrogate Modeling: Data-Driven Surrogates

Data-driven (also referred to as approximation-based) models are the most popular types of surrogates around. There are numerous modeling techniques available (e.g., Simpson *et al.*, 2001, Søndergaard, 2003; Forrester and Keane, 2009; Couckuyt, 2013; Lophaven *et al.*, 2002; Gorissen *et al.*, 2010a). Approximation models can be characterized as follows:

- They are data-driven (no problem-specific knowledge is required);
- They are generic and applicable to a wide range of problems;
- They are typically based on explicit analytical formulas;
- They are cheap to evaluate.

Perhaps, the only (yet important) disadvantage is large amount of training data necessary to ensure good predictive power of the model. Even more importantly, the number of training samples grows very quickly with the problem dimensionality, which is referred to as curse of dimensionality, (Forrester and Keane, 2009; Gorissen *et al.*, 2010b). In particular, due to highly nonlinear responses, data-driven modeling of antenna structures is essentially limited to a few (<5–6) dimensions unless the ranges of geometry parameters are considerably restricted. On the other hand, local approximation models can be quite useful as explained later in the book.

In this section, we outline the surrogate modeling flow as well as discuss popular approximation techniques and model validation methods.

5.2.1 *Surrogate Modeling Flow*

A surrogate is constructed by approximating sampled data from the high-fidelity model. The modeling process involves the following steps:

- Design of experiments (DOE), which is allocation of the training samples in the design space using a selected strategy. Given the computational budget, the number of samples is normally limited. Nowadays, space-filling DOEs are preferred (Simpson *et al.*, 2001; Forrester and Keane, 2009), cf. Sec. 5.2.2.
- Training data acquisition at the points allocated at the first stage.
- Surrogate model identification. The model parameters are typically determined by solving a suitably defined minimization problem (e.g., Kriging (Kleijnen, 2009), neural networks (Rayas-Sanchez, 2004)) or by solving an appropriate regression problem analytically (e.g., polynomial approximation, Queipo *et al.*, 2005).
- Model validation. It is carried out to estimate the generalization error of the surrogate, i.e., the predictive power at the designs not seen during the identification stage.

In many practical situations, the surrogate model construction is an iterative process so that the previous steps are repeated until required model accuracy is achieved (Koziel *et al.*, 2014c). In particular, a new set of samples and the corresponding high-fidelity model evaluations may be added to the training set upon validation, and utilized to re-identify the model (Bekasiewicz *et al.*, 2014a). The surrogate modeling process is shown in Fig. 5.2.

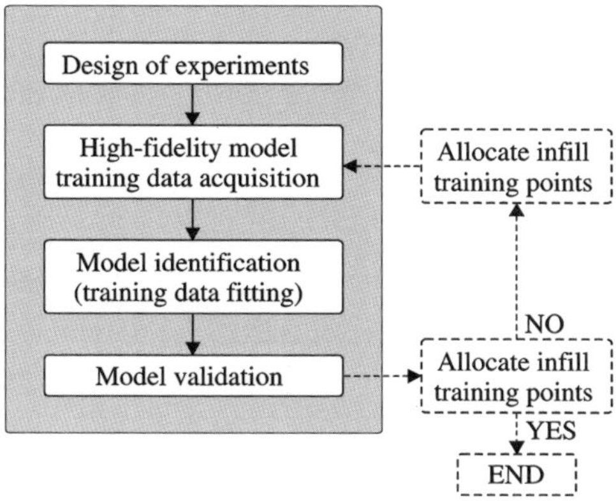

Fig. 5.2. A generic surrogate model construction flowchart. Optional iterative model enhancement is indicated using dashed lines.

5.2.2 *Design of Experiments*

The surrogate model is constructed using the high-fidelity model data that is acquired at the training points allocated in the design space by means of a specific DOE strategy (Giunta *et al.*, 2003; Santner *et al.*, 2003; Koehler and Owen, 1996). The training points may be selected beforehand or using the feedback about the model performance, e.g., by allocating more points in the regions where the system exhibits more nonlinear behavior (Couckuyt, 2013; Devebhaktuni *et al.*, 2001).

Figure 5.3(a) shows an example of a so-called factorial DOE, where the samples are allocated in the corners, edges and/or faces of the design space. This is a traditional approach that allows estimating the main effect and interactions between design variables without having too many samples. The most "economical" factorial DOE is a so-called star distribution (Fig. 5.3(a)) which is often used in conjunction with space mapping (Cheng *et al.*, 2006).

Majority of modern DOE approaches attempt to allocate the training points uniformly within the design space (Queipo *et al.*, 2005), which is useful for constructing surrogates when the knowledge about the system is limited.

Popular space-filing DOEs include pseudorandom sampling (Guinta *et al.*, 2003; Fig. 5.3(b)), uniform grid sampling (Fig. 5.3(c)),

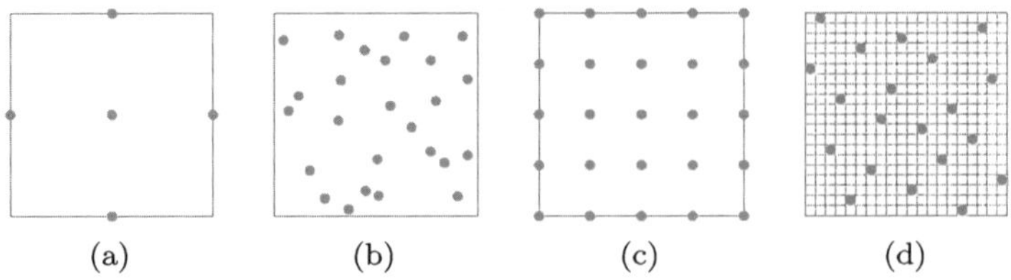

Fig. 5.3. Popular DOE techniques: (a) factorial designs (here, star distribution); (b) random sampling; (c) uniform grid sampling; and (d) Latin Hypercube Sampling (LHS).

and LHS (McKay *et al.*, 1979; Fig. 5.3(d)). Although simple, pseudo-random sampling does not ensure good uniformity. Grid sampling is only applicable for low-dimensional spaces. LHS and its improvements (e.g., Beachkofski and Grandhi, 2002; Leary *et al.*, 2003; Ye, 1998; Palmer and Tsui, 2001) offers good uniformity and flexibility. In order to allocate p samples with LHS, the range for each parameter is divided into p bins, which for n design variables, yields a total number of p^n bins in the design space. The samples are randomly selected in the design space so that (i) each sample is randomly placed inside a bin, and (ii) for all one-dimensional projections of the p samples and bins, there is exactly one sample in each bin. Figure 5.3(d) shows an example LHS realization of 20 samples in the two-dimensional space. Other uniform sampling techniques that are commonly used include orthogonal array sampling (Queipo *et al.*, 2005), quasi-Monte Carlo sampling (Guinta *et al.*, 2003), or Hammersley sampling (Hammersley, 1960). Uniform sample allocation can also be obtained through optimization (Leary *et al.*, 2003).

5.2.3 *Data-Driven Modeling Techniques*

Data-driven modeling techniques include a considerable number of approximation methods such as polynomial regression (Queipo *et al.*, 2005), radial-basis functions (Wild *et al.*, 2008), Kriging (Forrester and Keane, 2009), neural networks (Haykin, 1998), support vector regression (SVR) (Gunn, 1998), Gaussian Process Regression (Rasmussen and Williams, 2006), or multi-dimensional rational approximation (Shaker *et al.*, 2009). Here, we briefly outline some of these methods. More detailed information can be found in the references provided earlier or review works (e.g., Simpson *et al.*, 2001).

For the purpose of this section, $\{\boldsymbol{x}^{(i)}, f(\boldsymbol{x}^{(i)})\}, i = 1, \ldots, p$, will denote the training data points and the corresponding high-fidelity model evaluations.

5.2.3.1 *Polynomial Regression*

Polynomial regression is a simple yet popular approximation technique (Queipo *et al.*, 2005). The surrogate model is defined as

$$s(\boldsymbol{x}) = \sum_{j=1}^{K} \beta_j v_j(\boldsymbol{x}), \tag{5.3}$$

where v_j are the (polynomial) basis functions and β_j are the model parameters that can be found as a least-square solution to the linear system $\boldsymbol{F} = \boldsymbol{X}\boldsymbol{\beta}$, where $\boldsymbol{F} = [f(\boldsymbol{x}^{(1)})\ f(\boldsymbol{x}^{(2)}) \ldots f(\boldsymbol{x}^{(p)})]^T$, $\boldsymbol{X}$ is a $p \times K$ matrix containing the basis functions evaluated at the sample points, and $\boldsymbol{\beta} = [\beta_1\, \beta_2 \ldots \beta_{\mathrm{K}}]^T$. If $p \geq K$ and $\mathrm{rank}(\boldsymbol{X}) = K$, the model parameter can be determined in the least-squares sense through the pseudoinverse $\boldsymbol{X}^+$ of $\boldsymbol{X}$ (Gloub and Van Loan, 1996) as $\boldsymbol{\beta} = \boldsymbol{X}^+\boldsymbol{f} = (\boldsymbol{X}^T\boldsymbol{X})^{-1}\boldsymbol{X}^T\boldsymbol{f}$.

One of the simplest yet useful examples of a regression model is a second-order polynomial one defined as

$$s(\boldsymbol{x}) = s\left([x_1\ x_2\ \ldots\ x_n]^T\right) = \beta_0 + \sum_{j=1}^{n} \beta_j x_j + \sum_{i=1}^{n} \sum_{j \leq i}^{n} \beta_{ij} x_i x_j, \tag{5.4}$$

with the basis functions being monomials: 1, x_j, and $x_i x_j$.

5.2.3.2 *Radial Basis Functions*

Radial basis function model (Forrester and Keane, 2009; Wild *et al.*, 2008) is defined as a linear combination of radially symmetric functions ϕ

$$s(\boldsymbol{x}) = \sum_{j=1}^{K} \lambda_j \phi(\|\boldsymbol{x} - \boldsymbol{c}^{(j)}\|), \tag{5.5}$$

where $\boldsymbol{\lambda} = [\lambda_1\, \lambda_2 \ldots \lambda_K]^T$ is the vector of model parameters, and $\boldsymbol{c}^{(j)}, j = 1, \ldots, K$, are the (known) basis function centers. The parameters can be calculated as $\boldsymbol{\lambda} = \Phi^+ \boldsymbol{F} = (\Phi^T\Phi)^{-1}\Phi^T\boldsymbol{F}$, where $\boldsymbol{F} = [f(\boldsymbol{x}^{(1)})\; f(\boldsymbol{x}^{(2)}) \ldots f(\boldsymbol{x}^{(p)})]^T$, and the $p \times K$ matrix $\Phi = [\Phi_{kl}]_{k=1,\ldots,p;\; l=1,\ldots,K}$, with the entries defined as

$$\Phi_{kl} = \phi(||\boldsymbol{x}^{(k)} - \boldsymbol{c}^{(l)}||). \tag{5.6}$$

If $p = K$ and the centers of the basis functions coincide with the data points and are all different, Φ is a regular square matrix. Then $\boldsymbol{\lambda} = \Phi^{-1}\boldsymbol{F}$.

A popular basis function is a Gaussian one, $\phi(r) = \exp(-r^2/2\sigma^2)$, where σ is the scaling parameter.

5.2.3.3 *Kriging*

Kriging is a popular technique for interpolating deterministic noise-free data (Matheron, 1963; Journel and Huijbregts, 1981; Simpson *et al.*, 2001; Kleijnen, 2009; O'Hagan, 1978). In its basic formulation, kriging assumes that the function of interest is of the form,

$$f(\boldsymbol{x}) = \boldsymbol{g}(\boldsymbol{x})^T\boldsymbol{\beta} + Z(\boldsymbol{x}), \tag{5.7}$$

where $\boldsymbol{g}(\boldsymbol{x}) = [g_1(\boldsymbol{x})\, g_2(\boldsymbol{x}) \ldots g_K(\boldsymbol{x})]^T$ are known (e.g., constant) functions, $\boldsymbol{\beta} = [\beta_1\, \beta_2 \ldots \beta_K]^T$ are the unknown model parameters (hyperparameters); $Z(\boldsymbol{x})$ is a realization of a normally distributed Gaussian random process with zero mean and variance σ^2. The regression part $\boldsymbol{g}(\boldsymbol{x})^T\boldsymbol{\beta}$ is a trend function for f, and $Z(\boldsymbol{x})$ takes into account localized variations. The covariance matrix of $Z(\boldsymbol{x})$ is given as

$$Cov\left[Z\left(\boldsymbol{x}^{(i)}\right) Z\left(\boldsymbol{x}^{(j)}\right)\right] = \boldsymbol{\sigma}^2\boldsymbol{R}\left(\left[R\left(\boldsymbol{x}^{(i)}, \boldsymbol{x}^{(j)}\right)\right]\right), \tag{5.8}$$

where $\boldsymbol{R}$ is a $p \times p$ correlation matrix with $R_{ij} = R(\boldsymbol{x}^{(i)}, \boldsymbol{x}^{(j)})$. $R(\boldsymbol{x}^{(i)}, \boldsymbol{x}^{(j)})$ is the correlation function between the data points $\boldsymbol{x}^{(i)}$ and $\boldsymbol{x}^{(j)}$. Perhaps, the most popular correlation function is a Gaussian one,

$$R(\boldsymbol{x}, \boldsymbol{y}) = \exp\left[-\sum_{k=1}^{n} \theta_k |x_k - y_k|^2\right], \tag{5.9}$$

where θ_k are the correlation parameters to be determined; x_k and y_k are the kth components of the vectors $\boldsymbol{x}$ and $\boldsymbol{y}$, respectively. The Kriging predictor (Simpson *et al.*, 2001; Journel and Huijbregts, 1981) is defined as

$$s(\boldsymbol{x}) = \boldsymbol{g}(\boldsymbol{x})^T\boldsymbol{\beta} + \boldsymbol{r}^T(\boldsymbol{x})\boldsymbol{R}^{-1}(\boldsymbol{F} - \boldsymbol{G}\boldsymbol{\beta}), \tag{5.10}$$

where $\boldsymbol{r}(\boldsymbol{x}) = [R(\boldsymbol{x}, \boldsymbol{x}^{(1)}) \ldots R(\boldsymbol{x}, \boldsymbol{x}^{(p)})]^T$, $\boldsymbol{F} = [f(\boldsymbol{x}^{(1)})f(\boldsymbol{x}^{(2)}) \ldots f(\boldsymbol{x}^{(p)})]^T$, and $\boldsymbol{G}$ is a $p \times K$ matrix with $G_{ij} = g_j(\boldsymbol{x}^{(i)})$. We also have $\boldsymbol{\beta} = (\boldsymbol{G}^T\boldsymbol{R}^{-1}\boldsymbol{G})^{-1}\boldsymbol{G}^T\boldsymbol{R}^{-1}\boldsymbol{F}$. Model fitting is accomplished by maximum likelihood for θ_k (Journel and Huijbregts, 1981).

An important property of Kriging is that the random process $Z(\boldsymbol{x})$ gives information about the approximation error of the model. It can be used for improving the surrogate, e.g., by allocating additional training samples at the locations where the estimated model error is the highest (Forrester and Keane, 2009; Journel and Huijbregts, 1981). This feature is also utilized in various global optimization methods (e.g., Couckuyt, 2013, and references therein).

5.2.3.4 *Artificial Neural Networks (ANN)*

Despite their numerous applications, ANNs (Haykin, 1998) can be considered — in this context of this chapter — as nonlinear regression models. The key component of ANN is the neuron (or single-unit perceptron) (Haykin, 1998; Minsky and Papert, 1969). A neuron realizes a nonlinear operation (cf. Fig. 5.4(a)), where w_1–w_n are regression coefficients, $\boldsymbol{\beta}$ is the bias value of the neuron, and T is a user-defined slope parameter. The most common neural network architecture is the multi-layer feed-forward network shown in Fig. 5.4(b).

The construction of a neural network model involves selection of its architecture and the training, i.e., the assignment of the values to the regression parameters. The network training can be formulated as a nonlinear least-squares regression problem. A popular technique for solving this regression problem is the error back-propagation algorithm (Haykin, 1998; Simpson *et al.*, 2001). For large networks with complex architectures, the use of global optimization methods might be necessary (Alba and Marti, 2006).

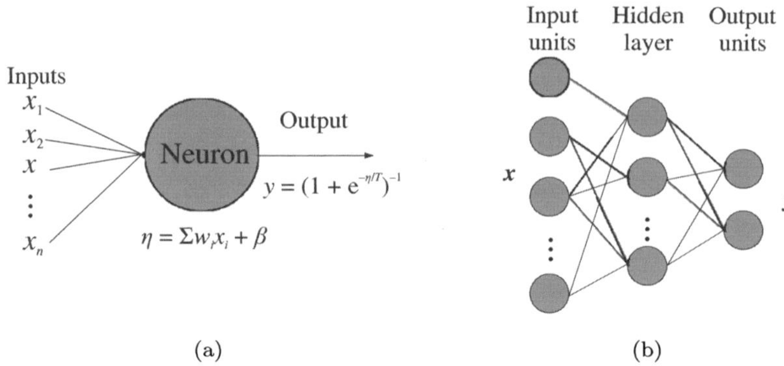

Fig. 5.4. Basic concepts of ANNs: (a) structure of a neuron; (b) two-layer feed-forward neural network architecture.

5.2.3.5 *Support Vector Regression (SVR)*

SVR (Gunn, 1998) is a relatively recent modeling approach that characterized good generalization capability (Angiulli *et al.*, 2007) and easy training by means of quadratic programming (Smola and Schölkopf, 2004). SVR gained popularity in various areas including electrical engineering and aerodynamic design (Pattipati *et al.*, 2011; Rojo-Alvarez *et al.*, 2005; Yang *et al.*, 2005; Meng and Xia, 2007; Xia *et al.*, 2007; Andrés *et al.*, 2012; Zhang and Han, 2013).

Here, we formulate SVR for vector-valued function $\boldsymbol{f}$. Let $r^k = [r_1^k\ r_2^k \ldots r_m^k]^T = \boldsymbol{f}(\boldsymbol{x}^k), k = 1, 2, \ldots, N$, denote the sampled high-fidelity model responses. We want to use SVR to approximate r^k at all base points $\boldsymbol{x}^k, k = 1, 2, \ldots, N$. For linear regression, we aim at approximating a training data set, here, the pairs $D_j = \{(\boldsymbol{x}^1, r_j^1), \ldots, (\boldsymbol{x}^N, r_j^N)\}, j = 1, 2, \ldots, m$, by a linear function $f_j(\boldsymbol{x}) = \boldsymbol{w}_j^T \boldsymbol{x} + b_j$. The optimal regression function is given by the minimum of the functional (Smola and Schölkopf, 2004)

$$\boldsymbol{\Phi}_j(\boldsymbol{w}, \xi) = \frac{1}{2}||\boldsymbol{w}_j||^2 + C_j \sum_{i=1}^{N} (\xi_{j.i}^+ + \xi_{j.i}^-). \tag{5.11}$$

Here, C_j is a user-defined value, and $\xi_{j.i}^+$ and $\xi_{j.i}^-$ are the slack variables representing upper and lower constraints on the output of the system. The typical cost function used in SVR is an ε-insensitive loss function defined as $L_\varepsilon(y) = 0$ for $|f_j(\boldsymbol{x}) - y| < \varepsilon$ and $L_\varepsilon(y) = |f_j(\boldsymbol{x}) - y|$

otherwise. The value of C_j determines the trade-off between the flatness of f_j and the amount up to which deviations larger than ε are tolerated (Gunn, 1998).

Here, we describe nonlinear regression employing the kernel approach, in which the linear function $\boldsymbol{w}_j^T\boldsymbol{x} + b_j$ is replaced by the $\Sigma_i \gamma_{j.i} K(\boldsymbol{x}^k, \boldsymbol{x}) + b_j$, where K is a kernel function. The SVR model is defined as

$$s(\boldsymbol{x}) = \left[\sum_{i=1}^{N} \gamma_{1.i} K(\boldsymbol{x}^i, \boldsymbol{x}) + b_1 \ldots \sum_{i=1}^{N} \gamma_{m.i} K(\boldsymbol{x}^i, \boldsymbol{x}) + b_m\right]^T, \qquad (5.12)$$

with the parameters $\gamma_{j.i}$ and $b_j, j = 1, \ldots, m, i = 1, \ldots, N$ obtained according to a general SVR methodology. In particular, the Gaussian kernels of the form $K(\boldsymbol{x}, \boldsymbol{y}) = \exp(-0.5 \cdot \|\boldsymbol{x} - \boldsymbol{y}\|^2/c^2)$ with $c > 0$ can be used, where c is the scaling parameter. Both c as well as parameters C_j and ε can be adjusted to minimize the generalization error calculated using cross-validation (Queipo *et al.*, 2005).

5.2.3.6 *Other Approximation Methods*

Although the approximation techniques outlined in Secs. 5.2.3.1–5.2.3.5 belong to the most popular ones, a number of other interesting methods exist. One of them is moving least squares (MLS) (Levin, 1998). In MLS, the error contribution from each training point $\boldsymbol{x}^{(i)}$ is multiplied by a weight ω_i that depends on the distance between $\boldsymbol{x}$ and $\boldsymbol{x}^{(i)}$. This improves the model flexibility and also increases computational complexity as the model has to be identified for each $\boldsymbol{x}$ separately. A typical choice for the weights is

$$\omega_i(||\boldsymbol{x} - \boldsymbol{x}^{(i)}||) = \exp(-||\boldsymbol{x} - \boldsymbol{x}^{(i)}||^2). \qquad (5.13)$$

Gaussian process regression (GPR) (Rasmussen and William, 2006) is a surrogate modeling technique that addresses the approximation problem from a stochastic point of view. From this perspective, and since Gaussian processes are mathematically tractable, it is relatively easy to compute error estimations for GPR-based surrogates in the form of uncertainty distributions. Under certain conditions, GPR models can be shown to be equivalent to large neural networks while requiring much less regression parameters than them (Rasmussen and William, 2006).

The last method mentioned here is co-Kriging (Forrester *et al.*, 2007; Toal and Keane, 2011). It is an extention of Kriging interpolation (cf. Sec. 5.2.3.3) that allows for combining information from computational models of various levels of fidelity. The fundamental benefit of such combination is that — by exploiting knowledge embedded in the lower-fidelity model — the surrogate can be created at much lower computational cost than for the models exclusively based on high-fidelity data. Co-Kriging is a rather recent method with relatively few applications in engineering (Toal and Keane, 2011; Huang and Gao, 2012; Laurenceau and Sagaut, 2008; Koziel *et al.*, 2013d; Koziel and Bekasiewicz, 2016g). It is also utilized in this book as one of the methods for Pareto front refinement (cf. Chapter 7).

5.2.4 *Model Validation*

An important stage of the modeling process is validation where the quality of the surrogate, in particular, its predictive power is assessed. Probably the most popular approach is a split sample method (Queipo *et al.*, 2005), where part of the available data set (the training set) is used to construct the surrogate, whereas the second part (the test set) serves purely for model validation. Unfortunately, the error estimation depends strongly on how the set of data samples is partitioned. More accurate estimation of the generalization error can be obtained using cross-validation (Queipo *et al.*, 2005; Geisser, 1993), where the available data set is divided into K subsets, and each of these subsets is sequentially used as testing set for a surrogate constructed on the other $K-1$ subsets. The prediction error can be estimated with all the K error measures obtained in this process. Cross-validation provides an error estimation that is less biased than with the split-sample method.

In practice, the process of model construction and validation is often iterated (with additional training samples added in each iteration) until the prescribed surrogate accuracy level is reached. The strategies of allocating the new samples (so-called infill criteria, Forrester and Keane, 2009) usually aim at improving the global accuracy of the model, i.e., inserting new samples at the locations where the estimated modeling error is the highest.

5.3 Surrogate Modeling: Physics-Based Surrogates

Despite their advantages highlighted in the previous section, applicability of data-driven models for antenna design is limited — especially in the SBO context — due to high computational cost of training data acquisition. Here, we discuss physics-based surrogates that seem to be more attractive this respect.

The key component of a physics-based surrogate is a low-fidelity model that is suitably corrected (or enhanced) in order to achieve good alignment with the high-fidelity model. As mentioned before, in the case of antennas, the low-fidelity model is typically obtained through coarse-discretization EM analysis (cf. Chapter 2).

The most important feature of physics-based surrogates is that only a limited amount of high-fidelity data is normally sufficient to ensure reasonable accuracy of the model. Good generalization capability of the model is a result of exploiting the knowledge about the system (here, antenna) under design embedded in the low-fidelity model. This normally translates into improved computational efficiency of the SBO process (Koziel *et al.*, 2011a; Koziel and Ogurtsov, 2014a).

Physics-based surrogate models are frequently used in multi-objective optimization algorithms presented later in this book. Here, we outline the main concepts of physics-based surrogate modeling. More detailed treatment of the subject can be found in the literature (Koziel *et al.*, 2011a; Koziel and Ogurtsov, 2014a; Bekasiewicz *et al.*, 2014a).

We will denote by $c(\boldsymbol{x})$ a low-fidelity model of the device or system of interest. For the sake of explaining the physics-based surrogate modeling concept, we discuss a simple case of multiplicative response correction, considered in the context of the SBO algorithm (5.1). From the point of view of ensuring convergence of a sequence $\{\boldsymbol{x}^{(i)}\}$ of approximate solutions to the original problem (3.1), produced by (5.1), it is important that the surrogate and the high-fidelity

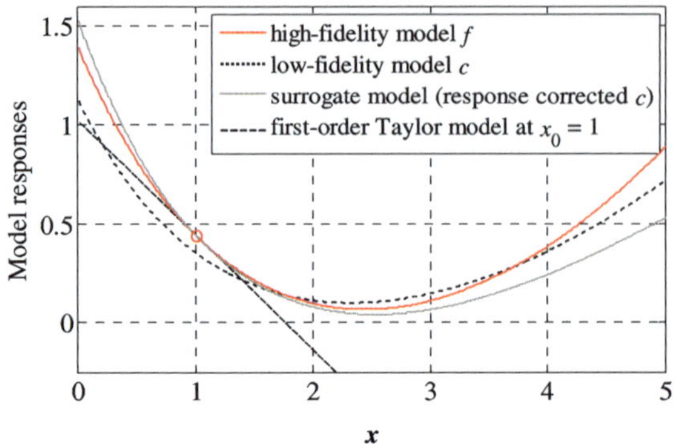

Fig. 5.5. Visualization of the response correction (5.14) for the example analytical functions c (low-fidelity model) and f (high-fidelity model). The correction is established at $\boldsymbol{x}_0 = 1$. Note that the surrogate exhibits good alignment with the high-fidelity model in relatively wide vicinity of $\boldsymbol{x}_0$, especially compared to the first-order Taylor model set up using the same data from f (the value and the gradient at $\boldsymbol{x}_0$).

model are well aligned at least locally. The surrogate $s^{(i)}(\boldsymbol{x})$ at iteration i can be constructed as

$$s^{(i)}(\boldsymbol{x}) = \beta_k(\boldsymbol{x})c(\boldsymbol{x}), \tag{5.14}$$

where $\beta_k(\boldsymbol{x}) = \beta_k(\boldsymbol{x}^{(i)}) + \nabla\beta(\boldsymbol{x}^{(i)})^T(\boldsymbol{x} - \boldsymbol{x}^{(i)})$, where $\beta(\boldsymbol{x}) = f(\boldsymbol{x})/c(\boldsymbol{x})$. This ensures that $s^{(i)}(\boldsymbol{x}^{(i)}) = f(\boldsymbol{x}^{(i)})$ (zero-order consistency) and $\nabla s^{(i)}(\boldsymbol{x}^{(i)}) = \nabla f(\boldsymbol{x}^{(i)})$ (first-order consistency; Alexandrov and Lewis, 2001) (see Fig. 5.5).

Another approach to low-fidelity model correction is to apply it at the level of the model domain. Perhaps, the most popular example of such a procedure is input space mapping (Bandler *et al.*, 2004), where the surrogate is created as

$$\boldsymbol{s}^{(i)}(\boldsymbol{x}) = \boldsymbol{c}(\boldsymbol{x} + \boldsymbol{q}^{(i)}), \tag{5.15}$$

where vector notation for the low-fidelity model $\boldsymbol{c}$ and the surrogate $\boldsymbol{s}$, respectively. The model parameters $\boldsymbol{q}^{(i)}$ are obtained by minimizing the misalignment between the surrogate and the high-fidelity model, i.e., $\|\boldsymbol{f}(\boldsymbol{x}^{(i)}) - \boldsymbol{c}(\boldsymbol{x}^{(i)} + \boldsymbol{q}^{(i)})\|$; $\boldsymbol{x}^{(i)}$ is a reference design (e.g., the most recent design encountered during the optimization run) at which the surrogate model is established. Benefits of using input

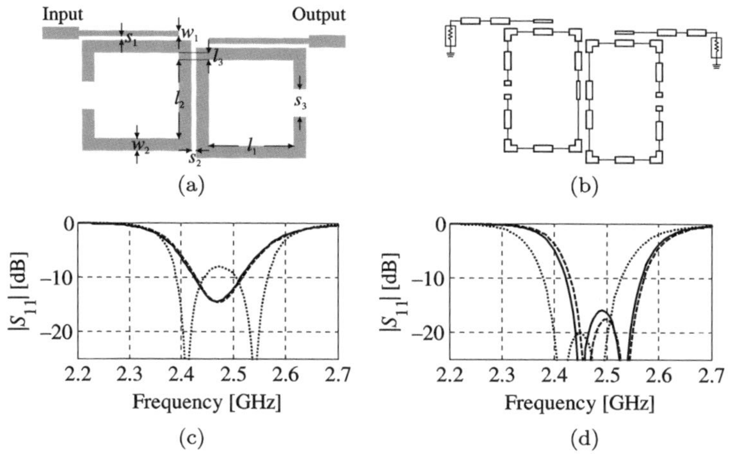

Fig. 5.6. Low-fidelity model correction through parameter shift (input space mapping): (a) microstrip filter geometry (high-fidelity model $\boldsymbol{f}$ evaluated using EM-simulation); (b) low-fidelity model $\boldsymbol{c}$ (equivalent circuit); (c) response of $\boldsymbol{f}$ (—) and $\boldsymbol{c}$ (····), as well as response of the surrogate model $\boldsymbol{s}$ (– –) created using input space mapping; (d) surrogate model verification at a different design (other than that at which the model was created) $\boldsymbol{f}$ (—), $\boldsymbol{c}$ (····), and $\boldsymbol{s}$ (– –). Good alignment indicates excellent generalization of the model.

space mapping for microwave filters is demonstrated in Fig. 5.6 (Hong and Lancaster, 2001); here, the high-fidelity model is evaluated using EM-simulation, whereas the low-fidelity model is an equivalent circuit; the response of interest is a so-called reflection coefficient $|S_{11}|$ as a function of frequency. For this particular example, input space mapping offers both good approximation and generalization capability (cf. Fig. 5.6(d)).

An alternative way of low-fidelity model correction is to exploit parameters that are normally fixed in the high-fidelity model (such as substrate height or dielectric permittivity in microstrip circuits, Koziel *et al.*, 2008a, 2004b). Because the surrogate model is just an auxiliary tool and it is not supposed to be built or measured, such parameters can be freely adjusted in the low-fidelity model for the sake of improving its alignment with the high-fidelity one. Implicit space mapping (ISM) (Bandler *et al.*, 2003, 2004b; Koziel *et al.*, 2010b) is a technique that utilizes this concept. More specifically,

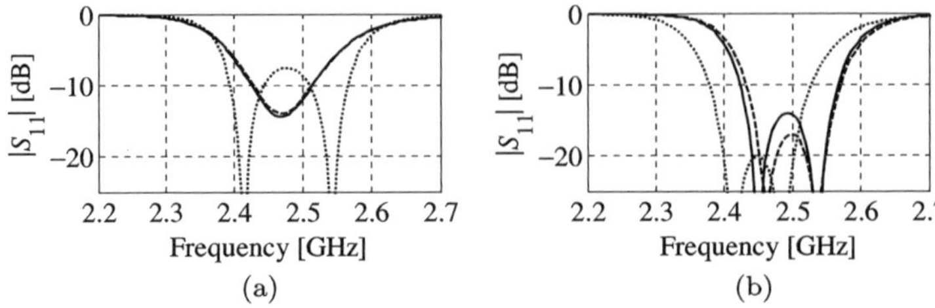

Fig. 5.7. Low-fidelity model correction through ISM applied to a microstrip filter of Fig. 5.6: (a) response of $\boldsymbol{f}$ (—) and $\boldsymbol{c}$ (····), as well as response of the surrogate model $\boldsymbol{s}$ (– –) created using ISM; (b) surrogate model verification at a different design (other than that at which the model was created) $\boldsymbol{f}$ (—), $\boldsymbol{c}$(····), and $\boldsymbol{s}$ (– –).

the surrogate is obtained as

$$\boldsymbol{s}^{(i)}(\boldsymbol{x}) = \boldsymbol{c}_I(\boldsymbol{x}, \boldsymbol{p}^{(i)}). \tag{5.16}$$

The symbol $\boldsymbol{p}$ in (5.16) denotes a vector of ISM parameters, whereas, $\boldsymbol{c}_I$ is the low-fidelity model with the explicit dependence on these parameters. In the context of design optimization, i.e., when the surrogate is established at the current design $\boldsymbol{x}^{(i)}$, the vector $\boldsymbol{p}^{(i)}$ is typically obtained by minimizing the norm-wise discrepancy between the models, i.e., $\|\boldsymbol{f}(\boldsymbol{x}^{(i)}) - \boldsymbol{c}_I(\boldsymbol{x}^{(i)}, \boldsymbol{p})\|$. Illustration of ISM for the filter of Figs. 5.6(a) and 5.6(b) is shown in Fig. 5.7. Here, the ISM parameters are thicknesses of the microstrip-line component substrates (rectangle elements in Fig. 5.6(b)).

Yet another possibility of low-fidelity model enhancement originates from the fact that, in many cases, the vector-valued responses of the system are actually evaluations of the same design but at different values of certain parameters such as the time, frequency (e.g., for microwave structures) or a specific geometry parameter (e.g., chord line coordinate for the airfoil profiles). This can be exploited by applying a linear or nonlinear scaling to this parameter which effectively changes the shape of the model response. Frequency scaling, popular in electrical engineering (Koziel *et al.*, 2006a; Koziel and Ogurtsov, 2014a), is a representative example of such a procedure. Let us assume that $\boldsymbol{f}(\boldsymbol{x}) = [f(\boldsymbol{x}, \omega_1)\ f(\boldsymbol{x}, \omega_2) \ldots f(\boldsymbol{x}, \omega_m)]^T$, where $f(\boldsymbol{x}, \omega_k)$ is the evaluation of the high-fidelity model at a frequency ω_k. Similarly, we have

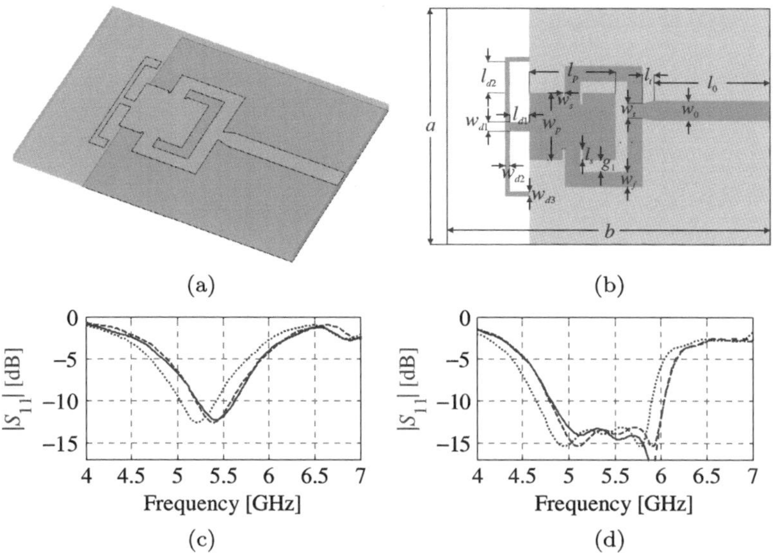

Fig. 5.8. Low-fidelity model correction through frequency scaling: (a) antenna visualization; (b) antenna geometry (both $\boldsymbol{f}$ and $\boldsymbol{c}$ evaluated using EM-simulation, coarse-discretization used for $\boldsymbol{c}$). Response of $\boldsymbol{f}$ (—) and $\boldsymbol{c}$ (····), as well as response of the surrogate model $\boldsymbol{s}$ (– – –) created using frequency scaling at: (c) a certain reference design; and (d) another (test) design. Note that the surrogate properly accounts for the frequency shifts between $\boldsymbol{c}$ and $\boldsymbol{f}$.

$\boldsymbol{c}(\boldsymbol{x}) = [c(\boldsymbol{x}, \omega_1)\ c(\boldsymbol{x}, \omega_2) \ldots c(\boldsymbol{x}, \omega_m)]^T$. The frequency-scaled (surrogate) model $\boldsymbol{s}_F(\boldsymbol{x})$ is then defined as

$$\boldsymbol{s}_F(\boldsymbol{x}, [F_0\ F_1]) = [c(\boldsymbol{x}, F_0 + F_1\omega_1) \ldots c(\boldsymbol{x}, F_0 + F_1\omega_m)]^T . \tag{5.17}$$

Here, F_0 and F_1 are scaling parameters obtained as $[F_0, F_1] = \text{argmin}\{[F_0, F_1] : \|\boldsymbol{f}(\boldsymbol{x}^{(j)}) - \boldsymbol{s}_F(\boldsymbol{x}^{(j)}, [F_0 F_1])\|\}$. Figure 5.8 shows an example of applying frequency scaling to the low-fidelity model of a substrate-integrated cavity antenna (Bekasiewicz and Koziel, 2016a). In this example, both the low- and high-fidelity models are evaluated using coarse- and fine-discretization EM-simulations, respectively.

Interested reader can find more extensive discussion on physics-based surrogates in the literature (e.g., Koziel and Ogurtsov, 2014a; Bandler *et al.*, 2004a; Koziel *et al.*, 2013e; Leifsson and Koziel, 2015a).

5.4 Optimization Using Data-Driven Surrogates

In this section, a brief outline of surrogate-assisted optimization exploiting data-driven models is provided. We discuss the response surface methodology and global optimization using Kriging surrogates. An important practical issue of balancing design space exploration and exploitation is also mentioned.

5.4.1 *Optimization Using Response Surfaces*

Optimization exploiting data-driven surrogates (also referred to as RSAs) is outlined here. The optimization process consists of the two major stages, that is, construction of the initial surrogate model, and iterative prediction-correction scheme in which the surrogate is used to identify the promising locations in the design space, and then updated using the high-fidelity model data evaluated at these candidate designs (Queipo *et al.*, 2005). The optimization flow is as follows (see also Fig. 5.9):

1. Construct an initial surrogate model using pre-sampled training data;
2. Identify a candidate design (i.e., an estimated high-fidelity optimum) by optimizing the surrogate model;
3. Evaluate the high-fidelity model at the design found in Step 2;
4. Update the surrogate by incorporating new high-fidelity data;
5. If the termination condition is not satisfied, go to 2.

Generating the new sample points by optimizing the surrogate is only one of the possible infill criteria (see Sec. 5.4.3). In case local optimization is the main goal, the SBO algorithm can be embedded in the trust-region-like framework (Forrester and Keane, 2009) for improved convergence.

In some situations, the surrogate model is constructed only once and optimized without further refinement. This one-shot approach normally requires considerable amount of training samples (to ensure sufficient accuracy) and it is typically preferred for constructing multiple-use models (Kabir *et al.*, 2008).

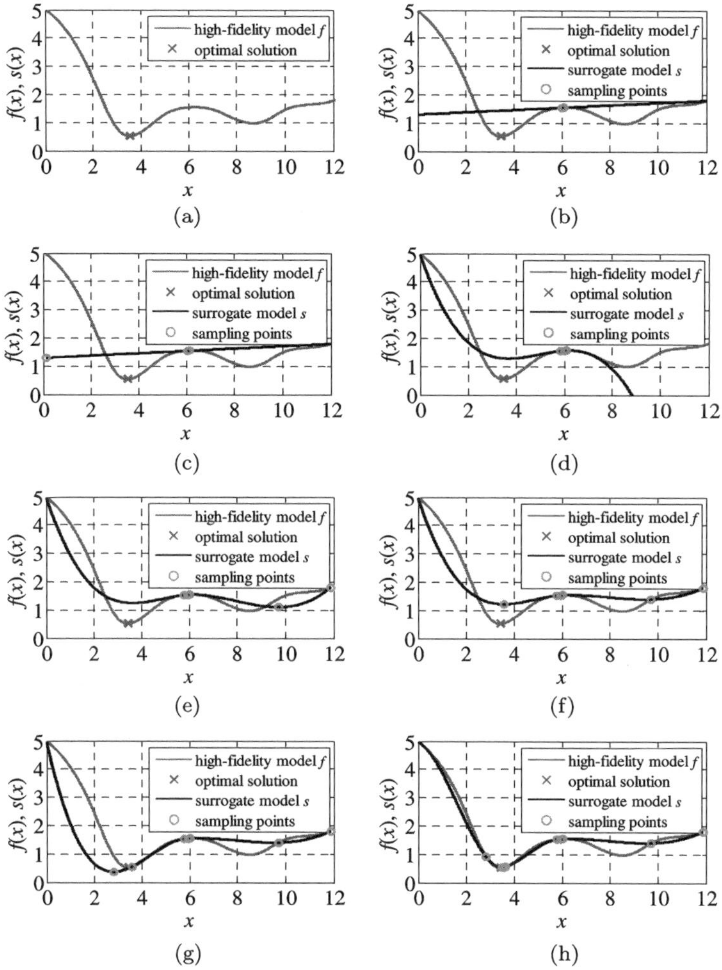

Fig. 5.9. RSA-based SBO illustration: (a) high-fidelity model $f(\cdots)$ to be minimized and the minimum of f ($\times$); (b) initial surrogate model (here, cubic splines) constructed using two data points; (c) a new data point obtained by optimizing the surrogate (evaluation of f at this point is used to update the surrogate); (d)–(h) after a few iterations, the surrogate model optimum adequately predicts the high-fidelity minimizer; global accuracy of the surrogate is not of primary concern.

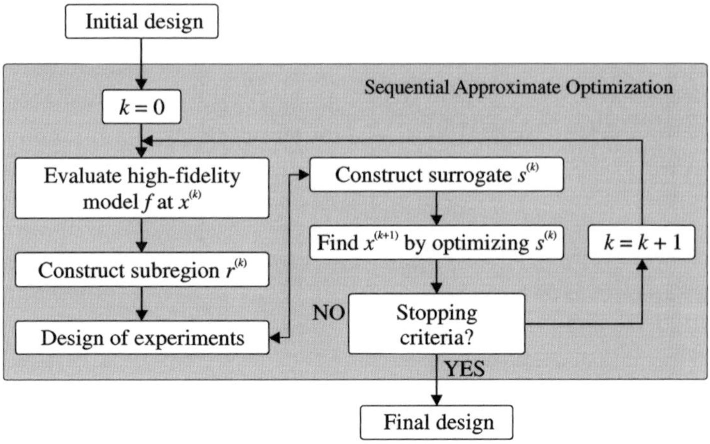

Fig. 5.10. Flowchart of sequential approximate optimization.

5.4.2 *Sequential Approximate Optimization*

SAO is one of the simplest realizations of SBO with data-driven surrogates (Jacobs *et al.*, 2004; Giunta and Eldred, 2000; Pérez *et al.*, 2002; Hosder *et al.*, 2001; Roux *et al.*, 1998; Sobieszczanski-Sobieski and Haftka, 1997; Giunta, 1997). In SAO, the search is restricted to a small region of the design space and a local approximation model (e.g., low-order polynomial) is used as a prediction tool. The search region is modified — upon optimizing the surrogate — using a chosen relocation strategy. The SAO flow is as follows (cf. Fig. 5.10):

1. Set $k = 0$ (iteration count);
2. Evaluate the high-fidelity model f at the design $\boldsymbol{x}^{(k)}$;
3. Construct a search region $r^{(k)}$ based on a relocation strategy;
4. Sample $r^{(k)}$ using a selected DOE plan;
5. Acquire high-fidelity data at the designs selected in Step 4;
6. Construct the surrogate $s^{(k)}$ using the data set of Step 5;
7. Find a new design $\boldsymbol{x}^{(k+1)}$ by optimizing the surrogate $s^{(k)}$ within $r^{(k)}$;
8. Terminate if the stopping criteria are met; otherwise set $k = k+1$ and go to 2.

Because of the use of simple surrogates, the number of training samples utilized in each iteration is small, which is an advantage of SAO. Obviously, only local search is possible this way.

Relocation strategy, i.e., adjustment of the location and the size of the new region before executing the next algorithm iteration (Step 3) is an important part of the process. A typical strategy is to allocate the center of the new region at the most current optimum (Jacobs *et al.*, 2004). In multi-point strategies, the last iteration optimum may become the corner point for the next search subregion (Toropov *et al.*, 1993). Adjustment of the subregion size may involve trust-region-like frameworks (Giunta and Eldred, 2000; Koziel and Ogurtsov, 2012b).

5.4.3 *Optimization with Kriging Surrogates: Exploration versus Exploitation*

Choosing appropriate infill criteria is an important aspect of surrogate-assisted optimization (Forrester and Keane, 2009). The two main goals for selecting the infill points are reduction of the objective function value and improvement of the global accuracy of the surrogate (Couckuyt, 2013).

As mentioned before, perhaps the simplest infill strategy is to allocate a single sample at the surrogate model optimum. Assuming that the optimization algorithm is embedded in the trust region framework (Forrester and Keane, 2009), and the surrogate model is first-order consistent with the high-fidelity model (Alexandrov *et al.*, 1998), this strategy is capable of finding at least a local minimum of the high-fidelity model.

In general, allocation of the new training point may be oriented towards global search or constructing a globally accurate surrogate. In this context, Kriging seems to be the most advantageous type of data-driven surrogate because it provides information about the expected model error (Kleijnen, 2008; Jones *et al.*, 1998; Gorissen *et al.*, 2010a). The following infill criteria based on this feature of Kriging models are commonly used:

1. Maximization of expected improvement, i.e., the improvement one expects to achieve at an untried point $\boldsymbol{x}$ (Jones *et al.*, 1998);

2. Minimization of the predicted objective function $\hat{y}(\boldsymbol{x})$, i.e., surrogate optimization mentioned already. A reasonable global accuracy of the surrogate has to be assumed (Liu *et al.*, 2012);
3. Minimization of the statistical lower bound, i.e., $LB(\boldsymbol{x}) = \hat{y}(\boldsymbol{x}) - As(\boldsymbol{x})$ (Forrester and Keane, 2009), where $\hat{y}(\boldsymbol{x})$ is the surrogate model prediction and $s^2(\boldsymbol{x})$ is the variance; A is a user-defined coefficient;
4. Maximization of the probability of improvement, i.e., identifying locations that give the highest change of improving the objective function value (Forrester and Keane, 2009);
5. Maximization of the mean square error, i.e., finding locations where the mean square error (predicted by the surrogate) is the highest (Liu *et al.*, 2012).

It should be emphasized that identifying the new samples according to the above infill criteria requires global optimization (Couckuyt, 2013). Some comments on constrained SBO optimization can be found in the works of (Forrester and Keane, 2009; Queipo *et al.*, 2005; Liu *et al.*, 2012).

The aforementioned methods are also referred to as adaptive sampling techniques. They may either aim at constructing globally accurate surrogates or carrying out global optimization. Putting more focus on design space exploitation usually leads to a reduced computational cost. Design space exploration normally results in higher cost but also global search capability (Forrester and Keane, 2009). On the other hand, global exploration is often impractical, especially for expensive functions with a medium/large number of optimization variables (more than a few tens). In the optimization context, there should be a balance between exploitation and exploration. Minimization of the statistical lower bound is an example of achieving such a balance controlled by the constant A (from pure exploitation, i.e., $\mathrm{LB}(\boldsymbol{x}) \to \hat{y}(\boldsymbol{x})$, for $A \to 0$ to pure exploration for $A \to \infty$). However, choosing a good value of A is a non-trivial task (Forrester and Keane, 2009).

SBO methods with adaptive sampling are often referred to as efficient global optimization (EGO) techniques. A detailed discussion of EGO is given in the works of Jones *et al.* (1998) or Forrester and Keane (2009). Kriging-based SBO can also be combined with pattern search methods (e.g., Surrogate Management Framework, SMF; Booker *et al.*, 1999). Several EGO-like methods enhanced by memetic operations (such as solution pre-screening) have been proposed recently (e.g., SADEA framework, Liu *et al.*, 2014a).

5.4.4 *Final Comments*

SBO algorithms with data-driven surrogates iteratively explore the design space in order to improve the model accuracy and search for the optimum design. Finding a proper balance between exploration of the design space and exploitation of the promising regions is one of the important issues for this type of method. Due to high cost of data acquisition, approximation-based surrogates may be appropriate solutions for problems with small or medium number of designable parameters. For more complex problems, the computational cost of creating the surrogate may be prohibitive.

5.5 Surrogate-Based Optimization Using Physics-Based Surrogates

One of the most attractive features of physics-based surrogates is their excellent generalization, which is a result of the knowledge about the system of interest embedded in the underlying low-fidelity models. Due to this, a reliable surrogate may be constructed with significantly less high-fidelity training points than for approximation-based models (Koziel *et al.*, 2011a). In this section, a few selected physics-based SBO techniques are outlined, in particular, some variations of space mapping (Bandler *et al.*, 2004a; Koziel *et al.*, 2006a, 2008a), approximation model management optimization (AMMO, Alexandrov and Lewis, 2001), MM (Echeverría and Hemker, 2005), shape-preserving response prediction (Koziel, 2010a), and adaptively adjusted design specifications (Koziel, 2010b).

5.5.1 *Space Mapping*

Space mapping (Bandler *et al.*, 1994, 2004a; Koziel *et al.*, 2006a) refers to a family of design optimization techniques originally developed for solving expensive problems in computational EMs, in particular, microwave engineering. Nowadays, the popularity of space mapping has spread across several engineering disciplines (Redhe and Nilsson, 2004; Bandler *et al.*, 2004a; Priess *et al.*, 2011; Marheineke *et al.*, 2012; Koziel and Leifsson, 2012; Tu *et al.*, 2013).

Initially, space mapping was exclusively based on the transformation of the low-fidelity model domain (ASM, input space mapping; Bandler *et al.*, 1994, 1995), which is sufficient for handling many engineering problems, especially in electrical engineering, where the ranges of both the low- and high-fidelity model responses are similar. Other space mapping variations were developed to handle situations where the low- and high-fidelity models are severely misaligned amplitude-wise (output space mapping; Bandler *et al.*, 2003, 2004b; Koziel *et al.*, 2006a, 2008a; Robinson *et al.*, 2008; Koziel *et al.*, 2014e).

In original space mapping, the existence of a mapping $\boldsymbol{P}$ between the high- and low-fidelity model domains is assumed (Bandler *et al.*, 1994), i.e., $\boldsymbol{x}_c = \boldsymbol{P}(\boldsymbol{x}_f)$ and so that $\boldsymbol{c}(\boldsymbol{P}(\boldsymbol{x}_f)) \approx \boldsymbol{f}(\boldsymbol{x}_f)$. Given $\boldsymbol{P}$, the direct solution of the original problem (3.1), can be replaced by finding $\boldsymbol{x}_f^{\#} = \boldsymbol{P}^{-1}(\boldsymbol{x}_c^*)$. Here, $\boldsymbol{x}_c^*$ is the optimal design of $\boldsymbol{c}$ defined as $\boldsymbol{x}_c^* = \arg\min\{\boldsymbol{x}_c\colon U(\boldsymbol{c}(\boldsymbol{x}_c))\}$; $\boldsymbol{x}_f^{\#}$ can be considered as a reasonable estimate of $\boldsymbol{x}_f^*$. In other words, the problem (3.1) can be reformulated as

$$\boldsymbol{x}_f^{\#} = \arg\min_{\boldsymbol{x}_f} U(\boldsymbol{c}(\boldsymbol{P}(\boldsymbol{x}_f))), \tag{5.18}$$

where $\boldsymbol{c}(\boldsymbol{P}(\boldsymbol{x}_f))$ is a surrogate model. However, $\boldsymbol{P}$ is not given explicitly: it can only be evaluated at any $\boldsymbol{x}_f$ by means of a parameter extraction (PE) procedure $\boldsymbol{P}(\boldsymbol{x}_f) = \arg\min\{\boldsymbol{x}_c\colon \|\boldsymbol{f}(\boldsymbol{x}_f) - \boldsymbol{c}(\boldsymbol{x}_c)\|$. One of the practical issues is possible non-uniqueness of the solution to (4.23) (Bandler *et al.*, 1995). Another issue is the assumption on the similarity of high- and low-fidelity model ranges, which is a very strong one (Alexandrov and Lewis, 2001). These and other issues led to numerous improvements, including parametric space mapping outlined later in this section.

Aggressive space mapping (ASM) (Bandler *et al.*, 1995) is one of the first versions of space mapping, still popular in microwave engineering (Sans *et al.*, 2014). Assuming uniqueness of $\boldsymbol{x}_c^*$, the solution to (5.18) is equivalent to reducing the residual vector $\boldsymbol{f} = \boldsymbol{f}(\boldsymbol{x}_f) = \boldsymbol{P}(\boldsymbol{x}_f) - \boldsymbol{x}_c^*$ to zero. The first step of the ASM algorithm is to find $\boldsymbol{x}_c^*$. Next, ASM iteratively solves the nonlinear system $\boldsymbol{f}(\boldsymbol{x}_f) = 0$ for $\boldsymbol{x}_f$. At the jth iteration, the calculation of the error vector $\boldsymbol{f}^{(j)}$ requires an evaluation of $\boldsymbol{P}^{(j)}(\boldsymbol{x}_f^{(j)}) = \boldsymbol{P}(\boldsymbol{x}_f^{(j)}) = \arg\min\{\boldsymbol{x}_c\colon \boldsymbol{f}(\boldsymbol{x}_f^{(j)}) - \boldsymbol{c}(\boldsymbol{x}_c)\|\}$. The quasi-Newton step in the high-fidelity model space is given by

$$\boldsymbol{B}^{(j)}\boldsymbol{h}^{(j)} = -\boldsymbol{f}^{(j)}, \tag{5.19}$$

where $\boldsymbol{B}^{(j)}$ is the approximation of the space mapping Jacobian $\boldsymbol{J}_P = \boldsymbol{J}_P(\boldsymbol{x}_f) = [\partial \boldsymbol{P}^T/\partial \boldsymbol{x}_f]^T = [\partial(\boldsymbol{x}_c^T)/\partial \boldsymbol{x}_f]^T$. Solving (5.19) for $\boldsymbol{h}^{(j)}$ gives the next iterate $\boldsymbol{x}_f^{(j+1)} = \boldsymbol{x}_f^{(j)} + \boldsymbol{h}^{(j)}$. The algorithm terminates if $\|\boldsymbol{f}^{(j)}\|$ is sufficiently small. The output of the algorithm is an approximation to $\boldsymbol{x}_f^{\#} = \boldsymbol{P}^{-1}(\boldsymbol{x}_c^*)$. A popular way of obtaining the matrix $\boldsymbol{B}$ is through a rank one Broyden update (Broyden, 1965) of the form $\boldsymbol{B}^{(j+1)} = \boldsymbol{B}^{(j)} + (\boldsymbol{f}^{(j+1)} - \boldsymbol{f}^{(j)} - \boldsymbol{B}^{(j)}\boldsymbol{h}^{(j)})\boldsymbol{h}^{(j)T}/\|\boldsymbol{h}^{(j)}\|^2$. Several improvements of the ASM algorithm have been proposed in the literature (Bakr *et al.*, 1998, 1999).

Parametric space mapping is a more recent and more generic type of space mapping, where the optimization algorithm is an iterative process (5.1), where the low-fidelity model correction has explicit analytical form. Simple examples of input and ISM were shown in Sec. 5.3. In general, the input space mapping surrogate model can take the form (Koziel *et al.*, 2006a)

$$\boldsymbol{s}^{(i)}(\boldsymbol{x}) = \boldsymbol{c}\left(\boldsymbol{B}^{(i)} \cdot \boldsymbol{x} + \boldsymbol{q}^{(i)}\right). \tag{5.20}$$

Here $\boldsymbol{B}^{(i)}$ and $\boldsymbol{q}^{(i)}$ are matrices obtained by minimizing misalignment between the surrogate and the high-fidelity model as

$$\left[\boldsymbol{B}^{(i)}, \boldsymbol{q}^{(i)}\right] = \arg\min_{[\boldsymbol{B},\boldsymbol{q}]} \sum_{k=0}^{i} w_{i.k} \left\|\boldsymbol{f}\left(\boldsymbol{x}^{(k)}\right) - \boldsymbol{c}\left(\boldsymbol{B} \cdot \boldsymbol{x}^{(k)} + \boldsymbol{q}\right)\right\|. \tag{5.21}$$

Here, $w_{i.k}$ are weighting factors; a common choice of $w_{i.k}$ is $w_{i.k} = 1$ for all i and all k (all previous designs contribute equally) or $w_{i.1} = 1$

and $w_{i.k} = 0$ for $k < i$ (the surrogate depends on the most recent design only).

In general, the space mapping surrogate model is constructed as follows:

$$\boldsymbol{s}^{(i)}(\boldsymbol{x}) = \bar{\boldsymbol{s}}(\boldsymbol{x}, \boldsymbol{p}^{(i)}), \tag{5.22}$$

where $\bar{\boldsymbol{s}}$ is a generic space mapping surrogate model, i.e., the low-fidelity model $\boldsymbol{c}$ composed with suitable (usually linear) transformations. The parameters $\boldsymbol{p}$ are obtained in the extraction process similar to (5.21). More information about specific space mapping surrogates can be found in the literature (e.g., Bandler *et al.*, 2004a; Koziel *et al.*, 2006a).

Some of the issues pertaining to application of space mapping for solving practical engineering problems involve ensuring sufficient quality of the low-fidelity model as well as a proper selection of the space mapping transformations (Koziel and Bandler, 2007a, 2007b), which may not be trivial (Koziel *et al.*, 2008b). On the other hand, space mapping has been found to be a very efficient tool for yielding satisfactory designs in various engineering disciplines e.g., (Redhe and Nilsson, 2004; Bandler *et al.*, 2004a; Marheineke *et al.*, 2012). A number of enhancements of space mapping algorithms have been suggested to alleviate some of the difficulties such as potential convergence problems (e.g., Koziel *et al.*, 2010a, 2010c).

5.5.2 *Approximation Model Management Optimization*

AMMO (Alexandrov and Lewis, 2001) is a simple framework with the surrogate model constructed through response correction. It exploits sensitivity data to ensure first-order consistency conditions, in particular, $s^{(i)}(\boldsymbol{x}^{(i)}) = f(\boldsymbol{x}^{(i)})$, and $\nabla s^{(i)}(\boldsymbol{x}^{(i)}) = \nabla f(\boldsymbol{x}^{(i)})$. Additionally, AMMO utilized the trust-region methodology (Conn *et al.*, 2000) to guarantee convergence of the optimization process to the high-fidelity model optimum. Assuming that $\beta(\boldsymbol{x}) = f(\boldsymbol{x})/c(\boldsymbol{x})$ is the correction function and

$$\beta_i(\boldsymbol{x}) = \beta(\boldsymbol{x}^{(i)}) + \nabla\beta(\boldsymbol{x}^{(i)})^T(\boldsymbol{x} - \boldsymbol{x}^{(i)}), \tag{5.23}$$

the surrogate model is defined as

$$s^{(i)}(\boldsymbol{x}) = \beta_i(\boldsymbol{x})c(\boldsymbol{x}), \tag{5.24}$$

and can be shown to satisfy the aforementioned consistency conditions. Obviously, (5.24) requires both the low- and high-fidelity derivatives.

5.5.3 *Manifold Mapping*

MM is essentially a response correction technique (Echeverría and Hemker, 2005, 2008) with the surrogate model set up using multiple high-fidelity data points as follows:

$$\boldsymbol{s}^{(i)}(\boldsymbol{x}) = \boldsymbol{f}\left(\boldsymbol{x}^{(i)}\right) + \boldsymbol{S}^{(i)}\left(\boldsymbol{c}(\boldsymbol{x}) - \boldsymbol{c}\left(\boldsymbol{x}^{(i)}\right)\right). \tag{5.25}$$

Here, $\boldsymbol{S}^{(i)} = \boldsymbol{\Delta F} \cdot \boldsymbol{\Delta C}^{\dagger}$ is the $m \times m$ correction matrix, where

$$\boldsymbol{\Delta F} = \left[\boldsymbol{f}\left(\boldsymbol{x}^{(i)}\right) - \boldsymbol{f}\left(\boldsymbol{x}^{(i-1)}\right) \ldots \boldsymbol{f}\left(\boldsymbol{x}^{(i)}\right) - \boldsymbol{f}\left(\boldsymbol{x}^{(\max\{i-n,0\})}\right)\right], \tag{5.26}$$

$$\boldsymbol{\Delta C} = \left[\boldsymbol{c}\left(\boldsymbol{x}^{(i)}\right) - \boldsymbol{c}\left(\boldsymbol{x}^{(i-1)}\right) \cdots \boldsymbol{c}\left(\boldsymbol{x}^{(i)}\right) - \boldsymbol{c}\left(\boldsymbol{x}^{(\max\{i-n,0\})}\right)\right]. \tag{5.27}$$

The pseudoinverse, denoted by †, is defined as $\boldsymbol{\Delta C}^{\dagger} = \boldsymbol{V}_{\boldsymbol{\Delta C}}\boldsymbol{\Sigma}^{\dagger}_{\boldsymbol{\Delta C}}\boldsymbol{U}^{T}_{\boldsymbol{\Delta C}}$, where $\boldsymbol{U}_{\boldsymbol{\Delta C}}, \sum_{\boldsymbol{\Delta C}}$, and $\boldsymbol{V}_{\boldsymbol{\Delta C}}$ are the factors in the singular value decomposition of the matrix $\Delta \boldsymbol{C}$. The matrix $\sum^{\dagger}_{\boldsymbol{\Delta C}}$ is the result of inverting the non-zero entries in $\sum_{\boldsymbol{\Delta C}}$, leaving the zeroes invariant. Upon convergence, the linear correction $\boldsymbol{S}^{*}$ (being the limit of $\boldsymbol{S}^{(i)}$ with $i \to \infty$) maps the point $\boldsymbol{c}(\boldsymbol{x}^{*})$ to $\boldsymbol{f}(\boldsymbol{x}^{*})$, and the tangent plane for $\boldsymbol{c}(\boldsymbol{x})$ at $\boldsymbol{c}(\boldsymbol{x}^{*})$ to the tangent plane for $\boldsymbol{f}(\boldsymbol{x})$ at $\boldsymbol{f}(\boldsymbol{x}^{*})$ (Echeverría and Hemker, 2008). If the sensitivity data is available for both the low- and high-fidelity model, the correction matrix $\boldsymbol{S}^{(i)}$ can be defined using the exact Jacobians of the models.

5.5.4 *Shape Preserving Response Prediction*

In Shape preserving response prediction (SPRP) (Koziel, 2010a, 2012), the surrogate model is constructed assuming that the change of the high-fidelity model response due to the adjustment of the design variables can be predicted using the actual response changes of the low-fidelity model. It is critically important for SPRP that

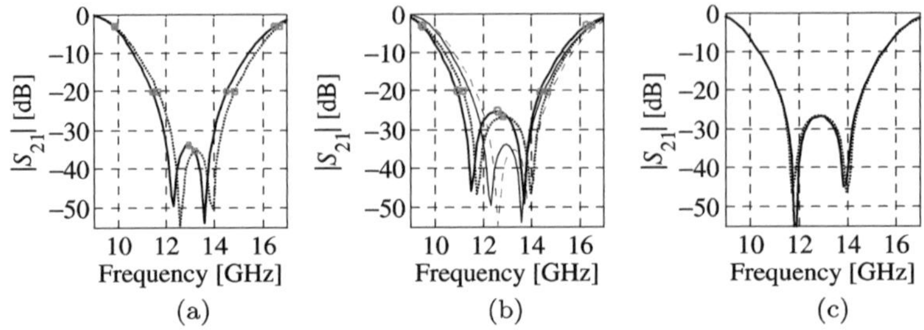

Fig. 5.11. SPRP concept (Koziel, 2010a): (a) Example low-fidelity model response at the design $\boldsymbol{x}^{(i)}, \boldsymbol{c}(\boldsymbol{x}^{(i)})$ (—), and at another design $\boldsymbol{x}, \boldsymbol{c}(\boldsymbol{x})$ (····), characteristic points of $\boldsymbol{c}(\boldsymbol{x}^{(i)})(\circ)$ and $\boldsymbol{c}(\boldsymbol{x})$, and the translation vectors (short lines); (b) High-fidelity model response at $\boldsymbol{x}^{(i)}, \boldsymbol{f}(\boldsymbol{x}^{(i)})$ (—) and the predicted fine model response at $\boldsymbol{x}$(····) obtained using SPRP based on characteristic points of Fig. 5.11(a); characteristic points of $\boldsymbol{f}(\boldsymbol{x}^{(i)})(\circ)$ and the translation vectors (short lines) were used to find the characteristic points of the predicted high-fidelity model response; low-fidelity model responses $\boldsymbol{c}(\boldsymbol{x}^{(i)})$ and $\boldsymbol{c}(\boldsymbol{x})$ are plotted using thin solid and dotted line, respectively; (c) predicted (····) and actual (—) high-fidelity model response at $\boldsymbol{x}$.

the low-fidelity model is physics-based, which ensures that the effect of the design parameter variations on the model response is similar for both models. The change of the low-fidelity model response is described by the translation vectors corresponding to certain (finite) number of characteristic points of the model's response. These translation vectors are subsequently used to predict the change of the high-fidelity model response with the actual response of $\boldsymbol{f}$ at the current iteration point, $\boldsymbol{f}(\boldsymbol{x}^{(i)})$, treated as a reference.

We explain the concept of SPRP using the example microstrip bandstop filter (Koziel, 2010c). Figure 5.11(a) shows the example of the low-fidelity model response $|S_{21}|$ (transmission coefficient) in the frequency range 8–18 GHz, at the design $\boldsymbol{x}^{(i)}$, as well as the low-fidelity model response at some other design $\boldsymbol{x}$. Circles denote five characteristic points of $\boldsymbol{c}(\boldsymbol{x}^{(i)})$, here, selected to represent $|S_{21}| = -3$ dB, $|S_{21}| = -20$ dB, and the local $|S_{21}|$ maximum (at about 13 GHz). Squares denote corresponding characteristic points for $\boldsymbol{c}(\boldsymbol{x})$, while small line segments represent the translation vectors that determine the "shift" of the characteristic points of $\boldsymbol{c}$ when changing the design variables from $\boldsymbol{x}^{(i)}$ to $\boldsymbol{x}$. Because the low-fidelity model is physics-based, the high-fidelity model response at the given design, here, $\boldsymbol{x}$, can be predicted using the same translation vectors applied to the corresponding characteristic points of the fine model response at $\boldsymbol{x}^{(i)}, \boldsymbol{f}(\boldsymbol{x}^{(i)})$, cf. Fig. 5.11(b). Figure 5.11(c) shows the predicted high-fidelity model response at the design $\boldsymbol{x}$, as well as the actual response, $\boldsymbol{f}(\boldsymbol{x})$, with a good agreement between both curves.

A rigorous formulation of SPRP can be found in the literature (Koziel, 2010a, 2010c). It should be mentioned that an important assumption of SPRP is that the overall shape of both the high- and low-fidelity model responses is similar. This means, in particular, that the characteristic points of responses of both the coarse model $\boldsymbol{c}$ and the fine model $\boldsymbol{f}$ are in one-to-one correspondence. If this assumption is not satisfied, the SPRP surrogate cannot be evaluated because the translation vectors are not well defined. Generalizations of SPRP that allow alleviating this difficulty in some cases can be found in the work of Koziel (2010c).

5.5.5 *Adaptively Adjusted Design Specifications*

All physics-based SBO algorithms outlined so far in this section exploit the idea of low-fidelity model correction in order to improve its alignment with the high-fidelity one. Adaptively adjusted design specifications (AADS) (Koziel, 2010b) proposes an alternative way of exploiting the system-specific knowledge embedded in the low-fidelity model by modifying the design specifications so as to account for the model discrepancies. AADS is not universally applicable, but it is extremely simple to implement as no changes of the low-fidelity model are required. AADS consists of the two basic steps:

1. Modify the design specifications of the original problem to account for the differences between the responses of the high-fidelity model $\boldsymbol{f}$ and the low-fidelity model $\boldsymbol{c}$ at their characteristic points.
2. Obtain a new design by optimizing the low-fidelity model with respect to the modified specifications.

AADS has been developed to handle minimax-type of specifications (Koziel, 2010b) so that the characteristic points of the

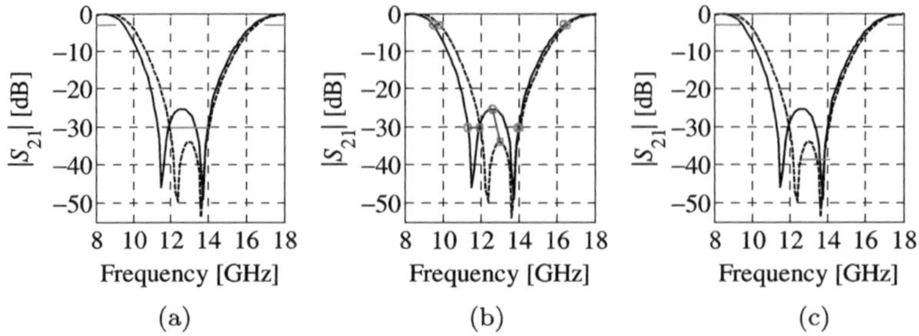

Fig. 5.12. AADS concept (responses of $\boldsymbol{f}$ and $\boldsymbol{c}$ are marked with solid and dashed line, respectively): (a) high- and low-fidelity model responses at the initial design (optimum of $\boldsymbol{c}$) as well as the original design specifications; (b) characteristic points of the responses corresponding to the specification levels (here, −3 dB and −30 dB) and to the local response maxima; (c) high- and low-fidelity model responses at the initial design and the modified design specifications.

responses should correspond to the relevant design specification levels. They may also include local maxima/minima of the respective responses at which the specifications may not be satisfied. Figure 5.12(a) shows the high- and low-fidelity model responses at the optimal design of $\boldsymbol{c}$, corresponding to the bandstop filter example (Koziel, 2010d); design specifications are indicated using horizontal lines (here, $|S_{21}| \leq -30$ dB for the frequencies 12–14 GHz, and $|S_{21}| \geq -3$ dB for the frequencies 8–9 GHz and 17–18 GHz). Figure 5.12(b) shows the characteristic points of $\boldsymbol{f}$ and $\boldsymbol{c}$, i.e., the points corresponding to −3 dB and −30 dB levels as well to the local maxima of the responses. In the first step of the AADS optimization procedure, the design specifications are modified so that the level of satisfying/violating the modified specifications by the low-fidelity model response corresponds to the satisfaction/violation levels of the original specifications by the high-fidelity model response.

In the example of Fig. 5.12, for each edge of the specification line, the edge frequency is shifted by the difference of the frequencies of the corresponding characteristic points, e.g., the left edge of the specification line of −30 dB is moved to the right by about 0.7 GHz, which is equal to the length of the line connecting the corresponding characteristic points in Fig. 5.12(b).

Similarly, the specification levels are shifted by the difference between the local maxima/minima values for the respective points, e.g., the –30 dB level is shifted down by about 8.5 dB because of the difference of the local maxima of the corresponding characteristic points of $\boldsymbol{f}$ and $\boldsymbol{c}$. Modified design specifications are shown in Fig. 5.12(c).

The low-fidelity model is subsequently optimized with respect to the modified specifications and the new design obtained this way is treated as an approximated solution to the original design problem. Steps 1 and 2 can be iterated if necessary. If the correlation between the low- and high-fidelity models is good, a substantial design improvement is typically observed after the first iteration, however, additional iterations may bring further enhancements (Koziel, 2010b). Figure 5.13 illustrates an AADS iteration applied

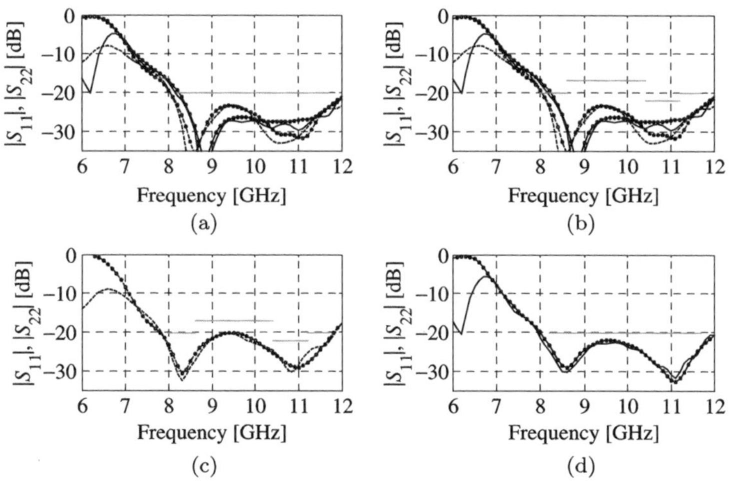

Fig. 5.13. AADS for optimization of microstrip-to-SIW transition (Ogurtsov and Koziel, 2011): high- and low-fidelity model response denoted as solid and dashed lines, respectively. $|S_{22}|$ distinguished from $|S_{11}|$ using circles. Design specs denoted by thick horizontal lines. (a) model responses at the beginning of the iteration and original design specifications; (b) model responses and modified design specifications that reflect the differences between the responses; (c) low-fidelity model optimized with respect to the modified specifications; (d) high-fidelity model at the low-fidelity model optimum shown versus original specifications.

to microstrip-to-SIW transition design (Ogurtsov and Koziel, 2011). Note that optimizing the low-fidelity model with respect to the modified specifications results in improving the high-fidelity model design with respect to the original specifications. Because the model discrepancies may change somehow from one design to another, a few iterations may be necessary to find an optimal high-fidelity design.

5.5.6 *Summary*

SBO algorithms with physics-based surrogates exploit the problem-specific knowledge embedded in the low-fidelity model, thereby permitting better computational efficiency compared to data-driven surrogates. On the other hand, they are typically more difficult in implementation and less transferrable between various classes of problems. Still, physics-based SBO methods seem to be the most promising approaches for handling complex engineering design problems involving expensive computational models with highly nonlinear responses. The techniques presented in this section are just selected examples. Information about other methods can be found in the literature, e.g., adaptive response correction (Koziel *et al.*, 2009), feature-based optimization (Koziel and Bandler, 2015), nested space mapping (Koziel *et al.*, 2014e), corrected space mapping (Robinson *et al.* , 2008), or tuning space mapping (Koziel *et al.*, 2011b).

Chapter 6

Multi-Objective Optimization

Majority of engineering design problems are of multi-objective nature, i.e., there are several performance figures that have to be controlled at the same time. In antenna design, typical figures include reflection response, gain, radiation pattern, axial ratio (for circular polarization antennas), as well as the structure size (for compact antennas) (Azaro, 2008; Gautam *et al.*, 2013; Ding *et al.*, 2008; Kuwahara, 2005; Bekasiewicz *et al.*, 2016b; Bai *et al.*, 2016; Smolders and Johannsen, 2011; Viani *et al.*, 2008). In most cases, either because the designer's priorities are known beforehand or just to simplify the design task, the problem can be reformulated into a single-objective one, which can be achieved by selecting the primary goals and handling the remaining objectives by means of constraints (Koziel and Bekasiewicz, 2015e). Another popular technique is aggregation of the objectives into a scalar cost function using a weighted sum approach (Kuwahara, 2005; Lizzi *et al.*, 2009) or penalty functions (Bekasiewicz and Koziel, 2016b). Nevertheless, in some situations, it might be important to acquire more comprehensive information about the antenna at hand, in particular, to identify the best possible trade-offs between the conflicting criteria. If this is the case, defaulting to proper multi-objective optimization becomes necessary. Multi-objective optimization leads to a set of alternative designs representing a so-called Pareto front (Fonseca, 1995). Subsequently, a decision making process allows for selecting one of these designs as a final solution.

Perhaps, the most popular multi-objective optimization approaches are population-based metaheuristics (Deb, 2001; Talbi, 2009). These algorithms are capable of yielding the entire Pareto set in one algorithm run, however, the computational cost of evolutionary optimization may be very high: thousands or even tens of thousands of objective function evaluations. In particular, direct multi-objective optimization of expensive electromagnetic (EM)-simulation models of antenna structures is normally prohibitive.

In this chapter, we discuss the formulation of multi-objective optimization problem, review popular solution approaches including a weighted sum approach, a goal attainment method, as well as multi-objective versions of several population-based metaheuristic algorithms. The latter will be used as one of the components of surrogate-assisted antenna optimization schemes presented later in the book. The algorithms for accelerated multi-objective design of antenna structures will be discussed in Chapters 7 and 8, whereas their applications for solving practical antenna and microwave design problems will be presented in Chapters 9–12.

6.1 Formulation of Multi-Objective Optimization Problem

As before, we denote by $\boldsymbol{f}(\boldsymbol{x})$ the high-fidelity simulation model of the antenna under design, where $\boldsymbol{x} \in X$ is a vector of designable parameters, whereas X is the high-fidelity model domain. We consider N_{obj} design objectives, $F_k(\boldsymbol{x}), k = 1, \ldots, N_{\text{obj}}$. The space of objectives consisting of vectors $\boldsymbol{f}(\boldsymbol{x}) = [F_1(\boldsymbol{x})\ F_2(\boldsymbol{x}) \ldots F_{N\text{obj}}(\boldsymbol{x})]^T$ is referred to as a feature space. In most practical cases, these objectives are conflicting with each other, so that improvement of one leads to degradation of the others.

Assuming that $N_{\text{obj}} > 1$, any two designs $\boldsymbol{x}^{(1)}$ and $\boldsymbol{x}^{(2)}$ for which $F_k(\boldsymbol{x}^{(1)}) < F_k(\boldsymbol{x}^{(2)})$ and $F_l(\boldsymbol{x}^{(2)}) < F_l(\boldsymbol{x}^{(1)})$ for at least one pair $k \neq l$, are not commensurable, i.e., none is better than the other in the multi-objective sense. For the sake of comparing the designs, a Pareto dominance relation $\prec$ can be used, which is defined as follows (Fonseca, 1995): for the two designs $\boldsymbol{x}$ and $\boldsymbol{y}$, we have $\boldsymbol{x} \prec \boldsymbol{y}$ ($\boldsymbol{x}$ dominates $\boldsymbol{y}$) if $F_k(\boldsymbol{x}) \leq F_k(\boldsymbol{y})$ for all $k = 1, \ldots, N_{\text{obj}}$, and $F_k(\boldsymbol{x}) < F_k(\boldsymbol{y})$ for at least one k. The goal of multi-objective optimization is to find a representation of a so-called Pareto front (or Pareto-optimal set) $\boldsymbol{f}(X_P)$ of the design space image through $\boldsymbol{f}(X)$, such that for any $\boldsymbol{x} \in X_P$, there is no $\boldsymbol{y} \in X$ for which $\boldsymbol{y} \prec \boldsymbol{x}$ (Fonseca, 1995). The concept of the Pareto dominance relation is explained in Fig. 6.1.

Figure 6.2 explains the concept of the Pareto front for the two-objective design problem. The part of the boundary of the design space image through $\boldsymbol{f}, \boldsymbol{f}(X)$, marked with the thick line, is the Pareto front, i.e., there are no designs in X that would dominate any design in X_P. In general, the Pareto front is at most an $N_{\text{obj}} - 1$

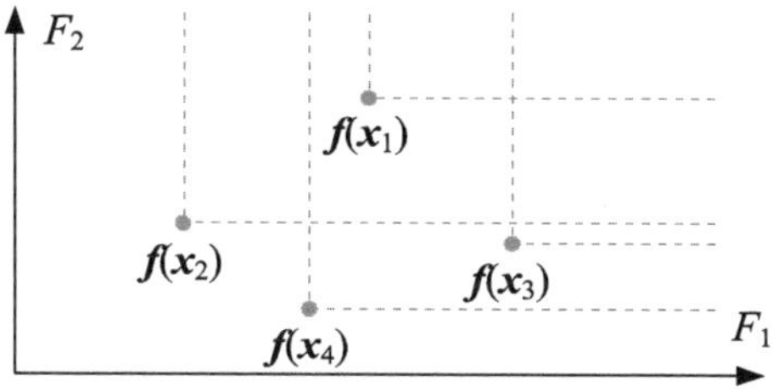

Fig. 6.1. Illustration of the Pareto dominance relation for a two-objective feature space. For the four designs $\boldsymbol{x}_k, k = 1, 2, 3, 4$, we have the following relations: $\boldsymbol{x}_2 \prec \boldsymbol{x}_1, \boldsymbol{x}_4 \prec \boldsymbol{x}_3$ and $\boldsymbol{x}_4 \prec \boldsymbol{x}_1$; the remaining pairs of designs are not in a dominance relation with each other.

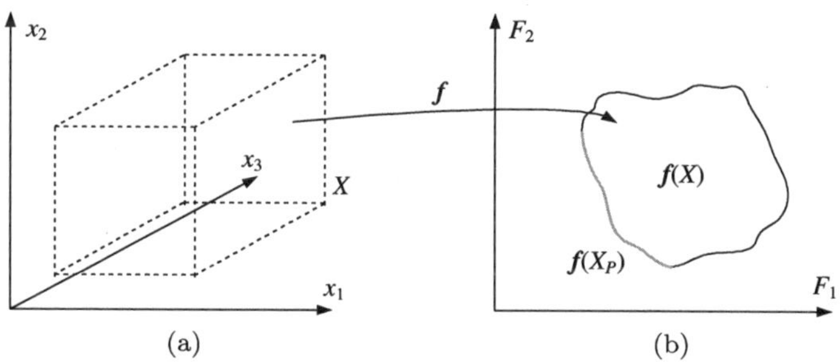

Fig. 6.2. Conceptual illustration of the Pareto front for a two-objective design problem in a three-dimensional design space: design space X (a), its image $\boldsymbol{f}(X)$ (b), and the image of the Pareto front $\boldsymbol{f}(X_P)$ marked using a thick line.

dimensional manifold in the feature space (e.g., a curve for $N_{\text{obj}} = 2$ and a surface for $N_{\text{obj}} = 3$, etc.).

Figure 6.3 illustrates a Pareto set for a ultra-wideband (UWB) dipole antenna (Koziel and Bekasiewicz, 2015f) optimized for two design objectives, reduction of the structure size and minimization of the maximum in-band reflection level. It can be observed that the antenna size changes from slightly over 700–1,200 mm^2, whereas the corresponding reflection level changes from -9 dB to -14 dB.

6.2 Solution Approaches

The relationship between design parameters and objectives, and, consequently, geometry of the Pareto front depends on various factors,

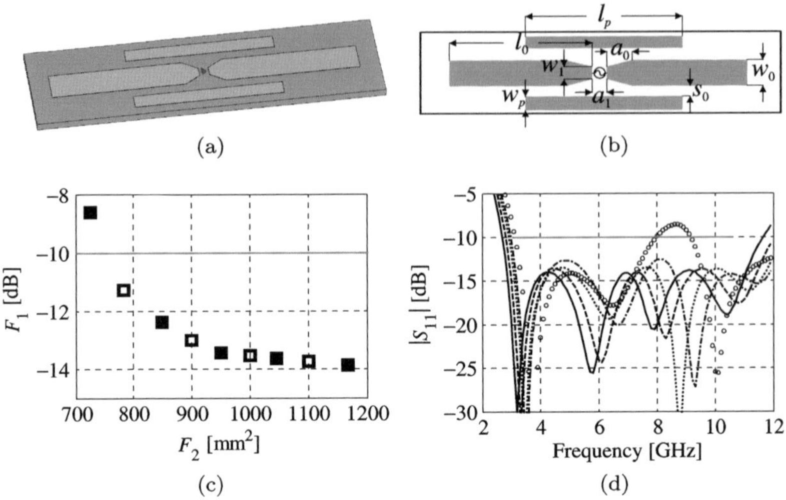

Fig. 6.3. Pareto front illustration for a UWB dipole antenna designed for two-objectives (Koziel and Bekasiewicz, 2015f), F_1 — size reduction and F_2 — minimization of the maximum in-band reflection: (a) visualization; (b) geometrical details, (c) Pareto set (designs 1, 3, 5, 7 and 9 marked black), (d) reflection characteristics of designs: 1 (ooo), 3 (–·), 5 (····), 7 (– –), 9 (—).

including complexity of the system under design. In general, there are three situations that are possible (Tan *et al.*, 2005):

- Totally conflicting objectives, i.e., there are no two designs $\boldsymbol{x}, \boldsymbol{y} \in X$ such that $\boldsymbol{x} \prec \boldsymbol{y}$ or $\boldsymbol{y} \prec \boldsymbol{x}$. In particular, no optimization is necessary for this class of problem because the feasible set X already represents the global Pareto optimal solutions, i.e., we have $X_P = X$.
- Non-conflicting objectives. In this case, for any two designs $\boldsymbol{x}, \boldsymbol{y} \in X$ we either have $\boldsymbol{x} \prec \boldsymbol{y}$ or $\boldsymbol{y} \prec \boldsymbol{x}$. This class of problems can be converted into single-objective task by considering only one objective (as improvement of any objective leads to improvement of all the others). For this class of problems, the Pareto set is usually a single point.
- Partially conflicting objectives. This case covers all other situations, i.e., when the objectives are neither totally conflicting nor non-conflicting.

For partially conflicting objectives, the image of the problem domain $\boldsymbol{f}(X)$ is a "thick" set and, consequently, the Pareto front is a proper subset of the domain, i.e., $X_P \subset X$ but $X_P \neq X$. Needless to say, the last situation is the most interesting and the most common one for majority of real-world engineering design problems, including those related to antenna design.

In the course of multi-objective optimization, a decision has to be made towards selecting a single final design. A decision making process is based upon the set of trade-off solutions identified in the process, but it can also be interleaved with the optimization itself. The following approaches can be distinguished (Coello Coello *et al.*, 2007):

- *A priori* preference articulation, where the multi-objective problem is transformed into a single-objective one (e.g., by objective aggregation) prior to optimization.
- Progressive preference articulation, where decision and optimization are intertwined; in particular, partial preference information is provided upon which optimization proceeds.
- Posteriori preference articulation, where a set of efficient candidate solutions is found using a specific algorithm and then a decision is made on the choice of the best solution.

If the relative importance of various objectives is clearly defined and known beforehand, *a priori* preference articulation seems to be a proper approach. If the knowledge of the problem is limited, the initial goals may be set for the objectives and subsequently modified during the optimization process, especially if it turns out that some objectives are difficult to be met (progressive preference articulation). In general, however, it might be preferable to acquire as much information about the Pareto optimal designs as possible. This would give an idea about available trade-offs between design objectives and allow for selecting the most suitable solution (posteriori preference articulation). As mentioned before, this approach might be very expensive in computational terms. For the purpose of this book, multi-objective optimization is merely understood as a procedure of Pareto set identification.

In the remaining part of this chapter, we discuss several multi-objective optimization approaches, including the weighted sum method, the goal attainment method, multi-objective evolutionary algorithms (MOEAs), as well as briefly mention multi-objective versions of several other population-based metaheuristics. Other methods (not covered here) include lexicographic ordering (Fourman, 1985), Chebyshev approach (Miettinen, 1999) or independent sampling (Srigiriraju, 2000). An excellent survey of preference articulation techniques can be found in the works of Coello Coello *et al.* (2007) and Deb (2001).

6.3 Weighted Sum Method

One of the simplest techniques for multi-objective optimization is a weighted sum method (Changkong and Haimes, 1983), where the problem is converted into a single-objective one by aggregating all objectives into a scalar cost function using a convex combination of the form,

$$\sum_{k=1}^{N_{\text{obj}}} a_k f_k(\boldsymbol{x}) \quad a_k > 0, \quad i = 1, 2, \ldots, N_{\text{obj}}, \quad \sum_{k=1}^{N_{\text{obj}}} a_k = 1. \qquad (6.1)$$

The weights a_k might be adjusted depending on relative importance of particular objectives. Minimizing the cost function (6.1) allows for finding a single point on the Pareto front. A location of that point depends on the values of the weighting factors as illustrated in Fig. 6.4.

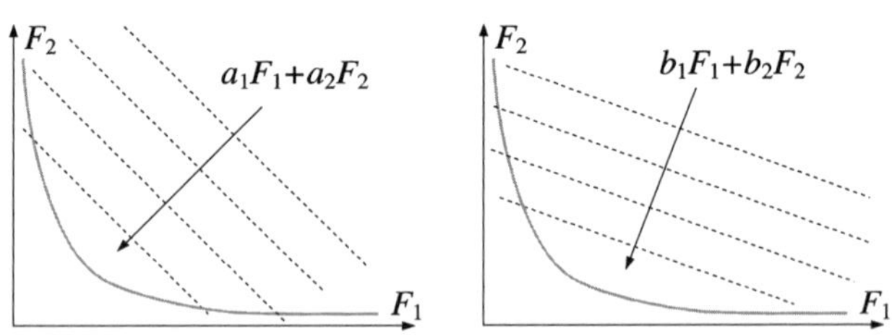

Fig. 6.4. Weighted sum method and dependence of the Pareto-optimal solution found on the weighting factor values. The dashed lines represent the contours of the aggregated cost function.

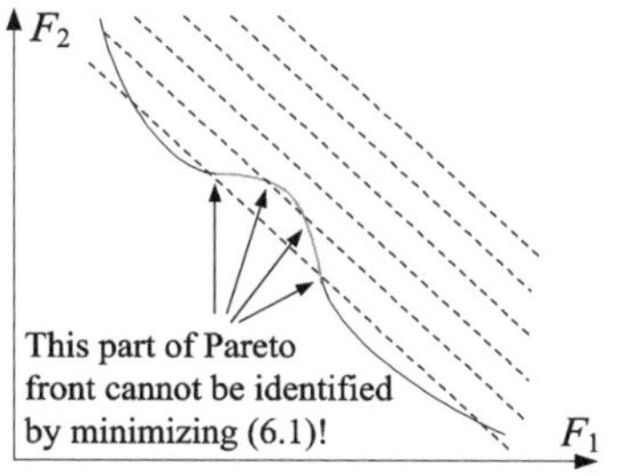

Fig. 6.5. Illustration of one of the disadvantages of the weighted sum method, that is, its inability to find Pareto-optimal designs on non-convex part of the Pareto front.

In order to find a representation of the Pareto front, multiple optimization runs of (6.1) are necessary for various setups of the weights a_k (using, e.g., a linear sweep). Obviously, the cost of such operation can be large, however, the process can be easily parallelized (if sufficient computational resources are available). Nevertheless, the weighted sum method has two serious disadvantages:

- It is difficult to generate a uniformly spread set of Pareto-optimal solutions because the geometrical properties of the Pareto front are not known beforehand. In particular, the designs obtained in various runs can cluster in certain parts of the front with no representation in other parts.
- Non-convex parts of the Pareto front are "invisible" for the method so that no designs in such regions can be found by minimizing the cost function (6.1). This property is illustrated in Fig. 6.5.

On the other hand, the advantage of this technique is its simplicity, including the fact that the aggregated cost function can be minimized using any available conventional optimization algorithms.

6.4 Goal Attainment Method

The goal attainment method (Gembicki, 1974) controls relative importance of the objectives by appropriately defined design constraints. Let $F^* = [F_1^* \; F_2^* \ldots F_{N\text{obj}}^*]^T$ denote a set of design goals associated with the objectives $F_k, k = 1, 2, \ldots, N_{\text{obj}}$. The relative

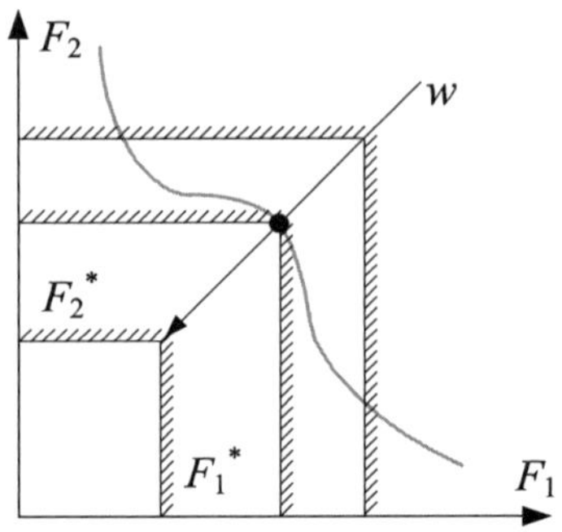

Fig. 6.6. Operation of the goal attainment method. Feasible region plots for various values of γ for a specific combination of the weighting factors. Adjusting the weights allows for identifying non-convex parts of the Pareto front.

degree of under- or over-achievement of the objectives is controlled by a vector of weighting coefficients $w = [w_1 \ldots w_m]^T$. The optimization problem is formulated as

$$\arg \min_{\gamma \in R,\, \boldsymbol{x} \in X} \gamma, \tag{6.2}$$

under the following set of constraints

$$F_k(\boldsymbol{x}) - w_k\gamma \leq F_k^*, \quad k = 1, \ldots, N_{\text{obj}}. \tag{6.3}$$

The weighting factors w_k allow for controlling the relative trade-offs between the objectives. In particular, decreasing w_k increases the importance of achieving the target value of F_k^* for the objective F_k. On the other hand, setting a particular weighting factor to zero introduces a hard constraint on that specific goal.

The advantage of the goal attainment method is the ability of locating non-convex areas of the Pareto front as explained in Fig. 6.6.

6.5 Multi-Objective Evolutionary Algorithms

By far, the most popular multi-objective optimization methods nowadays are population-based metaheuristics. For certain types of problems, especially those where the computational cost of objective evaluation is not of primary concern, they are also very effective and reliable. Their attractiveness is due to several features: ability of generating the entire Pareto front representation in a single algorithm run, ability of handling discontinuous and noisy objective functions, as well as straightforward parallelization. On the other hand, population-based methods are very expensive in terms of the required number of objective evaluation, which is a serious bottleneck for multi-objective EM-driven antenna design. The ways of alleviating these difficulties are discussed in Chapter 7.

In this section, we describe a general structure of MOEA, the mechanisms that allow for finding Pareto-optimal designs, as well as various practical issues that need to be taken care of to secure proper operation of the algorithm. A brief outline of multi-objective versions of other popular metaheuristics can be found in Sec. 6.6.

6.5.1 *Algorithm Structure and Search Mechanisms*

As shown in Fig. 6.7(a), a general structure of a MOEA is similar to that of a single-objective one. In particular, the new individuals are created using recombination and mutation operators similar to those described in Chapter 4 (floating-point representation is normally used). However, assessment of individuals, selection procedure as well as elitism, are all implemented differently. As indicated in Fig. 6.7(b) there are two types of incentives that have to be realized in order to obtain a uniform distribution of solutions along the Pareto front: (i) a so-called normal pressure P_n that pushes individuals towards the front, and (ii) the tangent pressure P_t that is applied in order to avoid solution clustering. The first one is enforced by a selection process where the preference is given to non-dominated individuals. The second can be implemented by means of fitness sharing that penalizes individuals that are clustered. In terms of elitism, in multi-objective algorithms, one needs to keep track of the entire set of non-dominated solutions instead of just one (as in single-objective algorithms), which allows for generating the final Pareto front representation and also for controlling the algorithm convergence.

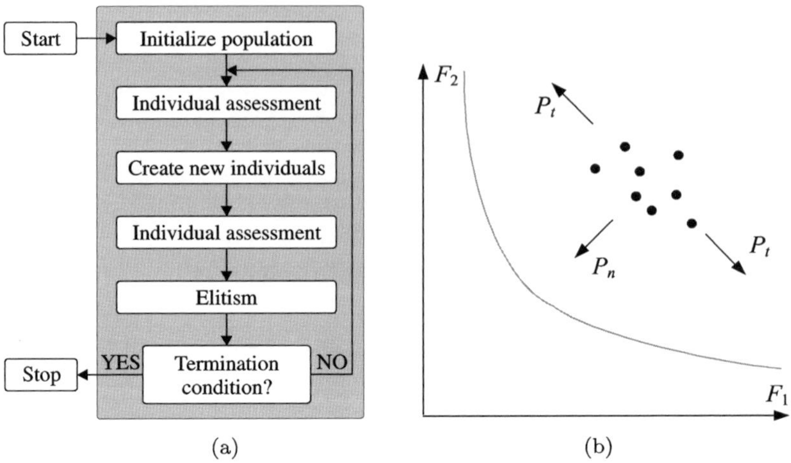

Fig. 6.7. MOEAs: (a) general structure, (b) normal pressure P_n and tangent pressure P_t implemented using appropriate mechanisms within the algorithm and necessary to obtain a uniform representation of the Pareto front (Tan *et al.*, 2005).

The aforementioned algorithm components are discussed based in subsequent paragraph on the MOEA: implementation presented in the work of Fonseca (1995).

6.5.2 *Assessment of Individuals*

Individuals in the population are assessed using Pareto dominance relation defined in Sec. 6.1. One of the possible methods is ranking where the rank r_k of the kth individual is given as (Coello Coello *et al.*, 2007)

$$r_k = 1 + q_k, \tag{6.4}$$

with q_k being the number of individuals that dominate over the kth individual. The fitness f_k of the kth individual is then computed as $f_k = 1/r_k$.

6.5.3 *Fitness Sharing*

In order to enforce individuals within the population to uniformly spread along the Pareto front, it is necessary to implement a fitness sharing that penalizes individuals in case they are too close to each other. The distance between individual k and j can be measured in the feature space as (Tan *et al.*, 2005)

$$d\left(\boldsymbol{f}\left(\boldsymbol{x}^{(k)}\right), \boldsymbol{f}\left(\boldsymbol{x}^{(j)}\right)\right) = \left\|\boldsymbol{f}\left(\boldsymbol{x}^{(i)}\right) - \boldsymbol{f}\left(\boldsymbol{x}^{(j)}\right)\right\|_2. \tag{6.5}$$

The sharing function can be defined as follows (Goldberg and Richardson, 1987):

$$SF(i,j) = \begin{cases} 1 - [d(\boldsymbol{f}(\boldsymbol{x}^{(i)}), & \text{if } d(\boldsymbol{f}(\boldsymbol{x}^{(i)}), \\ \quad \boldsymbol{f}(\boldsymbol{x}^{(j)}))/\sigma_{\text{share}}]^{\alpha} & \boldsymbol{f}(\boldsymbol{x}^{(j)})) < \sigma_{\text{share}}, \\ \quad 0 & \text{otherwise}, \end{cases} \tag{6.6}$$

where α is a parameter that controls the shape of the sharing function, whereas σ_{share} determines the extent of sharing allowed in terms of the radius distance. The shared fitness of the kth individual is then given as

$$f'_k = f_k \Big/ \sum_{j=1}^{N} SF(k,j). \tag{6.7}$$

As indicated in (6.7), the individuals in a "crowded neighborhood" have their fitness decreased, which promotes uniform spatial distribution of the individuals in the population.

Clearly, a critical issue is appropriate selection of the sharing distance σ_{share}. It should be chosen so that the size of the "niche" is in a proper relation to the number of individuals in the population and the size of the Pareto front. It is conceptually shown in Fig. 6.8. The value of σ_{share} should be selected carefully because if it is too large or too small, the regions of the feature space that are overcrowded cannot be detected.

Dynamic sharing (Tan *et al.*, 2005) allows for adaptive adjustment of σ_{share} by taking into account the estimated size (volume) of the Pareto front in relation to the population size. The sharing distance at iteration k of the algorithm is given as

$$\sigma_{\text{share}}^{(k)} = N^{1/(N_{\text{obj}}-1)} d^{(k)}/2, \tag{6.8}$$

where N is the population size, $d^{(k)} = (d_{\max}^{(k)} + d_{\min}^{(k)})/2$ in which $d_{\min}^{(k)} = [(d_1^{(k)})^2 + (d_2^{(k)})^2]^{1/2}$, $d_{\max}^{(k)} = d_1^{(k)} + d_2^{(k)}$, with $d_1^{(k)} = ||F_x^{(k)} - F_z^{(k)}||$, and

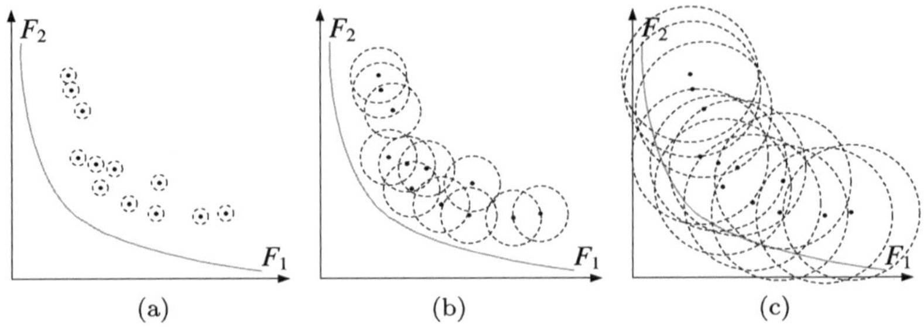

Fig. 6.8. Sharing distance σ_{share} and how it affects the fitness sharing efficiency: (a) σ_{share} too small: fitness is not penalized even at clusters, (b) σ_{share} of a proper value: fitness is penalized in crowded regions but not elsewhere, (c) σ_{share} too large: fitness is penalized in all regions.

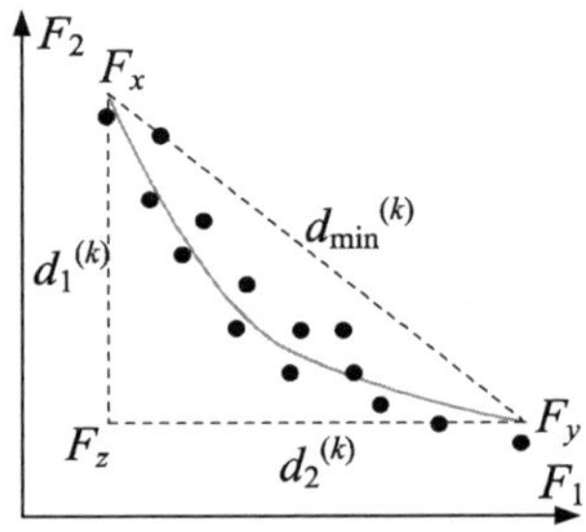

Fig. 6.9. Notation used in the definition of the dynamic sharing distance (6.8).

$d_2^{(k)} = ||F_y^{(k)} - F_z^{(k)}||$. The "extreme" estimated points of the Pareto front are defined as $||F_x^{(k)} - F_y^{(k)}|| = \max_{i,j\in\{1,\ldots,N_{\text{obj}}\},\, i\neq j}||\, \boldsymbol{f}(\boldsymbol{x}^{(i)}) - \boldsymbol{f}(\boldsymbol{x}^{(j)})||$, and $F_z^{(k)} = [f_{z.1}^{(k)} \ \ldots \ f_{z.N_{\text{obj}}}^{(k)}]^T$ with $f_{z.j}^{(k)} = \min_{j\in\{1,\ldots,N_{\text{obj}}\}} \{f_{x.j}^{(k)}, f_{y.j}^{(k)}\}$. The notation used in the above description has been explained in Fig. 6.9.

6.5.4 *Selection*

Selection is one of the most important components of the MOEA. In general, it can be based on any selection procedure used for single-objective algorithms, however, a Pareto-domination-based assessment with fitness sharing should be exploited to evaluate individuals. A specific example described here utilizes tournament selection.

Let c_1 and c_2 denote the two candidate individuals, C_{comp} denote a comparison set (randomly selected subset of the population), whereas s denotes an individual selected upon accomplishing the selection procedure. The selection process works as follows (Horn *et al.*, 1994):

1. Set $c_{1.\text{dominated}} = false$ and $c_{2.\text{dominated}} = false$;
2. If c_1 is dominated by any $c \in C_{\text{comp}}$ then set $c_{1.\text{dominated}} = true$;
3. If c_2 is dominated by any $c \in C_{\text{comp}}$ then set $c_{2.\text{dominated}} = true$;
4. If $c_{1.\text{dominated}} == c_{2.\text{dominated}}$ then
   Compute shared fitness f_1 of c_1;
   Compute shared fitness f_2 of c_2;
   Go to 6;
5. If $c_{1.\text{dominated}} == false$ then $s = c_1$, else $s = c_2$; go to 7;
6. If $f_1 > f_2$ then $s = c_1$, else $s = c_2$;
7. END.

In the procedure above, the two candidates are compared with individuals from the comparison set. If their domination flag is the same (both are either non-dominated by C_{comp} or both are dominated), the candidate featuring higher value of the shared fitness is selected. Otherwise, the non-dominated candidate is selected. The amount of selection pressure — in terms of the preference given to non-dominated solutions — can be controlled by changing the size of the comparison set (typically, about 10 percent of the population size).

6.5.5 *Elitism*

Similarly as for a single-objective case, a MOEA should implement elitism so as to keep track of a set of the best non-dominated individuals. This so-called elitist set eventually becomes the very outcome of the algorithm, i.e., a representation of the Pareto front. It should be mentioned that the elitist set can also be utilized in the reproduction process to speed up convergence of the algorithm. Clearly, such means have to be used with caution to avoid premature convergence due to over-influence of the best individuals (Deb, 2001).

Let Y denote a set that is a concatenation of the current elitist set and the newly created population. Let Z denote the new elitist set to be constructed. There are two popular implementations of the elitism that work as follows (Tan *et al.*, 2005):

Batch mode:

1. Evaluate all individuals in Y (i.e., compute shared fitness);
2. Select the subset of best individuals to fill Z.

Recurrence mode:

1. Evaluate all individuals in Y;
2. Remove the least promising individuals from Y and fill in Z;
3. If the size of Z is as required, terminate; else set $Y = Z$ and go to 1.

The recurrence mode is computationally more expensive, however, it has a higher tendency of avoiding the extinction of local individuals leading to discontinuity of the Pareto front.

6.5.6 *Mating Restrictions*

The mechanisms described so far, in particular, Pareto-dominance-based selection and dynamic sharing, implement forces that push the solutions towards the Pareto front and distribute them along it. On the other hand, recombination is one of the processes that may push the solutions away from the front. If the parent individuals are both Pareto-optimal yet allocated in different regions of the Pareto front, it is very likely that their offspring obtained through recombination will be poorly fitted in terms of Pareto dominance. This is because of the Pareto front geometry: recall that it is a manifold of the dimensionality of at most $N_{\text{obj}} - 1$, i.e., it is a "thin" subset of the feature space.

So-called mating restrictions alleviate this difficulty by ensuring that individuals too different from each other are generally less likely than similar individuals to produce fit offspring through crossover. In other words, mating of similar individuals is favored. Mating restrictions are implemented as follows (Deb and Goldberg, 1989):

1. Specify how close individuals should be in order to mate; set the corresponding parameter σ_{mate};
2. Upon selection, choose one individual from the mating pool and search for a mate within the distance σ_{mate}.

If such an individual can be found, the mating is performed; otherwise, a random individual is chosen. Typically σ_{mate} is related to σ_{share} used in dynamic sharing (cf. Sec. 6.5.4), e.g., $\sigma_{\text{mate}} = 3\sigma_{\text{share}}$.

6.5.7 *Stopping Criteria*

Similarly as for a single-objective case, the simplest termination criterion for a multi-objective algorithm is the maximum number of iterations. More sophisticated way of terminating the algorithm is to assess its advancement towards the Pareto front as formed by currently non-dominated individuals. A practical example of a convergence measure is (Tan *et al.*, 2005)

$$p^{(k)} = \frac{\bar{N}^{(k)}_{\text{non-dom}}}{N^{(k)}_{\text{non-dom}}}, \tag{6.9}$$

where $\bar{N}^{(k)}_{\text{non-dom}}$ is the number of non-dominated individuals at generation k that are dominating the non-dominated individuals at generation $k-1$, whereas $N^{(k)}_{\text{non-dom}}$ is the number of non-dominated individuals at generation k. Convergence of $p^{(k)}$ to zero indicates no further possibility of Pareto front improvement. However, in order to avoid irregularity (there might be significant fluctuations of $p^{(k)}$ from iteration to iteration), one may consider a moving average of $p^{(k)}$ of the form:

$$\bar{p}^{(k)} = \sum_{j=k+1-l}^{k} p^{(j)} \Bigg/ l, \tag{6.10}$$

where $l > 1$.

6.6 Other Multi-Objective Metaheuristics

As mentioned in Chapter 4, there is a considerable number of population-based metaheuristic algorithms available including well established methods (genetic/evolutionary algorithms, evolution strategies, particle swarm optimizers (PSO), differential evolution (DE), Horn *et al.*, 1994; Ding and Wang, 2013; Binh and Korn, 1997; Reyes-Sierra and Coello Coello, 2006; Robič and Filipič, 2005) and newer techniques (e.g., firefly algorithm, cuckoo search, harmony search and many others, Yang, 2010a; Geem *et al.*, 2001). Many of these algorithms have been generalized to permit multi-objective optimization (Yang, 2013; Yang and Deb, 2013; Sivasubramani and Swarup, 2011). Although the basic structures are inherited from the single-objective versions, certain modifications are introduced in order to achieve two goals: (i) enforce the population to move (or evolve) towards the Pareto front, and (ii) maximize the spread of the solutions to avoid clustering and obtain possibly uniform coverage of the Pareto front. Particular ways of realizing these goals are algorithm dependent. A detailed discussion of a selected variation of MOEA has been provided in Sec. 6.5. In this section, we briefly outline multi-objective PSO and DE.

6.6.1 *Multi-Objective Particle Swarm Optimization*

In single-objective PSO (see Sec. 4.5 for details), apart from other factors influencing relocation of the particles, there is a single globally best solution (the leader) that attracts the particles in the swarm. This is a mechanism that eventually leads to a convergence of the algorithm towards a single final solution. In a multi-objective case, the concept of a leader needs to be redefined (e.g., Parsopoulos *et al.*, 2004a; Srinivasan and Seow, 2003) and additional measures have to be taken to promote diversity in the population. Figure 6.10 outlines the flow of single- and multi-objective PSO. The major change is that an archive of non-dominated solutions is created and updated after each iteration. The leaders are selected separately for each particle so that the swarm may converge towards multiple locations. Diversity within the archive can be achieved using techniques similar to fitness sharing of MOEA (cf. Sec. 6.5.3). More information about multi-objective PSO and its various implementations can be found in the works of Reyes-Sierra and Coello Coello (2006).

(a)

1. Initialize the swarm;
2. Repeat for each particle:
   - Select leader
   - Update velocity
   - Update position
3. Update global best;
4. Go to 2.

(b)

1. Initialize the swarm and the archive (non-dominated solutions);
2. Repeat for each particle:
   - Select leader from the archive
   - Update velocity
   - Update position
3. Update the archive;
4. Go to 2.

Fig. 6.10. Algorithmic flow of PSO: (a) single-objective, (b) multi-objective.

6.6.2 *Multi-Objective Differential Evolution*

Similarly as for other metaheuristic algorithms, differential evolution (cf. Sec. 4.6) also has a variety of multi-objective versions. A comprehensive classification of these approaches, based on typical features to adopt DE for handling multi-objective problems, has been proposed by Chang *et al.* (1999). The following categories are distinguished (Mezura-Montes *et al.*, 2006b): (i) non-Pareto-based approaches, (ii) Pareto-based approaches (with the two sub-categories of methods using Pareto dominance and Pareto ranking), and (ii) combined approaches.

The non-Pareto-based approaches utilize concepts such as objective aggregation or controlling objectives through constraint functions. In the works of Babu and Jehan (2003), a multi-objective DE algorithm based on decomposition has been proposed using the weighted Chebyshev approach to decompose a multi-objective optimization problem into several scalar sub-problems.

The largest class of multi-objective DE variations is the one that utilizes the Pareto concepts. As mentioned before, there are two sub-classes here. The first one is based on Pareto dominance. For example, in the works of Chang *et al.* (1999), the authors utilized an external archive of non-dominated solutions as well as the fitness sharing concept to promote population diversity. In the works of Abbass *et al.* (2001), only non-dominated solutions are maintained in the population (all dominated ones are removed), and the offspring created using DE operations replaces the main parent only if it dominates over it. The second sub-class includes the approaches where a Pareto ranking procedure is exploited. Typically, it is in the form of a so-called $(\mu + \lambda)$ selection, where λ newly created solutions are added to the current population of μ individuals and the new population is selected from the aggregated set using Pareto ranking. In general, the mechanisms used in these versions of multi-objective DE are similar to those utilized by MOEAs (in particular, NSGA-II, Deb *et al.*, 2002). The example implementations can be found in the works of Madavan (2002); Sarker *et al.* (2002); Kukkonen and Lampinen (2004).

Combined approaches belong to the last class of multi-objective DE algorithms, where various concepts (e.g., Pareto domination, population-based mechanisms, or local search) are considered together. An example of such an approach is a parallel Vector Evaluated Differential Evolution (VEDE) algorithm (Parsopulous *et al.*, 2004b), where M sub-populations are considered in a ring topology, each being evaluated using one of the objective functions of the problem. Exchange of information between populations is arranged by means of migration of the best individuals. VEDE also utilizes a domination selection procedure to promote non-dominated individuals.

6.7 Summary

In this chapter, a brief introduction to multi-objective optimization has been presented along with the outline of popular solution approaches. Handling multiple design criteria is quantitatively different from single-objective design already at the level of comparing the solutions, the notion of optimality, and, consequently, design strategies. The central concept discussed here is Pareto dominance relation and Pareto front, a representation of which is a typical outcome of the multi-objective optimization process. Several solution techniques have been discussed including objective aggregation (the weighted sum method, the goal attainment method), but the main emphasis was put on MOEAs.

MOEAs and multi-objective variations of other population-based metaheuristics are currently the most popular techniques for handling this type of problems, yet, their applications for solving antenna design tasks is very limited due to high computational complexity. If the antenna is evaluated using full-wave EM analysis, which is the case for majority of contemporary structures (cf. Chapter 2), this cost may be prohibitive. The main focus of this book is on expedited multi-objective antenna design and the methods permitting such design are discussed in Chapters 7 and 8. MOEAs play their role in these approaches, however, they are executed at the level of fast approximation surrogates and within a restricted domain. On the other hand, metaheuristics are still attractive for solving antenna design problems when fast analytical models are available. Representative examples are array antennas where analytical array factor models (assuming isotropic radiators) provide reasonable approximation of the radiation pattern under the assumption of weak coupling between the antenna components.

Chapter 7

Multi-Objective Antenna Optimization Using Surrogate Models

Nowadays, the most popular approaches for solving multi-objective optimization problems are population-based metaheuristics. Their important advantage is the ability of generating the entire representation of the Pareto front in a single run of the algorithm. Other advantages include simplicity and availability of numerous (and often reliable) implementations. A brief exposition of these methods was provided in Chapter 6. On the other hand, a serious disadvantage of population-based algorithms is their considerable computational complexity. In a typical multi-objective setup, a large population size is utilized (from a 100 to a few hundreds of individuals) so that the overall number of objective function evaluations during the optimization run might be as high as a few thousands to tens of thousands. This becomes a fundamental bottleneck when applying metaheuristics for multi-objective design of contemporary antenna structures. As explained in Chapter 2, reliable performance evaluation of antennas requires full-wave electromagnetic (EM) analysis. Such analysis may be quite expensive particularly for realistic models (e.g., including connectors and installation fixtures) with typical simulation times of several minutes to a few hours per design. Clearly, straightforward optimization of EM models using population-based algorithms might be prohibitive.

In this chapter, we present several methods for surrogate-assisted multi-objective optimization of antenna structures. The aim of these techniques is to reduce the cost of the optimization process to practically acceptable levels of up to a few hundreds of high-fidelity EM-simulations of the antenna under design. The main tools utilized by the presented methods are variable-fidelity EM-simulations, local and global data-driven surrogates as well as response correction procedures. The first of the discussed techniques (Sec. 7.1) exploits metaheuristics for identification of the Pareto set; however, for the sake of computational efficiency, the search process is executed at the level of a fast surrogate constructed in a reduced design space (see Chapter 8 for discussion of the space reduction techniques). The other two methods are deterministic ones and are both based on exploring the Pareto front using local response surface approximation models (Secs 7.2 and 7.3). In this chapter, we only discuss the formulation of the aforementioned optimization algorithms. Their operation and performance are demonstrated in Chapter 9 for various antenna structures and in Chapter 12 for design problems in other engineering disciplines.

7.1 Optimization Using Response Surface Approximation Surrogates and Pareto Front Refinement

Given the high cost of full-wave EM analysis, necessary for reliable evaluation of antenna performance, direct optimization of EM-simulation models by means of population-based metaheuristics is not practical. In this section, we discuss a procedure for expedited multi-objective optimization of antenna structures using data-driven surrogates, evolutionary algorithms and design refinement (Koziel and Ogurtsov, 2013a). In this approach, metaheuristics are still utilized as primary tools for Pareto set identification. However, the optimization process is conducted at the level of a faster representation of the antenna structure, here, constructed by approximating sampled low-fidelity model data. As the low-fidelity model is obtained from coarse-discretization EM-simulations, the initial Pareto set has to be further refined in order to find Pareto-optimal designs at the high-fidelity EM model level.

In the remaining part of this section, we briefly describe the Kriging and co-Kriging interpolation models, a procedure for finding the initial Pareto set representation, as well as two alternative refinement procedures using response correction technique and co-Kriging. Similarly as in the previous chapters, the high-fidelity simulation model will be denoted as $\boldsymbol{f}$, whereas its low-fidelity counterpart will be denotes as $\boldsymbol{c}$. Normally, both models are evaluated using the same EM solver but with different discretization levels (see Chapter 2 for discussion of low-fidelity model setup). Although the low-fidelity model is considerably (from a few to up to 50 times) faster than the high-fidelity one, its direct optimization using metaheuristics is still too expensive.

7.1.1 *Kriging and Co-Kriging Interpolation*

Data-driven (or Response surface approximation (RSA)) surrogates are the key components of the optimization technique discussed here. In the first stage of the process, the initial approximation of the Pareto set is generated using a Kriging interpolation model constructed from sampled coarse-discretization model data. Kriging has been already discussed in Chapter 5 (Sec. 5.2.3.3). Here, we formulate it again using a slightly different notation so that it is consistent with the subsequent formulation of co-Kriging, utilized later on as one of the methods for Pareto front refinement.

Kriging is a popular technique to interpolate deterministic noise-free data (Jeong *et al.*, 2005; Koziel and Bandler, 2012; Tonkin and Larson, 2002; Rajaguru *et al.*, 2013). We denote by $X_{B.\mathrm{KR}} = \{\boldsymbol{x}_{\mathrm{KR}}^1, \boldsymbol{x}_{\mathrm{KR}}^2, \ldots, \boldsymbol{x}_{\mathrm{KR}}^{N\mathrm{KR}}\} \subset X_R$ the set of the training (base) designs and by $\boldsymbol{c}(X_{B.\mathrm{KR}})$ the associated low-fidelity model responses. The Kriging surrogate is defined by Koziel *et al.* (2013d) as

$$\boldsymbol{s}_{\mathrm{KR}}(\boldsymbol{x}) = M\alpha + r(\boldsymbol{x}) \cdot \Psi^{-1} \cdot (\boldsymbol{c}(X_{B.\mathrm{KR}}) - F\alpha), \qquad (7.1)$$

where M and F are Vandermonde matrices of the test point $\boldsymbol{x}$ and the base set $X_{B.\mathrm{KR}}$, respectively. The coefficient vector α is determined by Generalized Least Squares (GLS). $r(\boldsymbol{x})$ is an $1 \times N_{\mathrm{KR}}$ vector of correlations between the point $\boldsymbol{x}$ and the base set $X_{B.\mathrm{KR}}$, where the entries are $r_i(\boldsymbol{x}) = \psi(\boldsymbol{x}, \boldsymbol{x}_{\mathrm{KR}}^i)$, and Ψ is a $N_{\mathrm{KR}} \times N_{\mathrm{KR}}$ correlation matrix, with the entries given by $\Psi_{i,j} = \psi(\boldsymbol{x}_{\mathrm{KR}}^i, \boldsymbol{x}_{\mathrm{KR}}^j)$. Here, an exponential correlation function is used, i.e., $\psi(\boldsymbol{x}, \boldsymbol{x}') = \exp(\sum_{k=1,\ldots,n} -\theta_k |x_k - x'_k|)$, where the parameters $\theta_1, \ldots, \theta_n$ are identified by Maximum Likelihood Estimation (MLE). The regression function is chosen constant, $F = [1 \ldots 1]^T$ and $M = (1)$.

One of the strategies for refining the initial Pareto set considered in this chapter relies on combining information from low- and high-fidelity EM-simulations. This combination is realized using co-Kriging (Kennedy and O'Hagan, 2000). Co-Kriging is an extension of Kriging that allows for blending the low- and high-fidelity EM-simulation data into one surrogate by exploiting correlations between the models of various fidelities (Forrester *et al.*, 2008; Koziel *et al.*, 2013d). The co-Kriging model is set up by sequential construction of the two Kriging models: the first model $\boldsymbol{s}_{KRc}$ composed from the low-fidelity training samples $(X_{B.KRc}, \boldsymbol{c}(X_{B.KRc}))$, and the second $\boldsymbol{s}_{KRd}$ model generated on the residuals of the high- and low-fidelity samples $(X_{B.KRf}, \boldsymbol{d})$, where $\boldsymbol{d} = \boldsymbol{f}(X_{B.KRf}) - \rho \cdot \boldsymbol{c}(X_{B.KRf})$. The parameter ρ is a part of MLE of the second model. If $\boldsymbol{c}(X_{B.KRf})$ is not available, it can be approximated by the first model, i.e., as $\boldsymbol{c}(X_{B.KRf}) \approx \boldsymbol{c}_{KRc}(X_{B.KRf})$. Configuration of both models (i.e., the choice of the correlation function, regression function, etc.) can be adjusted separately for the low-fidelity data $\boldsymbol{c}$ and the residuals $\boldsymbol{d}$, respectively. Here, both models use the exponential correlation function together with constant regression function $F = [1\ 1 \ldots 1]^T$ and $M = (1)$.

The co-Kriging model $\boldsymbol{s}_{CO}(\boldsymbol{x})$ is defined similarly as in (7.1), i.e.,

$$\boldsymbol{s}_{CO}(\boldsymbol{x}) = M\alpha + r(\boldsymbol{x}) \cdot \Psi^{-1} \cdot (\boldsymbol{d} - F\alpha), \tag{7.2}$$

where the block matrices $M, F, r(\boldsymbol{x})$ and Ψ of (7.2) can be written as a function of the two underlying Kriging models $\boldsymbol{s}_{KRc}$ and $\boldsymbol{s}_{KRd}$:

$$r(\boldsymbol{x}) = [\rho \cdot \sigma_c^2 \cdot r_c(\boldsymbol{x}), \rho^2 \cdot \sigma_c^2 \cdot r_c(\boldsymbol{x}, X_{B.KR_f}) + \sigma_d^2 \cdot r_d(\boldsymbol{x})],$$

$$\Psi = \begin{bmatrix} \sigma_c^2 \Psi_c & \rho \cdot \sigma_c^2 \cdot \Psi_c(X_{B.KR_c}, X_{B.KR_f}) \\ 0 & \rho^2 \cdot \sigma_c^2 \cdot \Psi_c(X_{B.KR_f}, X_{B.KR_f}) + \sigma_d^2 \cdot \Psi_d \end{bmatrix}, \tag{7.3}$$

$$F = \begin{bmatrix} F_c & 0 \\ \rho \cdot F_d & F_d \end{bmatrix}, \quad M = [\rho \cdot M_c\ \ M_d],$$

where $(F_c, \sigma_c, \Psi_c, M_c)$ and $(F_d, \sigma_d, \Psi_d, M_d)$ of (7.3) are matrices obtained from the $\boldsymbol{s}_{KRc}$ and $\boldsymbol{s}_{KRd}$, respectively. Generally, σ_c^2 and σ_d^2 are process variances, while $\Psi_c(\cdot,\cdot)$ and $\Psi_d(\cdot,\cdot)$ stand for correlation matrices of two datasets with the optimized θ_k parameters and correlation function of $\boldsymbol{s}_{KRc}$ and $\boldsymbol{s}_{KRd}$, respectively.

A conceptual illustration of co-Kriging has been shown in Fig. 7.1 using a simple analytical function example. Typically, densely

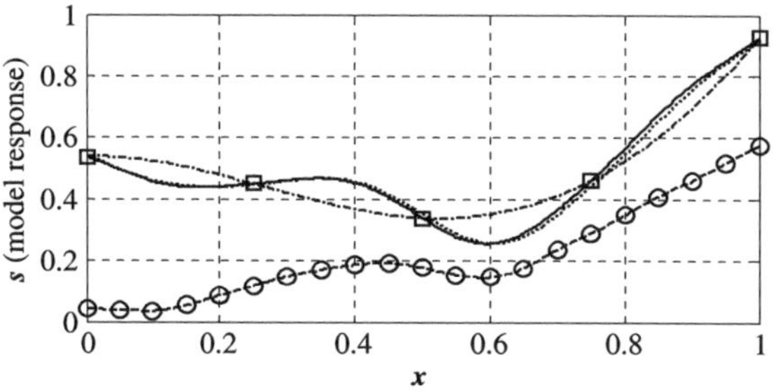

Fig. 7.1. Co-Kriging modeling concept (Koziel *et al.*, 2014): high-fidelity model (—), low-fidelity model (- - -), high-fidelity model samples (□), low-fidelity model samples (∘). Kriging interpolation of the high-fidelity model samples (- · -) is not an adequate representation of the high-fidelity model (due to the limited data set size). Co-Kriging interpolation (····) of blended low- and high-fidelity model data provides better accuracy at low computational cost.

sampled low-fidelity model data supplemented with a few samples of the high-fidelity model which allows for achieving very good accuracy when the two types of data are blended together (co-Kriging). More specifically, exploitation of the correlations between the low- and high-fidelity models results in enhancing the low-fidelity-based Kriging model so that its prediction power is good at the high-fidelity model level. At the same time, the accuracy of the Kriging model solely based on the high-fidelity data is quite limited.

7.1.2 *Construction of the Response Surface Approximation Surrogate: Obtaining Initial Pareto Set*

In order to maintain low computational cost of the design process, the initial approximation of the Pareto set is obtained by optimizing a fast surrogate model constructed using Kriging interpolation (cf. Sec 7.1.1) and based on the low-fidelity EM model data. Because no prior knowledge about the problem at hand is assumed, the training data is allocated uniformly (here, Latin hypercube sampling is utilized as a design of experiments technique, Beachkofski and Grandhi, 2002).

For most practical antenna design cases, it is neither possible nor necessary to construct the RSA model in the entire design space.

In particular, the number of training points necessary to ensure sufficient surrogate model accuracy may be impractically large for highly-dimensional spaces or if the ranges of antenna parameters are wide. The latter is a typical situation because the optimum parameter values can be rarely estimated beforehand; therefore, the parameter ranges are set wider rather than narrower to capture the entire Pareto front. In order to keep the number of training points at practically acceptable level, it might be necessary to perform the initial reduction of the design space. Specifically, we aim at identifying the subset of the design space containing the Pareto front, which is normally a very small part of the original design space (Koziel *et al.*, 2014c). More information about space reduction techniques as well as description of the specific methods can be found in Chapter 8 as well as in the literature of Lollini *et al.* (2009); Bekasiewicz *et al.* (2014a) and Koziel *et al.* (2014c).

Having constructed the kriging surrogate $\boldsymbol{s}_{\text{KR}}$, we apply a multi-objective evolutionary algorithm (MOEA) to find a set of designs representing Pareto-optimal solutions with respect to the objectives F_k of interest. Direct population-based optimization of the Kriging model is computationally feasible because — once established — the surrogate is very cheap to evaluate. Here, we use a standard MOEA with fitness sharing, Pareto-dominance tournament selection, and mating restrictions (Forrester, 1995; see also Chapter 6 for more details concerning the MOEA implementation).

The design optimization flow leading to identification of the initial Pareto set representation is as follows:

1. (Optional) Perform design space reduction (see Chapter 8);
2. Sample the design space and acquire low-fidelity model data;
3. Construct the Kriging interpolation model $\boldsymbol{s}_{\text{KR}}$;
4. (Optional) Correct the Kriging model $\boldsymbol{s}_{\text{KR}}$ using space mapping (SM);
5. Obtain the Pareto front by optimizing $\boldsymbol{s}_{\text{KR}}$ using MOEA;

It should be emphasized that the high-fidelity model $\boldsymbol{f}$ is not evaluated in the above procedure. The two methods of refining the initial Pareto set so that it can be elevated to the high-fidelity model level are described in Secs. 7.1.3 and 7.1.4. Step 3 can be executed in case of considerable discrepancy between $\boldsymbol{s}_{\mathrm{KR}}$ and $\boldsymbol{f}$. If this is the case, before finding the Pareto set, the kriging model is enhanced by aligning it with the high-fidelity model at certain (usually small) number of designs using space mapping. In antenna design, output space mapping (OSM) and frequency scaling are typically preferred (Koziel and Bekasiewicz, 2015b; Bekasiewicz and Koziel, 2016b).

7.1.3 *Pareto Set Refinement Using Response Correction*

The first refinement approach relies on sequential construction of the high-fidelity model Pareto set representation. The starting points are the designs sampled along the initial Pareto set obtained using the algorithm of Sec. 7.1.2. The latter consists of the Pareto optimal solutions of the surrogate, which, because of the discrepancies between the low-fidelity model $\boldsymbol{c}$ and the high-fidelity model $\boldsymbol{f}$, have to be corrected to adequately represent the high-fidelity model.

We denote by $\boldsymbol{x}_s^{(k)}, k = 1, \ldots, K$, the selected elements of the Pareto front found by the MOEA. The refinement stage exploits the OSM (Koziel *et al.*, 2005) process of the following form:

$$\boldsymbol{x}_f^{(k.i+1)} = \arg\min_{\boldsymbol{x}} F_1\left(\boldsymbol{s}(\boldsymbol{x}) + \left[\boldsymbol{f}\left(\boldsymbol{x}_s^{(k.i)}\right) - \boldsymbol{s}\left(\boldsymbol{x}_s^{(k.i)}\right)\right]\right) \tag{7.4}$$

under the following constraints $F_k(\boldsymbol{x}) \leq F_k(\boldsymbol{x}^{(k.i)})$, for $k = 2, 3, \ldots, N_{\mathrm{obj}}$. In other words, the optimization process (7.4) is constrained not to increase the remaining objectives (other than F_1) as compared to $\boldsymbol{x}_s^{(k)}$. The surrogate model $\boldsymbol{s}$ is corrected using the OSM term $\boldsymbol{f}(\boldsymbol{x}_s^{(k.i)}) - \boldsymbol{s}(\boldsymbol{x}_s^{(k.i)})$ (here, $\boldsymbol{x}_f^{(k.0)} = \boldsymbol{x}_s^{(k)}$), so that the corrected surrogate model coincides with $\boldsymbol{f}$ at the beginning of each iteration. In practice, two or three iterations of (7.4) are sufficient to find a refined high-fidelity model design $\boldsymbol{x}_f^{(k)}$. After completing this stage, we create a set of Pareto-optimal high-fidelity model designs. This set is the final outcome of our multi-objective optimization process.

7.1.4 *Pareto Set Refinement Using Co-Kriging*

An alternative approach to Pareto set refinement is by using co-Kriging surrogates (cf. Sec. 7.1.4). In this case, the high-fidelity models evaluated at the designs sampled from the initial Pareto set obtained by optimizing the Kriging model $\boldsymbol{s}_{\mathrm{KR}}$ are not individually refined as described in the previous section but they are included altogether into the surrogate model so that it becomes more and more accurate representation of $\boldsymbol{f}$ in the vicinity of the Pareto front.

The design algorithm flow is as follows:

1. Evaluate the high-fidelity model $\boldsymbol{f}$ at selected locations along the current Pareto front representation;
2. Update the co-Kriging surrogate $\boldsymbol{s}_{\mathrm{CO}}$ (cf. (7.2));
3. Update Pareto set by optimizing $\boldsymbol{s}_{\mathrm{CO}}$ using MOEA;
4. If the termination condition is not satisfied go to 2; else END.

When executing Step 1 for the first time, the current Pareto front representation is a Pareto set obtained using the algorithm of Sec. 7.1.2. Typically, about 10 high-fidelity model evaluations are used in Step 1, and the number of iterations necessary to converge is two to three. Our convergence criterion is the maximum distance between the Pareto front estimated in Step 3 and the sampled $\boldsymbol{f}$ data (here, we use 0.5 dB for reflection objective).

There are two practical differences between both refinement approaches presented here. The first one is simpler to implement; however, it only generates a relatively sparse representation of the Pareto front (typically, 10 designs or so). The co-Kriging-based procedure is more complex but — upon convergence — the entire Pareto set generated optimizing the co-Kriging surrogate becomes a reliable representation of the high-fidelity Pareto set.

7.1.5 *Optimization Flow Summary*

Figure 7.2 shows the flow diagram of the entire surrogate-assisted procedure for multi-objective design optimization presented in this section. The Pareto front refinement presented in the figure corresponds to the procedure of Sec. 7.1.3 (response correction).

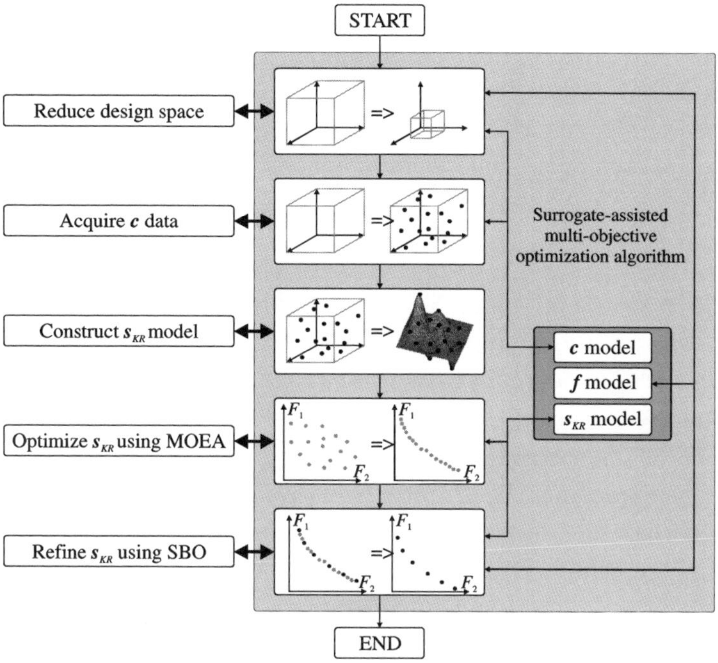

Fig. 7.2. Flowchart of the surrogate-assisted procedure for computationally efficient multi-objective design optimization of antenna structures.

7.2 Optimization by Means of Pareto Front Exploration

The multi-objective optimization technique presented in Sec. 7.1 was still relying on population-based metaheuristics. Here, we discuss another method that is fully deterministic and identifies the trade-off designs by moving along the Pareto front using surrogate-based optimization (SBO) techniques (Koziel *et al.*, 2015b; Koziel and Bekasiewicz, 2015f). This eliminates the need for using evolutionary methods for Pareto front determination as well as construction of global response surface approximation models.

7.2.1 *Optimization Algorithm*

In the following description, two design objectives are assumed. Generalization for a larger number of objectives is rather straightforward, yet two-objective case is the most important one, particularly for compact antennas where the trade-offs between the structure size and its electrical performance are sought. The algorithm produces a sequence of designs $\boldsymbol{x}^{(k)}, k = 1, 2, \ldots$, where $\boldsymbol{x}^{(1)}$ is a solution to the single-objective optimization problem of the form

$$\boldsymbol{x}^{(1)} = \arg\min_{\boldsymbol{x}} F_1\left(\boldsymbol{f}(\boldsymbol{x})\right), \tag{7.5}$$

which gives the best possible value of the first objective. Solution to (7.5) is obtained using SBO with the coarse-discretization EM model $\boldsymbol{c}$ as an underlying low-fidelity model and output space mapping as a model correction method (Koziel *et al.*, 2005; Koziel and Bekasiewicz, 2015f). Let $F_2^{(1)} = F_2(\boldsymbol{f}(\boldsymbol{x}^{(1)}))$ denote the corresponding value of the second objective. The subsequent designs are determined by exploring the Pareto front point by point. Clearly, it is assumed that the Pareto front is a connected set, which is a limitation of the method. On the other hand, this assumption is normally satisfied for antenna design problems.

The process is realized as a series of constrained single-objective optimizations. The optimal design obtained in previous iteration is utilized as a starting point for the current one. The procedure can be formulated as follows. Let $F_2^{(k)}$ be the threshold value for the second objective. Then, we have

$$\boldsymbol{x}^{(k)} = \arg\min_{\boldsymbol{x},\ F_2(\boldsymbol{f}(\boldsymbol{x})) \le F_2^{(k)}} F_1\left(\boldsymbol{f}(\boldsymbol{x})\right). \tag{7.6}$$

Here, $\boldsymbol{x}^{(k)}$ is the kth element of the Pareto set. The process is continued until $F_1(\boldsymbol{f}(\boldsymbol{x}^{(k)}))$ is still satisfactory from the point of view of given design specifications. The thresholds $F_2^{(k)}$ can be set in any convenient way, for example, as $F_2^{(k)} = \alpha \cdot F_2^{(k-1)}$ with $\alpha < 1$, or $F_2^{(k)} = F_2^{(k-1)} - \beta$ with $\beta > 0$. Conceptual illustration of the algorithm is shown in Fig. 7.3.

7.2.2 *Pareto Front Exploration Using Local Response Surface Approximation Models*

In practical situations, it is reasonable to assume that the design $\boldsymbol{x}^{(k)}$ is allocated close to the preceding one, i.e., $\boldsymbol{x}^{(k-1)}$. Having this in mind, the solution to the problem (7.6) can be obtained iteratively

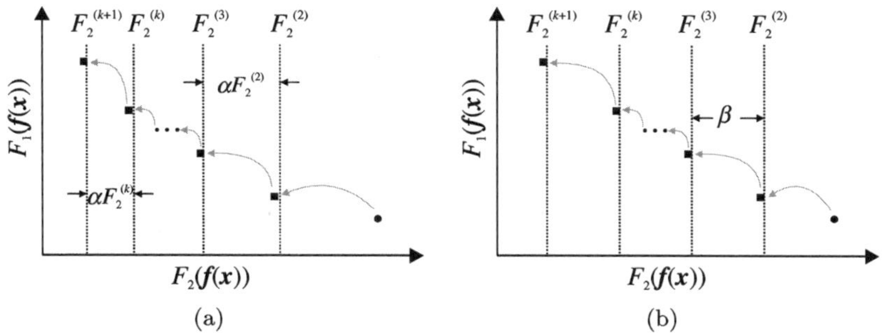

Fig. 7.3. Conceptual illustration of point-by-point exploration of the Pareto front. The initial design (black circle) is determined using (7.5). Subsequently, consecutive Pareto optimal designs (black squares) are obtained sequentially using (7.6). Vertical dotted lines represent thresholds along objective F_2: (a) multiplicative threshold $F_2^{(k)} = \alpha F_2^{(k-1)}$; and (b) additive threshold $F_2^{(k)} = F_2^{(k-1)} - \beta$.

as a sequence $\boldsymbol{x}^{(k.j)}, j = 0, 1, \ldots,$ with $\boldsymbol{x}^{(k.0)} = \boldsymbol{x}^{(k-1)}$, as follows

$$\boldsymbol{x}^{(k.j+1)} = \arg \min_{\boldsymbol{x},\ F_2(\boldsymbol{s}(\boldsymbol{x})) \le F_2^{(k)}} F_1\left(\boldsymbol{s}^{(k.j)}(\boldsymbol{x})\right), \tag{7.7}$$

where the surrogate $\boldsymbol{s}^{(k.j)}$ is defined as

$$\boldsymbol{s}^{(k.j)}(\boldsymbol{x}) = \boldsymbol{q}^{(k.j)}(\boldsymbol{x}) + \left[\boldsymbol{f}(\boldsymbol{x}^{(k.j)}) - \boldsymbol{q}^{(k.j)}(\boldsymbol{x}^{(k.j)})\right], \tag{7.8}$$

whereas $\boldsymbol{q}^{(k.j)}$ is a local RSA model of $\boldsymbol{c}$, constructed in the vicinity of the current design $\boldsymbol{x}^{(k.j)}$ defined as an interval $[\boldsymbol{x}^{(k.j)} - \boldsymbol{d}, \boldsymbol{x}^{(k.j)} + \boldsymbol{d}]$. The solution of the problem (7.7) is constrained to this region. Its size vector $\boldsymbol{d}$ is determined by the initial sensitivity analysis of the antenna. The surrogate itself is a quadratic model without mixed terms (Koziel, 2010e), based on $2n + 1$ evaluations (where n is the number of design parameters) of $\boldsymbol{c}$ allocated using star distribution design of experiments (see Fig. 7.4(a)) (Bandler *et al.*, 2004a; Koziel *et al.*, 2011a). More specifically, the surrogate $\boldsymbol{q}^{(k.j)}$ is defined as

$$\begin{aligned}\boldsymbol{q}^{(k.j)}(\boldsymbol{x}) &= \boldsymbol{q}^{(k.j)}\left([x_1 \ldots x_n]^T\right) \\ &= \lambda_0 + \sum_{l=1}^{n} \lambda_l \left(x_l - x_l^{(k.j)}\right) + \sum_{l=1}^{n} \lambda_{n+l} \left(x_l - x_l^{(k.j)}\right)^2\end{aligned} \tag{7.9}$$

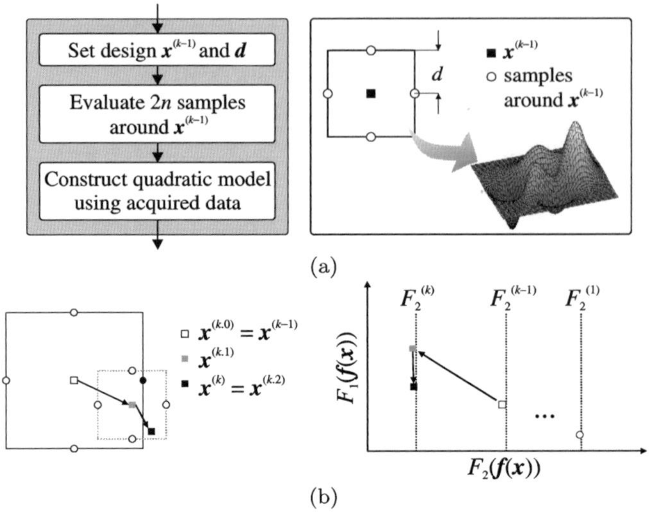

Fig. 7.4. Optimization using local RSA models: (a) construction of the local RSA model using star-distribution design of experiments; and (b) approximation of $\boldsymbol{x}^{(k)}$ in two iterations of (7.7).

with the coefficients λ_l found by solving linear regression problems

$$\boldsymbol{q}^{(k.j)}\left(\boldsymbol{x}^{(b.l)}\right) = \boldsymbol{c}\left(\boldsymbol{x}^{(b.l)}\right), \tag{7.10}$$

for $l = 1, \ldots, 2n+1$, where $\boldsymbol{x}^{(b.l)}$ are the star-distributed training points mentioned earlier.

The vicinity size $\boldsymbol{d}$ is reduced after each iteration of (7.7) as $\boldsymbol{d} \leftarrow \boldsymbol{d}/m$ ($m = 2$ is used in the numerical experiments presented later in the book). In practice, up to three iterations of (7.7) are sufficient to obtain a good approximation of $\boldsymbol{x}^{(k)}$. The correction term in (7.8) ensures zero-order consistency between the surrogate and the high-fidelity model at $\boldsymbol{x}^{(k.j)}$, i.e., $\boldsymbol{q}^{(k.j)}(\boldsymbol{x}^{(k.j)}) = \boldsymbol{f}(\boldsymbol{x}^{(k.j)})$ (Alexandrov and Lewis, 2001; Koziel and Ogurtsov, 2013a). Figure 7.4 shows the process of constructing and optimizing the RSA model. More detailed discussion on surrogate-assisted optimization of local RSA models can be found in the literature of Koziel and Ogurtsov (2014a) and Conn *et al.* (2009).

It should be noted that the computational cost of each iteration of (7.7) is only $2n + 1$ evaluations of the coarse-discretization model

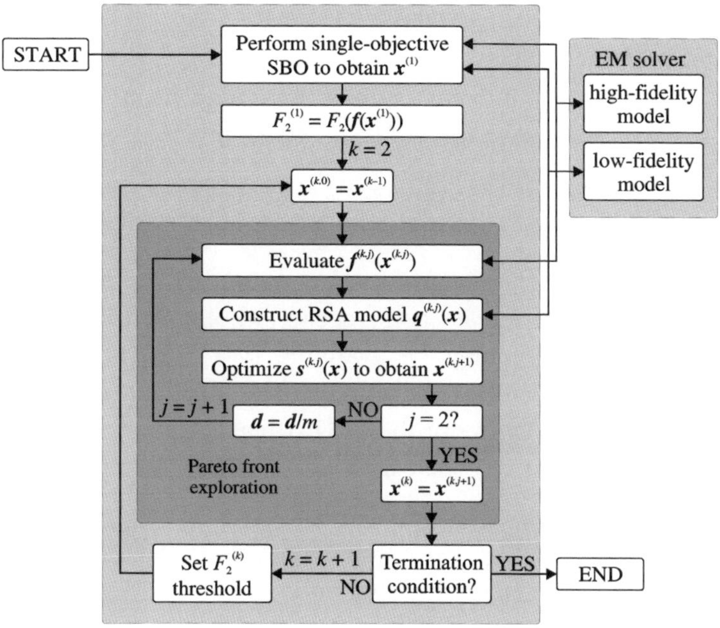

Fig. 7.5. Flow diagram of the multi-objective antenna optimization using Pareto front exploration.

$\boldsymbol{c}$ and a single evaluation of the high-fidelity model $\boldsymbol{f}$. Consequently, the overall cost of finding one Pareto-optimal design corresponds to just a few evaluations of the high-fidelity model.

7.2.3 *Optimization Flow*

The flow of the multi-objective optimization procedure has been illustrated in Fig. 7.5. It can be summarized as follows:

(1) Obtain $\boldsymbol{x}^{(1)}$ by performing single-objective SBO;
(2) Set $F_2^{(1)} = F_2(\boldsymbol{f}(\boldsymbol{x}^{(1)}))$ and set $k = 2$;
(3) Set $\boldsymbol{x}^{(k.0)} = \boldsymbol{x}^{(k-1)}$;
(4) Evaluate $\boldsymbol{f}^{(k.j)}(\boldsymbol{x}^{(k.j)})$;
(5) Construct local RSA model $\boldsymbol{q}^{(k.j)}(\boldsymbol{x})$ within the range region defined as $[\boldsymbol{x}^{(k.j)} - \boldsymbol{d}, \boldsymbol{x}^{(k.j)} + \boldsymbol{d}]$;
(6) Use (7.7) to obtain $\boldsymbol{x}^{(k.j+1)}$ by optimizing surrogate model $\boldsymbol{s}^{(k.j)}(\boldsymbol{x})$;

(7) If $j = 2$ set $\boldsymbol{x}^{(k)} = \boldsymbol{x}^{(k.j+1)}$ and go to 8; else set $\boldsymbol{d} = \boldsymbol{d}/m$, $j = j + 1$ and go to 4;

(8) If termination condition is met end algorithm; otherwise set $k = k + 1$, set threshold $F_2^{(k)}$ and go to 3.

7.2.4 *Alternative Exploration Methods*

Exploration of the Pareto front using local RSA models is one of possible options. In some cases, cheap adjoint sensitivities of the EM-simulation model may be available (CST, 2013; HFSS, 2012) which allows for further acceleration of the optimization process. In particular, the solution to problem (7.6) can be found iteratively as a sequence $\boldsymbol{x}^{(k.j)}, j = 0, 1, \ldots,$ with $\boldsymbol{x}^{(k.0)} = \boldsymbol{x}^{(k-1)}$, as follows

$$\boldsymbol{x}^{(k.j+1)} = \arg \min_{\substack{\boldsymbol{x},\ F_2(\boldsymbol{G}^{(k.j)}(\boldsymbol{x})) \le F_2^{(k)} \\ ||\boldsymbol{x} - \boldsymbol{x}^{(k.j)}|| \le \delta^{(k.j)}}} F_1\left(\boldsymbol{G}^{(k.j)}(\boldsymbol{x})\right), \tag{7.11}$$

where $\boldsymbol{G}^{(k.j)}$ is a linear expansion model of the high-fidelity model $\boldsymbol{f}$ at $\boldsymbol{x}^{(k.j)}$ defined as

$$\boldsymbol{G}^{(k.j)}(\boldsymbol{x}) = \boldsymbol{f}\left(\boldsymbol{x}^{(k.j)}\right) + \boldsymbol{J_f}\left(\boldsymbol{x}^{(k.j)}\right) \cdot \left(\boldsymbol{x} - \boldsymbol{x}^{(k.j)}\right). \tag{7.12}$$

Here, $\boldsymbol{J_f}(\boldsymbol{x}^{(k.j)})$ is a Jacobian of the high-fidelity EM model $\boldsymbol{f}$ at $\boldsymbol{x}^{(k.j)}$ evaluated using adjoint sensitivities; $\delta^{(k.j)}$ is the trust region radius updated using conventional rules of Conn *et al.* (2000). The candidate design produced by (7.11) is accepted if it leads to the improvement of the objective function at the level of the EM-simulation model, i.e., when $F_1(\boldsymbol{f}(\boldsymbol{x}^{(k.j+1)}) < F_1(\boldsymbol{f}(\boldsymbol{x}^{(k.j)}))$. In principle, any local optimization method can be utilized for solving (7.6), except perhaps, direct optimization of the high-fidelity model which would increase the cost of the entire process.

7.3 Optimization Using Sequential Domain Patching

The last technique presented here is multi-objective optimization using sequential domain patching (SDP). It is, again, a deterministic procedure that exploits variable-fidelity simulation models. In the first stage of the process, a path connecting the "extreme" Pareto-optimal designs is identified and the subset of non-dominated designs is further refined using a response correction technique as described in Sec. 7.1.3. The computational cost of the process can be predetermined beforehand because it depends on the size of patches used to cover the patch found by SDP.

7.3.1 *Optimization Flow*

The optimization procedure starts from identifying the "extreme" points of the Pareto front. These are obtained through single-objective optimization runs, one objective at a time, and correspond to the best possible designs with respect to particular objectives. For the sake of computational efficiency, the extreme points are obtained at the level of a coarse-discretization EM model $\boldsymbol{c}$ by solving

$$\boldsymbol{x}_k^* = \arg\min_{\boldsymbol{x}} F_k\left(\boldsymbol{c}(\boldsymbol{x})\right) \tag{7.13}$$

for $k = 1, \ldots, N_{\text{obj}}$ (Koziel *et al.*, 2014c). In practice, the problem (7.13) can be solved directly using, e.g., the pattern search algorithm (Kolda *et al.*, 2003) which is immune to the numerical noise (inherent to coarse-discretization simulation models).

In the next stage of the process, an initial representation of the Pareto front is found using a SDP algorithm. It is conceptually illustrated in Fig. 7.6. Hereby, two objectives are assumed, which is the most typical practical situation. As mentioned before, in case of compact antennas, one of the objectives would be minimization of the antenna size, whereas the other objective is normally related to electrical performance (e.g., minimization of maximum in-band reflection level for ultra-wideband (UWB) antennas).

Having the initial approximation of the Pareto front, the designs selected from the non-dominated part of the set produced by SDP are subjected to the refinement process using the method already described in Sec. 7.1.3. The high-fidelity Pareto-optimal designs obtained this way are the final product of the entire procedure.

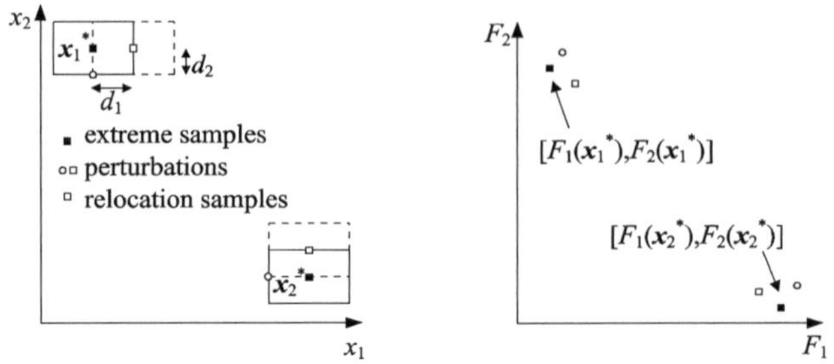

(a)

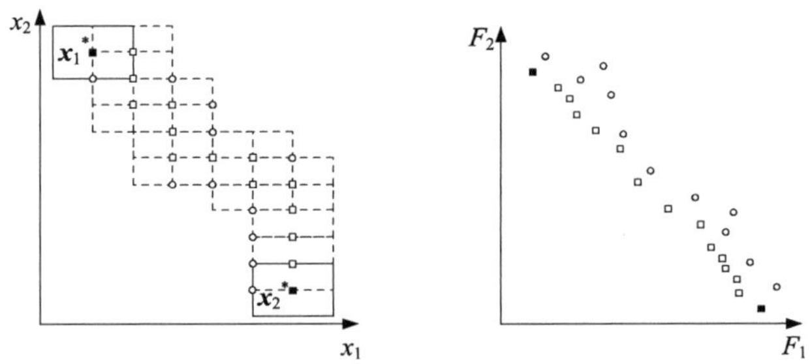

(b)

Fig. 7.6. Conceptual illustration of the Pareto set estimation using SDP (Koziel and Bekasiewicz, 2016c): (a) initialization by generating perturbations (○, □) around initial (extreme Pareto solutions obtained using (7.13)) designs (■) and construction of the new patches based on the objective function evaluations; relocation points (□) at the best perturbations that become centers of the next patches along the path; (b) final path and the corresponding initial Pareto set representation (■, □) together with the dominated designs (○) in the feature space. The design space and the feature space are shown on the left and the right panels, respectively.

7.3.2 *Sequential Domain Patching Algorithm*

The algorithm operation can be rigorously formulated as follows (Koziel and Bekasiewicz, 2016c).

1. Determine the patch size $\boldsymbol{d} = [d_1 \ldots d_n]^T$ using the procedure of Sec. 7.3.3;
2. Initialize the locations of the current path points $\boldsymbol{x}_{c1}$ and $\boldsymbol{x}_{c2}$ as $\boldsymbol{x}_{c1} = \boldsymbol{x}_1^*$ and $\boldsymbol{x}_{c2} = \boldsymbol{x}_2^*$;
3. Evaluate n perturbations of the size $\boldsymbol{d}$ around $\boldsymbol{x}_{c1}^*$ (towards $\boldsymbol{x}_{c2}^*$ only) and select the one that brings the largest improvement with respect to the second objective F_2;

4. Relocate the patch so that it is centered at the best perturbation selected in Step 2; update $\boldsymbol{x}_{c1}$;
5. Evaluate n perturbations of the size $\boldsymbol{d}$ around $\boldsymbol{x}_{c2}^*$ (towards $\boldsymbol{x}_{c1}^*$ only) and select the one that brings the largest improvement with respect to the second objective F_1;
6. Relocate the patch so that it is centered at the best perturbation selected in Step 5; update $\boldsymbol{x}_{c2}$;
7. If the path between $\boldsymbol{x}_1^*$ and $\boldsymbol{x}_2^*$ is not complete, go to 3.

The purpose of the algorithm is to yield the initial approximation of the set of Pareto-optimal solutions, as well as a set of patches covering the part of the design space that contains these solutions. The EM-simulation data gathered along the way will be further used for Pareto front refinement. It should be noted that not only the algorithm is deterministic but also its computational cost is predetermined by the design space dimension n and the total number of patches. Its upper bound is (excluding the cost of solving (7.13)) $(M-1)\cdot(n-1)$, where $M = \sum_{k=1,\ldots,n} m_k$, and is m_k the number of intervals in the direction j. In practice, the cost is lower as certain perturbations are not evaluated due to the imposed constraints.

7.3.3 *Automated Determination of Patch Sizes*

The numbers of intervals m_k along each dimension of the design space are the control parameters of the SDP algorithm. This number should be set appropriately in order to account for the fact that sensitivity of the antenna responses is generally different for various parameters. Although the number of patches in all dimensions can be made equal, this would lead to unnecessary increase of the computational cost of the procedure. In the following text, a technique for automatic sensitivity-based determination of the parameters m_k is described.

The patch size is selected so that its components split the distance between $\boldsymbol{x}_1^*$ and $\boldsymbol{x}_2^*$ into integer-valued number of intervals. The number of intervals in each direction should be determined so that the change of the antenna responses is (e.g., norm-wise) similar for patch-size perturbations with respect to each direction. Here, the numbers m_k are assigned using the following procedure (Koziel and

Bekasiewicz, 2016h), where we use the notation $\boldsymbol{x}_1^* = [x_{1.1}^* \ldots x_{1.n}^*]^T$ (similarly for $\boldsymbol{x}_2^*$):

1. Evaluate $\boldsymbol{c}$ at n points $\boldsymbol{x}_{1-2}^{(k)} = [x_{1.1}^* \ldots x_{1.k-1}^* x_{2.k}^* x_{1.k+1}^* \ldots x_{1.n}^*]^T$, $k = 1, \ldots, n$;
2. Calculate $E_{1.k} = ||\boldsymbol{c}(\boldsymbol{x}_{1-2}^{(k)}) - \boldsymbol{c}(\boldsymbol{x}_1^*)||/||\boldsymbol{c}(\boldsymbol{x}_1^*)||, k = 1, \ldots, n$;
3. Evaluate $\boldsymbol{c}$ at n points $\boldsymbol{x}_{2-1}^{(k)} = [x_{2.1}^* \ldots x_{2.k-1}^* x_{1.k}^* x_{2.k+1}^* \ldots x_{2.n}^*]^T$, $k = 1, \ldots, n$;
4. Calculate $E_{2.k} = ||\boldsymbol{c}(\boldsymbol{x}_{2-1}^{(k)}) - \boldsymbol{c}(\boldsymbol{x}_2^*)||/||\boldsymbol{c}(\boldsymbol{x}_2^*)||,\ k = 1, \ldots, n$;
5. Set $E_k = (E_{1.k} + E_{2.k})/2$;
6. Normalize $E_k = E_k / \max\{E_j : j = 1, \ldots, n\}$;
7. Set $m_k = \max\{2, \lceil m_{\max} \cdot E_k \rceil\},\ k = 1, \ldots, n$.

The factors $E_{1.k}$ are relative (norm-wise) response changes when varying the kth components of the design $\boldsymbol{x}_1^*$ towards $\boldsymbol{x}_2^*$ (similarly for $E_{2.k}$). They give an idea of how much the antenna responses change when moving from one extreme Pareto point to another along a specific geometry dimension. The factors E_k are the average values. Clearly, the number of intervals should be larger when the value of E_k is higher. Here, the number of intervals is made proportional to E_k (with the minimum number of intervals equal 2, and rounded up to the nearest integer).

The maximum number of intervals per direction, $m_{\max}$, is the only user-defined parameter. It can be set based on the maximum allowed relative response change $E_{\max}$ as follows: $m_{\max} = \lceil \max\{E_k : k = 1, \ldots, n\}/E_{\max} \rceil$ (calculated for un-normalized E_k factors). For example, if $E_{\max} = 0.05$ (we allow maximum 5 percent relative response changes within the patch), and $\max\{E_k\} = 0.72$, we get $m_{\max} = 15$. The value of $m_{\max}$ can also be calculated having in mind the maximum computational budget for the SDP algorithm (and calculated using the upper bound for the cost as given in Sec. 7.3.2).

7.3.4 *Pareto Set Refinement*

All operations up to this point, including generation of the initial Pareto set, have been performed using the low-fidelity model $\boldsymbol{c}$.

In order to find the high-fidelity Pareto-optimal designs, a refinement process is necessary which is applied for the designs $\boldsymbol{x}_c^{(k)}$, $k = 1, \ldots, K$, selected from the initial front (typically, in a uniform way with respect to one of the objectives). As mentioned before, the refinement procedure is similar to that used by the first technique considered in this chapter. The details can be found in Sec. 7.1.3. One important difference is that the surrogate model used in (7.4) is now a local response surface model, specifically, a simplified (without mixed terms) second-order polynomial approximation of the low-fidelity model $\boldsymbol{c}$ set up using the evaluation of $\boldsymbol{c}$ at $\boldsymbol{x}_c^{(k)}$ and the perturbed designs corresponding to the patch centered around $\boldsymbol{x}_c^{(k)}$. Note that at least half of the necessary data is already available from the patching process.

7.4 Summary

In this chapter, several surrogate-assisted techniques for computationally efficient multi-objective optimization of antenna structures have been presented. All methods involve variable-fidelity EM-simulation models as well as identification of the extreme points of the Pareto front (normally found at the level of the low-fidelity EM model). The first method exploits a quasi-global data-driven surrogate set up in a reduced design space and population-based metaheuristics to generate the initial approximation of the Pareto front. Selected designs are further refined to find a representation of the Pareto set at the level of the accuracy of the high-fidelity EM model of the antenna structure at hand. Two refinement strategies based on response correction and co-Kriging have been discussed. The remaining two methods are fully deterministic and work by exploring the Pareto front either directly (the method of Sec. 7.2) or indirectly (i.e., at the level of the low-fidelity model with subsequent Pareto set refinement, Sec. 7.3). The advantage of the first technique is its generality because no particular assumptions are made concerning the geometry of the Pareto set. A disadvantage is the necessity of constructing a global approximation model which has to be preceded by reduction of the design space in order to allocate the training samples only in the region containing Pareto-optimal designs. Such a reduction is mandatory for higher-dimensional cases (see Chapter 8 for exposition of the reduction techniques).

The deterministic methods are generally more efficient in computational terms; however, the implicit assumption is that the Pareto front is a connected set. Although this is not the case in general, the assumption normally holds for majority of practical antenna design problems. Another issue is that the specific formulation of the deterministic techniques discussed here only permits handling two design objectives.

All three methods will be comprehensively demonstrated using real-world antenna design cases in the remaining part of the book, that is, in Chapters 9–12 (the last illustrating applications for solving design problems in other engineering disciplines). More detailed discussion and practical recommendations will be provided in Chapter 13.

Chapter 8

Design Space Reduction Methods

Multi-objective optimization of contemporary antennas is a very challenging task not only due to the necessity of evaluating the structures using computationally expensive full-wave electromagnetic (EM) analysis but also because a typical search space is large both in terms of the number of (usually geometry) parameters to be adjusted and its wide ranges. Performing the optimization process in such a space is a waste of resources because vast majority of the designs are non-optimal and the region of the space containing Pareto-optimal solutions is normally a very small subset of the original space. Therefore, appropriate reduction of the space may be critical for carrying out multi-objective optimization in a computationally feasible manner. In Chapter 7, several surrogate-assisted optimization techniques for multi-objective design optimization of antenna structures have been described, all involving variable-fidelity EM-simulation models. The first and the most generic of these methods exploits population-based metaheuristics (here, a multi-objective evolutionary algorithm, MOEA) for generating the initial approximation of the Pareto front. Because neither low- nor high-fidelity EM model can be directly handled by MOEA, the optimization process is conducted using an auxiliary data-driven surrogate constructed by Kriging interpolation of sampled low-fidelity model data. Construction of the Kriging model in the original space is impossible if the antenna structure is described by more than a few parameters.

In this chapter, we discuss several ways of reducing the design space to permit construction of the data-driven model setup even for relatively high-dimensional cases of 15 to over 20 parameters. The presented methods aim at reducing the ranges of the parameters rather than dimensionality of the space. This is because parameter redundancy is lower in antenna design than for many other engineering problems (e.g., airfoil design in aerospace engineering) in the sense that most of the parameters are relevant and their number cannot be significantly reduced without considerable degradation of the structure flexibility (to the point where it may not be possible to fulfill given performance requirements). All of the considered space reduction methods are based on identification of the extreme points of the Pareto front.

8.1 Design Space Reduction for Antenna Design

Constructing a reliable approximation model is an important step of one of the multi-objective optimization methods considered in Chapter 7 (cf. Sec. 7.2). Although the surrogate model is obtained from coarse-discretization EM-simulation data (computationally much cheaper than the high-fidelity ones), the data acquisition is still a considerable challenge in computational terms. In practical problems, the number of the degrees of freedom in antenna structures tends to be large (cases with 10 to over 20 parameters are not uncommon). Moreover, the ranges of parameters are wide because initial guesses about optimal dimensions are very difficult. On the other hand, the solutions of interest from the multi-objective-optimization standpoint normally occupy a very small region of the design space. A relevant example is shown in Fig. 8.1 using an example of a three-parameter ultra-wideband (UWB) monocone antenna (Bekasiewicz *et al.*, 2014a). It can be observed that even in a low-dimensional case, the region containing the interesting part of the Pareto front (here, corresponding to the designs for which the maximum in-band reflection level is −10 dB or less) is a very small fraction of the original design space. For higher-dimensional spaces, that fraction may be many orders of magnitude smaller (volume-wise) (Koziel *et al.*, 2014c; Bekasiewicz *et al.*, 2016b; Koziel and Bekasiewicz, 2016d).

Clearly, sampling the original space and subsequent training data acquisition may be prohibitively expensive even for relatively low-dimensional design cases. Also, as mentioned in the previous paragraph, vast majority of the space is of no interest. Consequently, the key point is to limit the ranges of the antenna parameters as much as possible but in such a way that the reduced space still contains a possibly large part of the Pareto front.

In the remaining part of this chapter we discuss several methods that are based on identification of the extreme points of the Pareto front, obtained by means of single-objective optimization

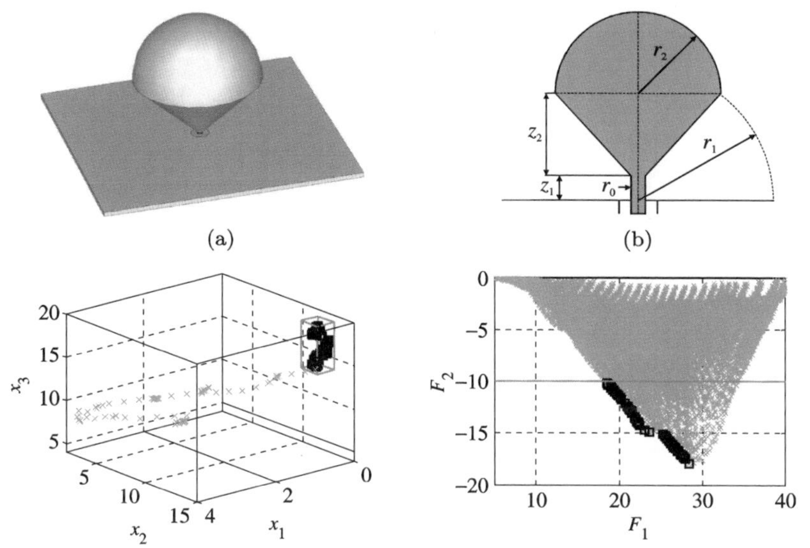

Fig. 8.1. UWB monocone antenna (Koziel and Ogurtsov, 2013a): (a) three-dimensional view, (b) cut view, (c) Pareto set inside design variable space and (d) entire search space mapped to the feature space (×). The acceptable Pareto-optimal designs with $F_1 \leq -10$ dB (□) are located in a subset, whose boundaries are determined by a small cuboid. Note that the relevant design space subset is a very small fraction of the entire space, here, even for a low-dimensional case.

(one objective at a time). It should be mentioned — although none of these techniques is discussed here — that, in general, design space reduction can be realized in many various ways. The most rudimentary and frequently used approach is variable pre-screening, where sensitivity analysis can be used to eliminate the least significant variables from the optimization process, thus leading to reduction of the problem dimensionality (Dean and Lewis, 2006). Unfortunately, in case of antennas, structure parameterization is rarely redundant and most of geometry parameters are of importance (Koziel and Bekasiewicz, 2016d). Another popular technique for dimensionality reduction is principal component analysis (PCA) (Jolliffe, 2002), where a reduced latent space is spanned by a limited number of principal components found by analyzing parameter correlations (normally, based on sensitivity data acquired for a set of samples allocated within the original space). Because PCA requires gradient information (e.g., from adjoint sensitivities), it is not a universal solution for antenna designers. Also, as already mentioned before, the major issue is to reduce parameter ranges rather than the dimensionality of the space.

8.2 Space Reduction Using Extreme Pareto-Optimal Designs

The first design space reduction method is very straightforward (Koziel *et al.*, 2014c). Let $\boldsymbol{l}$ and $\boldsymbol{u}$ be the initially defined lower/upper bounds for the design parameters. Let

$$\boldsymbol{x}^{*(k)} = \arg \min_{\boldsymbol{l} \leq \boldsymbol{x} \leq \boldsymbol{u}} F_k\left(\boldsymbol{c}(\boldsymbol{x})\right), \tag{8.1}$$

$k = 1, \ldots, N_{\text{obj}}$ be an optimum design of the low-fidelity model $\boldsymbol{c}$ obtained with respect to the kth objective F_k.

The bounds of the reduced design space are then defined as

$$\boldsymbol{l}^* = \min\{\boldsymbol{x}^{*(1)}, \ldots, \boldsymbol{x}^{*(N_{\text{obj}})}\} \tag{8.2}$$

and

$$\boldsymbol{u}^* = \max\{\boldsymbol{x}^{*(1)}, \ldots, \boldsymbol{x}^{*(N_{\text{obj}})}\}. \tag{8.3}$$

The min/max in the above statements is understood componentwise. The single-objective optimization runs are normally executed having in mind satisfaction of other objectives at reasonable levels, which can be ensured by imposing appropriate constraints or employing suitable penalty functions (Koziel and Bekasiewicz, 2016a). Conceptual illustration of the space reduction scheme is shown in Fig. 8.2.

In general, not all Pareto-optimal designs would reside in the reduced space, however, for antenna design problems that are characterized by relatively simple geometry of the Pareto front, majority of the optimal designs will be accounted for. In particular, the extreme points of the Pareto front are included by the very definition of the reduced space.

In practice, the space reduction is executed at the level of the low-fidelity EM model of the antenna structure at hand. However, in case of considerable discrepancies between the models, it might be

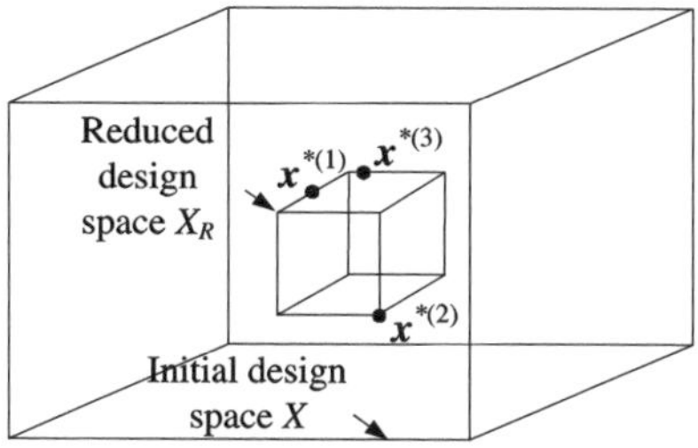

Fig. 8.2. Conceptual illustration of the original and reduced design space for three design objectives and three-dimensional design space.

advantageous to also include the extreme high-fidelity Pareto-optimal designs $\boldsymbol{x}_f^{*(k)}$. These can be obtained by solving the problem (8.2) for the high-fidelity model $\boldsymbol{f}$ (typically, using surrogate-assisted methods; Koziel *et al.*, 2014c). The reduced space would then be defined as

$$\bar{\boldsymbol{l}}^* = \min\{\boldsymbol{x}^{*(1)}, \ldots, \boldsymbol{x}^{*(N_{\text{obj}})}, \boldsymbol{x}_f^{*(1)}, \ldots, \boldsymbol{x}_f^{*(N_{\text{obj}})}\} \tag{8.4}$$

and

$$\bar{\boldsymbol{u}}^* = \max\{\boldsymbol{x}^{*(1)}, \ldots, \boldsymbol{x}^{*(N_{\text{obj}})}, \boldsymbol{x}_f^{*(1)}, \ldots, \boldsymbol{x}_f^{*(N_{\text{obj}})}\}. \tag{8.5}$$

8.3 Rotational Design Space Reduction Algorithm

In many situations, setting up a data-driven model may be expensive even in the design space reduced as shown in Sec. 8.2 (Koziel *et al.*, 2014c). As indicated in the literature (e.g., Koziel and Bekasiewicz, 2016g; Bekasiewicz *et al.*, 2014b), for antenna structures described by a dozen or more adjustable parameters, the number of training samples required to ensure sufficient surrogate model accuracy (i.e., relative least-square error lower than a few percent) is 1,000 or more.

Further confinement of the region containing the Pareto set is based on the observation that the majority of the Pareto-optimal solutions are — in practical antenna design cases — allocated close to the diagonal of the initially reduced space connecting the extreme points $\boldsymbol{x}^{*(1)}$ and $\boldsymbol{x}^{*(2)}$, cf. Fig. 8.3(a). The "rotational" space reduction procedure works as follows:

Step 1: Execute the reduction procedure of Sec. 8.2 to obtain the initially reduced space (the "box");

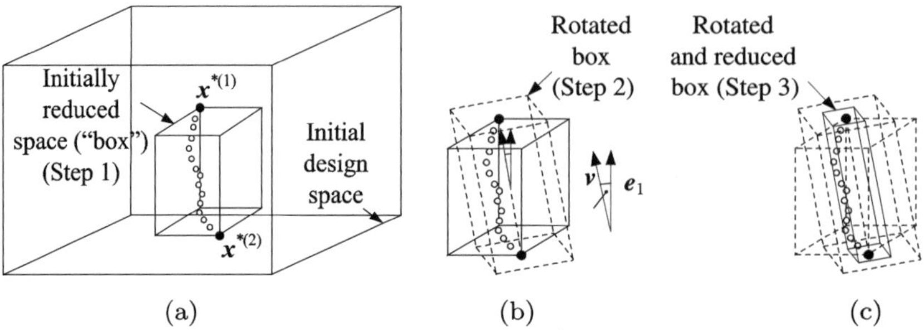

Fig. 8.3. Design space reduction procedure (Koziel and Bekasiewicz, 2016d): (a) initial design space and Step 1 of the reduction process (finding the hypercube encompassing extreme (•) Pareto front points $\boldsymbol{x}^{*(1)}$ and $\boldsymbol{x}^{*(2)}$), (b) Step 2: box rotation (here, $\boldsymbol{v}$ is a unit vector defining the box diagonal, $\boldsymbol{e}_1$ is a unit vector along the largest box dimension and θ represents rotation angle), (c) Step 3: reduction (linear scaling) of the rotated box; (∘) denote Pareto-optimal designs.

Step 2: Perform the box rotation (cf. Fig. 8.3(b));
Step 3: Perform the box reduction (cf. Fig. 8.3(c)).

Step 2 is realized as follows (Koziel and Bekasiewicz, 2016d). Let $\boldsymbol{x}_0 = (\boldsymbol{x}^{*(1)} + \boldsymbol{x}^{*(2)})/2$ be the box center. Let $\boldsymbol{v}$ be a unit vector $\boldsymbol{v} = (\boldsymbol{x}^{*(1)} - \boldsymbol{x}^{*(2)})/||\boldsymbol{x}^{*(1)} - \boldsymbol{x}^{*(2)}||$, and $\boldsymbol{e}_1$ be a unit vector of the standard basis $\{\boldsymbol{e}_k\}_{k=1,\ldots,n}$, associated with the largest of the box dimensions. The box is rotated around $\boldsymbol{x}_0$ with respect to a two-dimensional subspace M spanned by vectors $\boldsymbol{v}$ and $\boldsymbol{e}_1$, so that the points $\boldsymbol{x}^{*(1)}$ and $\boldsymbol{x}^{*(2)}$ become the centers of its two faces as shown in Fig. 8.3(b). Let $\boldsymbol{v}_1 = (\boldsymbol{v} - \boldsymbol{e}_1\boldsymbol{e}_1^T\boldsymbol{v})/||\boldsymbol{v} - \boldsymbol{e}_1\boldsymbol{e}_1^T\boldsymbol{v}||$ be the unit vector orthogonal to $\boldsymbol{e}_1$. The subspace M is also spanned by $\boldsymbol{e}_1$ and $\boldsymbol{v}_1$. The rotation angle θ is given by $\cos\theta = \boldsymbol{v}^T\boldsymbol{e}_1$. We are interested in finding out the rotated base vectors $\boldsymbol{e}_k$, $k = 1,\ldots,n$. Let $\boldsymbol{e}_k^p = \boldsymbol{e}_1\boldsymbol{e}_1^T\boldsymbol{e}_k + \boldsymbol{v}_1\boldsymbol{v}_1^T\boldsymbol{e}_k$ (note that $\boldsymbol{e}_1\boldsymbol{e}_k = 0$ for $k > 1$) be the projection of $\boldsymbol{e}_k$ onto M, and $\boldsymbol{e}_k^v = \boldsymbol{e}_k - \boldsymbol{e}_k^p$ be the corresponding normal vector. The rotation is only applied to $\boldsymbol{e}_k^p(\boldsymbol{e}_k^v$ stays intact). The representation of $\boldsymbol{e}_k^p$ in M is $\boldsymbol{e}_k^p = [\boldsymbol{e}_1^T\boldsymbol{e}_k\boldsymbol{v}_1^T\boldsymbol{e}_k]^T$. Note that $\boldsymbol{e}_k^p = [0\ \boldsymbol{v}_1^T\boldsymbol{e}_k]^T$ for $k > 1$, which is because $\boldsymbol{e}_k^p$ is orthogonal to $\boldsymbol{e}_1$ so that the entire vector $\boldsymbol{e}_k^p$ is actually parallel to $\boldsymbol{v}_1$ for $k > 1$. The rotation matrix in M is given by

$$\boldsymbol{R} = \begin{bmatrix} \cos\theta & -\sin\theta \\ \sin\theta & \cos\theta \end{bmatrix}. \tag{8.6}$$

The rotated vector $\boldsymbol{e}_k^p$ (in the M subspace representation) is then given as

$$\boldsymbol{R}\cdot\boldsymbol{e}_k^p=\begin{bmatrix}\cos\theta & -\sin\theta\\ \sin\theta & \cos\theta\end{bmatrix}\cdot\begin{bmatrix}\boldsymbol{e}_1^T\boldsymbol{e}_k\\ \boldsymbol{v}_1^T\boldsymbol{e}_k\end{bmatrix}=\begin{bmatrix}\boldsymbol{e}_1^T\boldsymbol{e}_k\cos\theta-\boldsymbol{v}_1^T\boldsymbol{e}_k\sin\theta\\ \boldsymbol{e}_1^T\boldsymbol{e}_k\sin\theta+\boldsymbol{v}_1^T\boldsymbol{e}_k\cos\theta\end{bmatrix}. \tag{8.7}$$

The rotated vector $\boldsymbol{e}_k^p$ in the original space is given as

$$\boldsymbol{e}_k^{\mathrm{rot}.p}=\boldsymbol{e}_1\left(\boldsymbol{e}_1^T\boldsymbol{e}_k\cos\theta-\boldsymbol{v}_1^T\boldsymbol{e}_k\sin\theta\right)+\boldsymbol{v}_1\left(\boldsymbol{e}_1^T\boldsymbol{e}_k\sin\theta+\boldsymbol{v}_1^T\boldsymbol{e}_k\cos\theta\right). \tag{8.8}$$

The entire vector $\boldsymbol{e}_k$ upon rotation is given by (Koziel and Bekasiewicz, 2016d)

$$\begin{aligned}\boldsymbol{e}_k^{\mathrm{rot}}&=\boldsymbol{e}_k^v+\boldsymbol{e}_k^{\mathrm{rot}.p}=\boldsymbol{e}_k-\boldsymbol{e}_1\boldsymbol{e}_1^T\boldsymbol{e}_k-\boldsymbol{v}_1\boldsymbol{v}_1^T\boldsymbol{e}_k\\&\quad+\boldsymbol{e}_1\left(\boldsymbol{e}_1^T\boldsymbol{e}_k\cos\theta-\boldsymbol{v}_1^T\boldsymbol{e}_k\sin\theta\right)\\&\quad+\boldsymbol{v}_1\left(\boldsymbol{e}_1^T\boldsymbol{e}_k\sin\theta+\boldsymbol{v}_1^T\boldsymbol{e}_k\cos\theta\right)\boldsymbol{e}_1\\&=\left[\boldsymbol{I}+\boldsymbol{e}_1\left[\boldsymbol{e}_1^T(\cos\theta-1)-\boldsymbol{v}_1^T\sin\theta\right]\right.\\&\quad\left.+\boldsymbol{v}_1\left[\boldsymbol{e}_1^T\sin\theta+\boldsymbol{v}_1^T(\cos\theta-1)\right]\right]\boldsymbol{e}_k,\end{aligned} \tag{8.9}$$

for $k=2,\ldots,n$. Here, $\boldsymbol{I}$ is the identity matrix. Thus,

$$\bar{\boldsymbol{R}}=\boldsymbol{I}+\boldsymbol{e}_1\left[\boldsymbol{e}_1^T(\cos\theta-1)-\boldsymbol{v}_1^T\sin\theta\right]+\boldsymbol{v}_1\left[\boldsymbol{e}_1^T\sin\theta+\boldsymbol{v}_1^T(\cos\theta-1)\right] \tag{8.10}$$

can be considered the overall rotation matrix. Note that $\boldsymbol{e}_1^{\mathrm{rot}}=\boldsymbol{v}/||\boldsymbol{v}||$.

The size of the original box is given as $\boldsymbol{d}=[d_1\ldots d_n]^T=(\boldsymbol{u}-\boldsymbol{l})/2$ (i.e., box dimensions with respect to its center $\boldsymbol{x}_0$). The size $\boldsymbol{d}'$ of the rotated box is the same, except the first dimension (spanned by the vector $\boldsymbol{v}$), which is equal to $||\boldsymbol{v}||/2$; thus, we have $\boldsymbol{d}'=[d_1'\ldots d_n']^T=[||\boldsymbol{v}||/2d_2\ldots d_n]^T$. In other words, the rotated box is a convex hull of the vectors $d_k'\boldsymbol{e}_k^{\mathrm{rot}}$ shifted by $\boldsymbol{x}_0$.

In Step 3 of the process, the rotated box is linearly reduced in all dimensions except the "diagonal" one. One way of finding the appropriate reduction rate is an iterative one, i.e., identification of the Pareto sets for the increasing reduction rates, and looking for the front "saturation". The rule of thumb developed based on numerous experiments with real-world antenna structures (Koziel

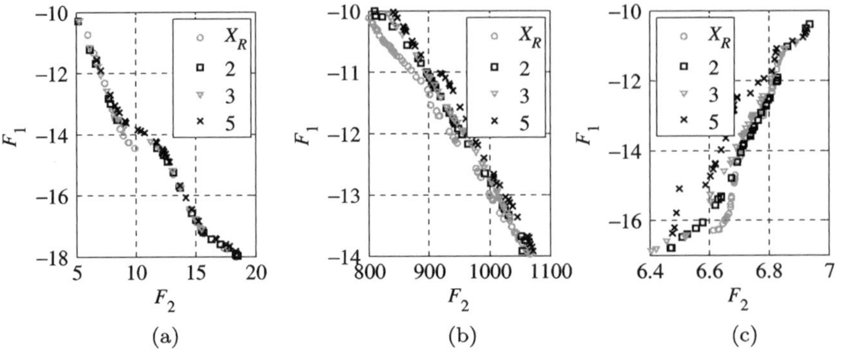

Fig. 8.4. Testing rotated box reduction factors (Koziel and Bekasiewicz, 2016d). Pareto sets generated using the reduction factors of 2, 3 and 5 for: (a) monocone antenna of Fig. 8.1; (b) UWB dipole antenna of Fig. 6.3 and (c) quasi Yagi–Uda antenna of Fig. 9.46. The Pareto set obtained within the initially reduced design space is marked as X_R (see Fig. 8.2).

and Bekasiewicz, 2016d) recommends the factor of 3. Note that this reduction rate results in approximate (assuming relatively small rotation angle θ) 3^{n-1} reduction of the design space volume compared to that obtained at Step 1. For $n = 10$ dimensions, this additional reduction is over 10^4 (volume-wise).

Clearly, space reduction may affect the shape of the Pareto-front and thus finding the appropriate scaling factor amounts to seeking for a trade-off between the cost of surrogate model training data acquisition and reducing the distance between the Pareto-optimal set contained in the reduced space and the actual Pareto front. In order to illustrate the effect of the reduction factor value, consider Fig. 8.4 that shows three antenna structures, a monocone antenna, UWB dipole and a quasi Yagi–Uda antenna. Geometries of these structures can be found in Figs. 8.1, 6.3 and 9.46, respectively. Figure 8.4 shows the Pareto sets obtained using the method of Sec. 7.1 with the data-driven surrogate set up in a design space reduced by means of the "rotational" method considered here. The Pareto sets are compared with the ones obtained within the initially reduced space (Sec. 8.2). The results indicate that the reduction factor of 3 ensures sufficient accuracy of the Pareto front identification within about 0.5 dB tolerance with respect to the true front, which is considered acceptable. Simultaneously, the cost of surrogate model preparation is considerably lower than within the initially reduced space (Koziel and Bekasiewicz, 2016d).

It is also possible to obtain more accurate (non-heuristic) estimation of the scaling factor. In particular, a Pareto-optimal design corresponding to the middle part of the front can be obtained through local optimization (e.g., starting from the design that is a center of the diagonal spanned by the two extreme Pareto-optimal designs). The distance of this design from the diagonal can be directly used to calculate the scaling factor.

8.4 Design Space Confinement

The last reduction method discussed here, design space confinement (Koziel and Bekasiewicz, 2015b), has been developed to identify a subset of the design space that contains a relevant part of the Pareto front, e.g., containing the designs that are acceptable with respect to the reflection response.

For example, in case of narrow-band antennas, although the extreme points of the Pareto set, found by (8.1), are obtained to ensure satisfaction of the design specifications with respect to the reflection response, majority of the designs in the initial Pareto set may not satisfy the specifications, mostly because of frequency shifts of $|S_{11}|$ when moving along the Pareto front. This situation is illustrated in Fig. 8.5(a) for a dielectric resonator antenna considered in (Koziel and Bekasiewicz, 2015b). Still, the feasible part of the initial front can be utilized to further confine the design space region that contains the Pareto subset of interest (i.e., the design satisfying specifications for reflection, such as $|S_{11}| \leq -10\,\text{dB}$ in the frequency range of interest). Subsequently, the response surface approximation (RSA) model is re-set in that region, and re-optimized using, e.g., an evolutionary algorithm, to find the final Pareto set representation.

The confinement method aims at identifying the geometry of the design space region of interest (as defined in the previous paragraph) by finding the orthogonal basis that spans the smallest hypercube containing the feasible Pareto-optimal solutions generated in the first

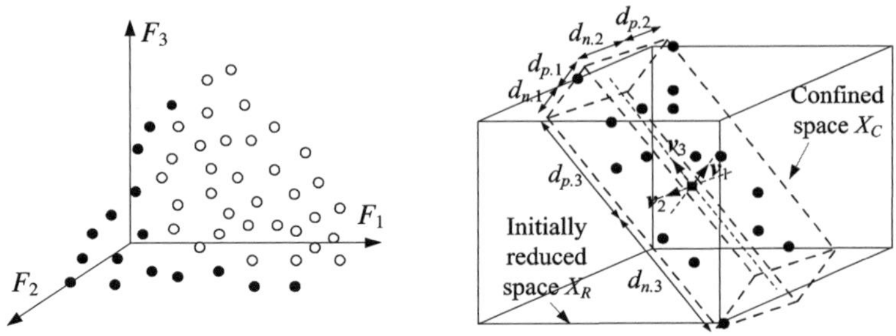

(a) (b)

Fig. 8.5. Conceptual illustration of the design space confinement. (a) Feasible (·) and infeasible (o) designs of the initial Pareto set for three-objective problem. For narrow-band antennas, majority of the Pareto designs will be infeasible from the point of view of the reflection response (due to the frequency shift). (b) Identification of the minimum (volume-wise) subset containing all feasible elements of the initial Pareto set (Koziel and Bekasiewicz, 2016d). Center point $\boldsymbol{x}^c$ is denoted by (■).

stage of the optimization process. Let $X_F = \{\boldsymbol{x}_f^{(k)}\}_{k=1,\ldots,Nf}$, denote the feasible subset of the initial Pareto set. The confinement procedure consists of identifying a set of vectors $\boldsymbol{v}_k, k = 1,\ldots,n$, and dimensions $d_{p.k}, d_{n.k}, k = 1,\ldots,n$ (see Fig. 8.5(b) for symbol explanation) defining the confined space X_C, with respect to the center point $\boldsymbol{x}^c = (1/N_f)\sum_{k=1,\ldots,Nf}\boldsymbol{x}_f^{(k)}$ (X_C center).

Let us assume that the vectors $\boldsymbol{v}_1,\ldots,\boldsymbol{v}_{k-1}$ are already known. Let $M_k = R^n\backslash\text{span}(\{\boldsymbol{v}_1,\ldots,\boldsymbol{v}_{k-1}\})$ be the orthogonal complement of the n-dimensional Euclidean space R^n and the subspace spanned by $\boldsymbol{v}_1$ through $\boldsymbol{v}_{k-1}$; we also have $M^k = R^n$ for $k = 1$. The vector $\boldsymbol{v}_k$ is found as the direction at which the diameter of the orthogonal projection of $X_F - \{\boldsymbol{x}^c\}$ onto $M_k, P_k(X_F - \{\boldsymbol{x}^c\})$ reaches its minimum, i.e.,

$$\boldsymbol{v}_k = \arg\min_{\boldsymbol{v}\in M_k} D(P_k(X_F - \{\boldsymbol{x}^c\}), \boldsymbol{v}), \tag{8.11}$$

where the diameter of a set Y in the direction of $\boldsymbol{v}$, $D(Y,\boldsymbol{v})$, is defined as

$$D(Y,\boldsymbol{v}) = \max_{\boldsymbol{y}\in Y}\{\boldsymbol{v}^T\boldsymbol{y}\} - \min_{\boldsymbol{y}\in Y}\{\boldsymbol{v}^T\boldsymbol{y}\}. \tag{8.12}$$

Having the vectors $\boldsymbol{v}_k$, the sizes of the confined space are found as

$$d_{p.k} = \max_{\boldsymbol{x} \in X_F} \left\{ \boldsymbol{v}_k^T \boldsymbol{x} \right\} \quad d_{n.k} = - \min_{\boldsymbol{x} \in X_F} \left\{ \boldsymbol{v}_k^T \boldsymbol{x} \right\}, \tag{8.13}$$

for $k = 1, \ldots, n$. More specifically, these dimensions determine the maximum distance between the center point $\boldsymbol{x}^c$ and the points from X_F (both "positive" and "negative" as $\boldsymbol{x}^c$ is not necessarily the confined space center) along the directions of the vectors $\boldsymbol{v}_k$.

The confined design space X_C is normally significantly smaller (volume-wise) than the initially reduced space X_R and the reduction rate will increase with the number of the design variables. Consequently, the number of training samples necessary for creating a reliable RSA model in X_C is expected to be smaller than for X_R. A specific design example illustrating utilization of the confinement method will be discussed in Chapter 9.

8.5 Summary

In this chapter, several design space reduction techniques have been discussed. The methods aim at identifying the region of the space that contains the Pareto-optimal solutions (or a feasible part of it) so as to facilitate further steps of the multi-objective optimization process. Typically, these would involve construction of the auxiliary data-driven surrogate that is subsequently optimized using metaheuristic algorithms (cf. Sec. 7.1). The techniques described here aim at reducing the ranges of the antenna geometry parameters rather than reducing the design space dimensionality. It should be mentioned, however, that in some cases, reducing the variable ranges may lead to dimensionality reduction as well. This is because the ranges of certain parameters may be narrowed down to very low values, which justifies their elimination from the design process (Koziel and Bekasiewicz, 2016g,i).

Considerable reduction of the ranges for some of the parameters also results in a "flattening" effect that improves the data-driven model scalability properties (with respect to the problem dimensionality) as elaborated in the following text. Figure 8.6(a) shows

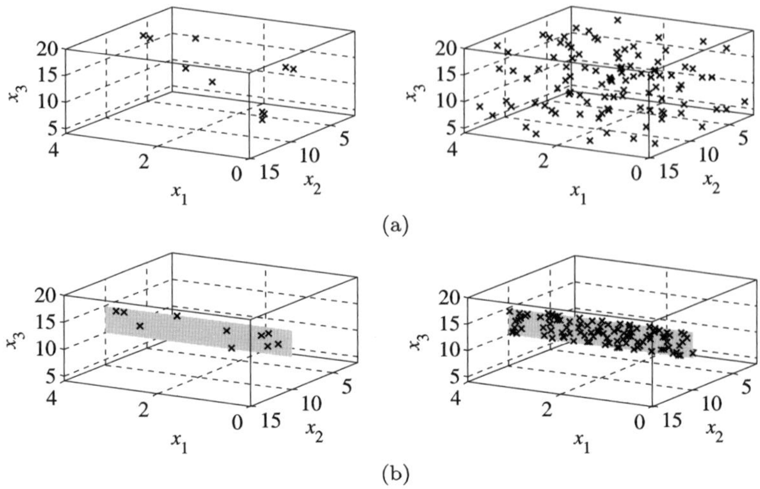

Fig. 8.6. Sample allocation within a three-dimensional design space. An increase of the number of samples from 10 to 100 in: (a) "thick" design space; (b) "flat" design space (gray area). In the latter case, the increase of the number of training points results in larger reduction of the average minimum distance between the points along the "thick" dimension leading to better scalability of the data-driven model in terms of its predictive power versus the training data set size.

a typical "thick" design space where increasing the number of training samples results in only slight reduction of the average distance between the training points (proportional to $1/N^{1/n}$ with n being dimension of the design space). The modeling error, on the other hand, is more or less proportional to that average distance. For "thin" space (Fig. 8.6(b)), the average minimum distance along "thick" (i.e., critical) dimensions decreases much faster than $1/N^{1/n}$ so that increasing the number of training points has more effect on the surrogate model error. Clearly, rigorous assessment of these effects is not possible in general because they are dependent on "flattening" effects and nonlinearity of the antenna responses along specific dimensions (both problem dependent).

The "flattening" effect can be illustrated using a UWB monopole antenna shown in Fig. 8.7(a) (Koziel *et al.*, 2014c). The antenna contains 13 geometry parameters. The initially reduced design space (cf. Fig. 8.7(b)) is about 14 orders of magnitude smaller

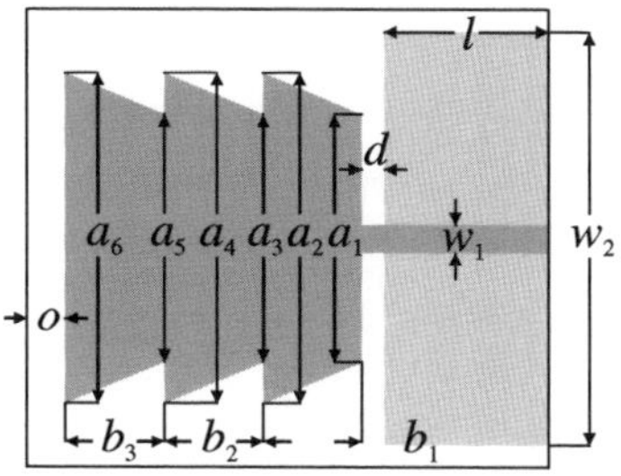

(a)

Parameter	Design bounds				Parameter	Design bounds			
	Initial		Reduced			Initial		Reduced	
	l	u	l^*	u^*		l	u	l^*	u^*
a_1	5	25	10.07	11.30	b_2	1	15	3.80	4.00
a_2	5	25	21.63	21.96	b_3	1	15	12.32	13.08
a_3	5	25	22.20	24.30	w_2	0.2	2	0.57	0.74
a_4	5	25	21.00	24.15	l	8	15	8.30	11.20
a_5	5	25	20.80	21.27	d	20	40	22.07	39.35
a_6	5	25	22.70	24.60	o	5	10	5.00	5.75
b_1	1	15	3.20	3.90					

(b)

Fig. 8.7. Illustration of the "flattening" effect: (a) Geometry of 13-variable planar UWB monopole antenna (Koziel *et al.*, 2014c), (b) initial parameter ranges $\boldsymbol{l}$ and $\boldsymbol{u}$, and the reduced ranges, $\boldsymbol{l}^*$ and $\boldsymbol{u}^*$, using the technique of Sec. 8.2.

(volume-wise) than the original space. More importantly, the ranges for many of the parameters are narrowed down considerably (to small fractions of the original ranges) which dramatically improves the surrogate model scalability.

The average relative error of the Kriging interpolation model constructed using 500 low-fidelity model samples is 6 percent (Koziel *et al.*, 2013c), whereas the relative error of the model constructed with 1,500 samples is 3.5 percent. Thus, the accuracy improvement of the second model compared to the first one, is 1.72, is much better than $3^{1/13} = 1.08$, which is due to the "flattening" of the design space.

The importance of design space reduction will be illustrated using real-world antenna design cases presented in the remaining parts of this book, especially in Chapter 9.

Chapter 9

Multi-Objective Optimization of Antenna Structures: Application Case Studies

The purpose of this chapter is to demonstrate the operation and performance of the surrogate-assisted multi-objective optimization procedures formulated in Chapters 7 and 8. The numerical studies included here are very comprehensive and contain nine antenna structures of various levels of complexity. For some of the test cases, numerical results are supported by experimental data. The number of geometry parameters ranges from five to 24. The considered design objectives include reflection response, antenna size (footprint area or volume), antenna gain, as well as radiation pattern. In vast majority of cases, two-objective problems are solved, which is the most typical situation, especially for compact antennas, where the trade-offs between the antenna size and its electrical performance (specifically, the maximum in-band reflection level) are sought. There are several antenna types considered in this study, including ultra-wideband (UWB) monopoles, planar Yagi antennas, a dielectric resonator antenna, as well as a UWB multi-input multi-output (MIMO) structure.

The presented test cases are arranged with respect to increasing level of the optimization problem complexity. The decisive factor is primarily design space dimensionality as it affects the computational cost of the design process and also enforces utilization of more involved tools (such as more sophisticated design space reduction techniques). For most antenna structures, one selected optimization method is demonstrated. For some of the structures, a comparison of several techniques is provided, which allows for both qualitative and quantitative assessment of the algorithms and discussion of their advantages and limitations.

The chapter is divided into 10 sections. Sections 9.1–9.9 contain descriptions of the considered antenna structures and their computational models, formulations of the optimization tasks, as well as numerical (and, in some cases, experimental) results. The chapter is concluded in Sec. 9.10, which contains a brief discussion of the presented methods and qualitative analysis of the results.

9.1 Design of Planar Yagi Antenna Using Decomposition

In this section, we consider multi-objective optimization of a planar Yagi antenna including its feeding structure (Koziel and Ogurtsov, 2013a). A multi-objective evolutionary algorithm (MOEA)-based optimization procedure of Sec. 7.1 is utilized along with the structure decomposition to ensure computationally feasible construction of the Kriging surrogate model.

9.1.1 *Antenna Geometry and Electromagnetic Models*

The antenna structure is shown in Fig. 9.1 (Qian *et al.*, 1998; Koziel and Ogurtsov, 2013a). It comprises a driven element fed by a coplanar strip-line, director, and a microstrip balun. The dielectric substrate is a 0.635 mm thick Rogers RT6010 ($\varepsilon_r = 10.2$, $\tan\delta = 0.0023$). The antenna is fed through a 50 Ohm microstrip line. Design variables are $\boldsymbol{x} = [s_1\ s_2\ v_1\ v_2\ u_1\ u_2\ u_3\ u_4]^T$. Parameters $w_1 = w_3 = w_4 = 0.6$, $w_2 = 1.2$, $u_5 = 1.5$, $s_3 = 3.0$ and $v_3 = 17.5$ remain fixed during the optimization process. All dimensions are in mm.

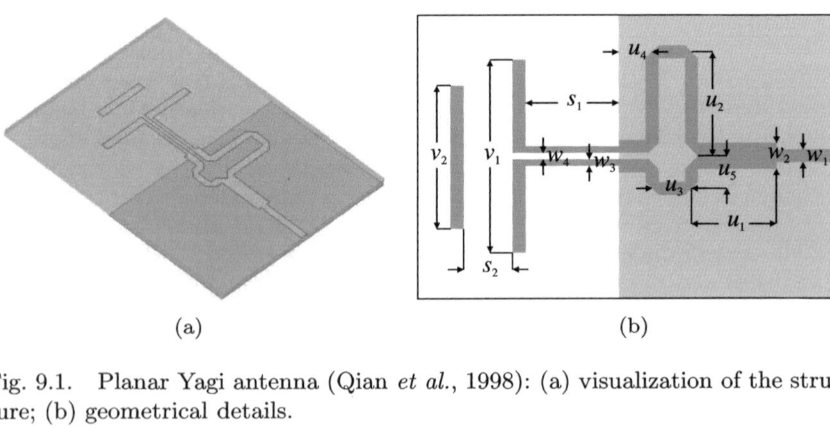

(a) (b)

Fig. 9.1. Planar Yagi antenna (Qian *et al.*, 1998): (a) visualization of the structure; (b) geometrical details.

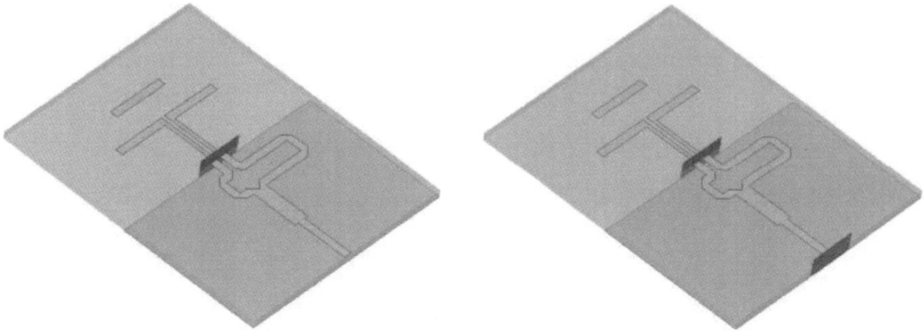

(a) (b)

Fig. 9.2. EM models with excitation ports shown: (a) coarse-discretization model $\boldsymbol{c}_a$ with excitation applied directly to the coplanar line of the driven element; (b) coarse-discretization model of the balun $\boldsymbol{c}_b$ with two ports shown. For visualization of the high-fidelity model, see Fig. 9.1(a).

The eight-dimensional design space is too large to construct a globally accurate Kriging interpolation model. On the other hand, in this initial example, we do not use any design space reduction. Instead, we demonstrate how dimensionality problem can be handled using the structure decomposition. We decompose the structure into two parts: the radiator (Fig. 9.2(a)), and the balun (Fig. 9.2(b)). In order to build the Kriging surrogate, the radiator part of the antenna is described by the coarse-discretization model $\boldsymbol{c}_a$ which contains only a half of the design variables, $\boldsymbol{x}_a = [s_1\ s_2\ v_1\ v_2]^T$. At the same time, the balun component is described using the coarse-discretization model $\boldsymbol{c}_b$ comprising the remaining variables $\boldsymbol{x}_b = [u_1\ u_2\ u_3\ u_4]^T$. The high-fidelity model $\boldsymbol{f}$ (for visualization, see Fig. 9.1(a)) contains the complete set of eight variables $\boldsymbol{x} = [\boldsymbol{x}_a\, \boldsymbol{x}_b]^T$. Each coarse-discretization model is defined with its complement being inactive, as shown in Figs. 9.2(a)–9.2(b), using the internal excitation which is available with the transient solver of CST Microwave Studio (CST, 2013).

The high-fidelity model is densely discretized and takes into account the material losses. Both coarse models utilize lossless materials and sparse meshes. In addition, they run with a more relaxed residual energy termination condition (−25 dB for $\boldsymbol{c}_a$ and −30 dB for $\boldsymbol{c}_b$). The high-fidelity model $\boldsymbol{f}$ contains ~1,400,000 hexahedral mesh cells. Its average simulation time is about 36 min. The coarse model $\boldsymbol{c}_a$ consists of about 85,000 mesh cells and its simulation time is 76 s. The coarse model $\boldsymbol{c}_b$ contains ~85,000 cells (simulation time 134 s).

The low-fidelity model of the entire structure is set as follows: the radiation response is configured from that of model $\boldsymbol{c}_a$ (shown in Fig. 9.2(a)) as a response surface approximation (RSA) surrogate model; the reflection response at a particular frequency point is obtained as

$$S_{11} = S_{11,b} + \frac{S_{12,b}S_{21,b}S_{11,a}}{1 - S_{22,b}S_{11,a}}, \tag{9.1}$$

where $S_{ij,b}$ are the S-parameters of the surrogate model $\boldsymbol{c}_b$ and $S_{11,a}$ is the reflection coefficient of the surrogate model $\boldsymbol{s}_a$.

For the Yagi antenna considered here, we are going to study objectives related to the antenna reflection and radiation responses using the MOEA and targeting the best matching (objective F_1) and the highest average end-fire gain (objective F_2) within the 10–11 GHz bandwidth.

9.1.2 *Surrogate Models*

To find the Pareto front, the surrogate model $\boldsymbol{s}$ of the Yagi antenna is constructed as follows. In the first stage, we obtain the Kriging model $\boldsymbol{s}_a$ of the coarse-discretization model of the radiator $\boldsymbol{c}_a$. The design space is defined using the following lower and upper bounds: $\boldsymbol{l}_a = [3.8\ 2.8\ 8.0\ 4.0]^T$ and $\boldsymbol{u}_a = [4.4\ 4.4\ 9.8\ 5.2]^T$. The Kriging model is constructed using 256 samples allocated on a uniform rectangular grid in the literature of Koziel *et al.* (2011a).

In the second stage, the Kriging model $\boldsymbol{s}_b$ of the balun is obtained. Here, the design space is defined by: $\boldsymbol{l}_b = [3.0\ 4.5\ 1.8\ 1.3]^T$ and $\boldsymbol{u}_b = [4.2\ 5.2\ 2.6\ 1.8]^T$, respectively. The Kriging model is constructed using 625 samples allocated on a uniform rectangular grid.

The initial surrogate model is obtained from the two aforementioned Kriging models so that the antenna gain is directly modeled by $\boldsymbol{s}_a$, whereas its reflection response is obtained from both $\boldsymbol{s}_a$ and $\boldsymbol{s}_b$

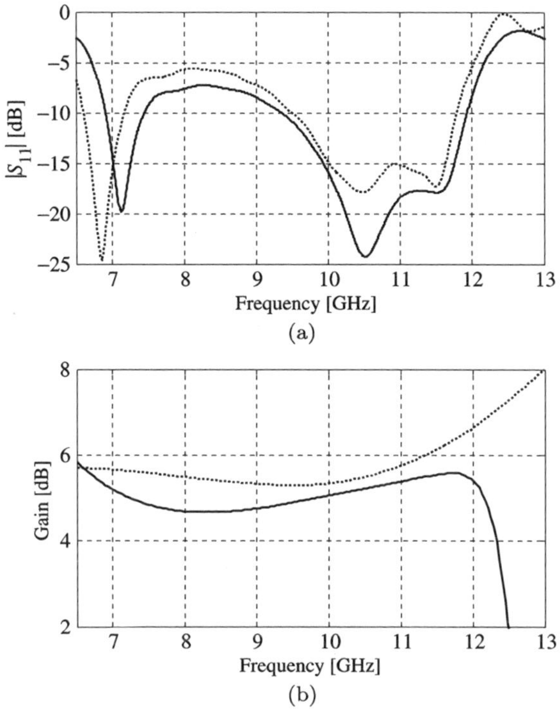

Fig. 9.3. Model responses at the initial design (Koziel and Ogurtsov, 2013a): (a) reflection with $\boldsymbol{f}$ (—) and reflection obtained from S-parameters of the coarse models $\boldsymbol{c}_a$ and $\boldsymbol{c}_b(\cdots)$; (b) end-fire gain with $\boldsymbol{f}$ (—) and with $\boldsymbol{c}_a(\cdots)$.

using (9.1). Because of discrepancy between $\boldsymbol{s}$ and the high-fidelity model (see Fig. 9.3), $\boldsymbol{s}$ is further enhanced using space mapping alignment with $2n + 1$ (n being the number of design variables, here, 8) high-fidelity training samples allocated using so-called star distribution design of experiment (Koziel *et al.*, 2011a). The space mapping used here is a parameter shift combined with multiplicative and additive response correction (Bandler *et al.*, 2004a). It should be reiterated that the major reason for decomposing the structure into two parts is that such decomposition allows alleviate the curse of dimensionality problem (Forrester and Keane, 2009). Otherwise, an accurate response surface model of the entire structure would require thousands of training samples (Bekasiewicz *et al.*, 2014a). The total

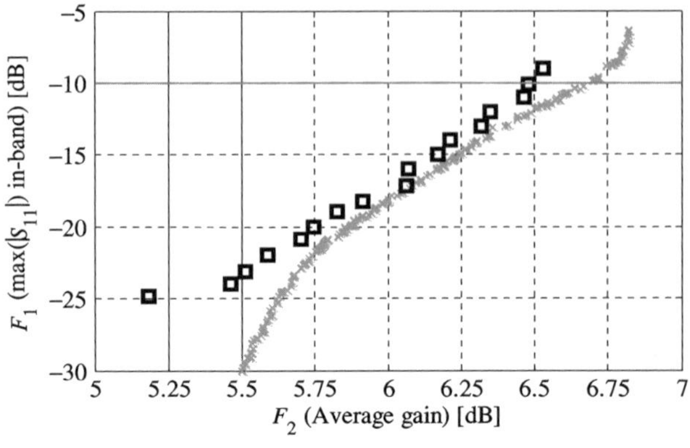

Fig. 9.4. Multi-objective optimization of the planar Yagi antenna: comparison of the low-fidelity Pareto set generated by MOEA optimization of the RSA model (×) and the high-fidelity designs obtained after the refinement stage (□).

cost of creating the surrogate, including the space mapping correction corresponds to about 60 evaluations of the high-fidelity model.

9.1.3 *Numerical Results*

Using the surrogate model created as described in Sec. 9.1.2, the initial Pareto front was generated by means of the MOEA as shown in Fig. 9.4.

It can be observed that the level of antenna reflection within the frequency band 10–11 GHz can change from around −9 dB to about −30 dB, whereas the average end-fire gain can change between 5.2 and 6.8 dB. In the next step, 17 designs out of Pareto front have been refined. These refined designs are listed in Table 9.1, whereas a comparison of the high- and low-fidelity Pareto sets is shown in Fig. 9.4. Responses of the selected designs are shown in Figs. 9.5 and 9.6.

The results indicate the capability of the considered antenna structure in terms of matching and gain, as well as the possible trade-offs between these quantities. At instance, design 16 can be attractive if the highest directivity and end-fire gain (see Figs. 9.6(a) and 9.6(b)) are required in for a particular application. This design, however, shows no benefits in terms of the realized end-fire gain compared to design 11 as it is seen from Fig. 9.6(c), while its maximal

Table 9.1 Yagi antenna optimization results

No	Design variables [mm] s_1	s_2	v_1	v_2	u_1	u_2	u_3	u_4	max $\|S_{11}\|$ [dB]	Gain* [dB]
1	4.13	3.23	9.55	4.65	3.92	4.72	2.19	1.68	−24.8	5.2
2	4.16	3.38	9.34	4.67	3.90	4.76	2.20	1.72	−24.0	5.5
3	4.15	3.45	9.38	4.70	3.82	4.86	2.16	1.74	−23.2	5.5
4	4.15	3.55	9.29	4.68	3.87	4.80	2.21	1.73	−22.0	5.6
5	4.26	3.75	9.13	4.66	3.91	4.80	2.23	1.68	−20.9	5.7
6	4.22	3.82	9.13	4.68	3.87	4.83	2.22	1.68	−20.0	5.7
7	4.20	4.02	9.08	4.65	3.85	4.87	2.22	1.70	−18.9	5.8
8	4.33	4.21	8.77	4.65	3.81	4.86	2.16	1.62	−18.3	5.9
9	4.36	4.28	8.70	4.79	3.76	4.83	2.28	1.74	−17.2	6.1
10	4.30	4.25	8.76	4.77	3.88	4.84	2.21	1.65	−16.0	6.1
11	4.31	4.26	8.60	4.82	3.86	4.81	2.17	1.67	−15.0	6.2
12	4.31	4.26	8.57	4.87	3.90	4.82	2.17	1.64	−14.0	6.2
13	4.37	4.26	8.47	4.98	3.84	4.85	2.23	1.65	−13.0	6.3
14	4.36	4.22	8.46	5.03	3.85	4.86	2.20	1.63	−12.0	6.3
15	4.35	4.29	8.38	5.10	3.82	4.85	2.22	1.63	−11.0	6.5
16	4.34	4.25	8.26	5.12	3.92	4.75	2.20	1.65	−10.1	6.5
17	4.33	4.26	8.21	5.16	3.91	4.74	2.16	1.58	−9.0	6.5

*End-fire gain averaged over the 10–11 GHz bandwidth.

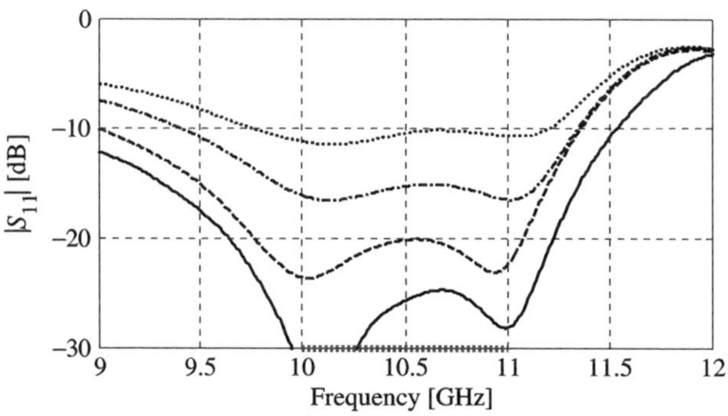

Fig. 9.5. Reflection responses of selected Pareto-optimal designs obtained in the refinement stage (Koziel and Ogurtsov, 2013a): design 1 (—), design 6 (– –), design 11 (· — ·), design 16 (· · ·).

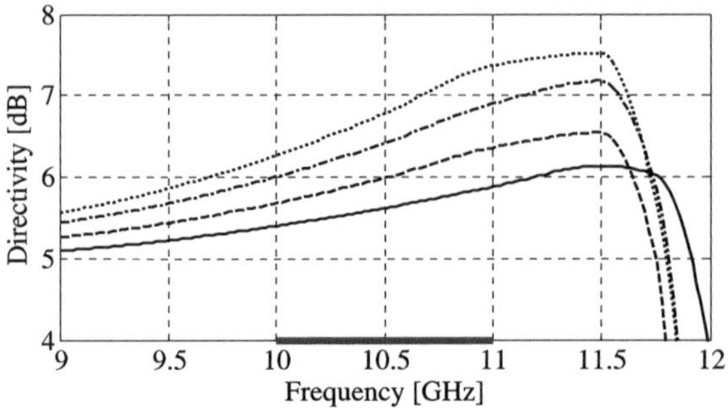

(a)

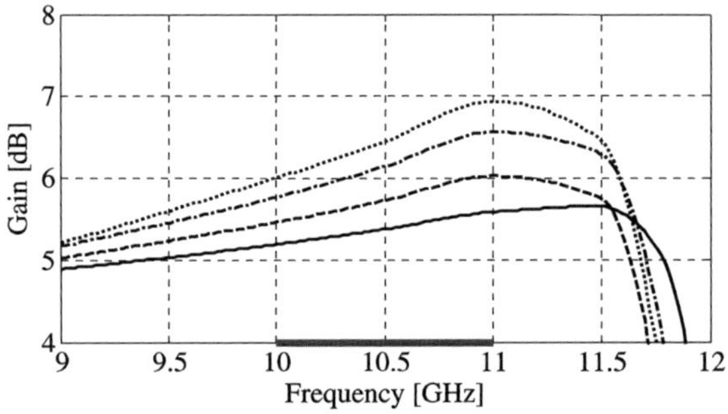

(b)

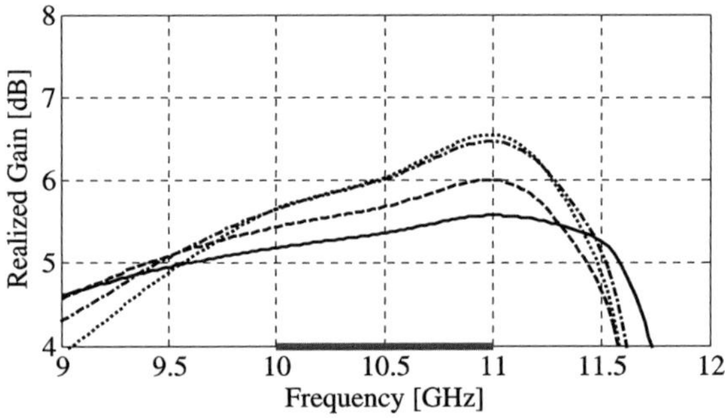

(c)

Fig. 9.6. End-fire radiation responses of selected Pareto-optimal designs (Fig. 9.5) obtained in the refinement stage, design 1 (—), design 6 (– –), design 11 (· — ·) and design 16 (· · ·): (a) directivity; (b) gain; (c) realized gain (Koziel and Ogurtsov, 2013a).

reflection coefficient in the bandwidth is higher in about 5 dB. Moreover, designs 11–16 have essentially the same end-fire realized gain values but they can be substantially different in reflection and other radiation and figures as indicated with Figs. 9.5, 9.6(a) and 9.6(b) (also see Table 9.1).

The computational cost of creating the Kriging surrogate corresponds to about 60 evaluations of the high-fidelity model $\boldsymbol{f}$. Multi-objective optimization of $\boldsymbol{s}$ takes about 20 min. Design refinement takes, on average, two iterations of the refinement procedure (i.e., two evaluations of $\boldsymbol{f}$). Thus, the total design cost (obtaining Pareto front and 17 antenna designs as in Table 9.1) corresponds to only about 80 evaluations of the high-fidelity model.

9.2 Design of Ultra-Wideband Monopole Antenna Using Multi-Objective Evolutionary Algorithm and Co-Kriging

In this section, we demonstrate the multi-objective optimization algorithm of Sec. 7.1 with co-Kriging-based Pareto front refinement, using an example of a UWB monopole antenna (Koziel *et al.*, 2014b). In this case, dimensionality of the design space is relatively low (five variables) so that, similarly as for the previous case, we do not use design space reduction.

9.2.1 *Antenna Geometry and Design Objectives*

Antenna geometry is shown in Fig. 9.7. The structure is composed of a driven element in the form of a trapezoid and a rectangle, stacked together into a radiator and fed through a 50 Ohm microstrip line. The antenna is implemented on a 0.762 mm thick Taconic RF-35 dielectric substrate ($\varepsilon_r = 3.5$, $\tan\delta = 0.0018$). Vector of design parameters is $\boldsymbol{x} = [a_1\ a_2\ b_1\ l\ w_2]^T$. Variables $w_1 = 1.7$, $d = 0.6$ and $o = 5$ remain fixed, whereas $b = 3b_1$ (all dimensions are expressed in mm). The design space is defined using the following lower and upper bounds: $\boldsymbol{l} = [10\ 23\ 5\ 11.5\ 28]^T$ and $\boldsymbol{u} = [11\ 24\ 7\ 12\ 38]^T$.

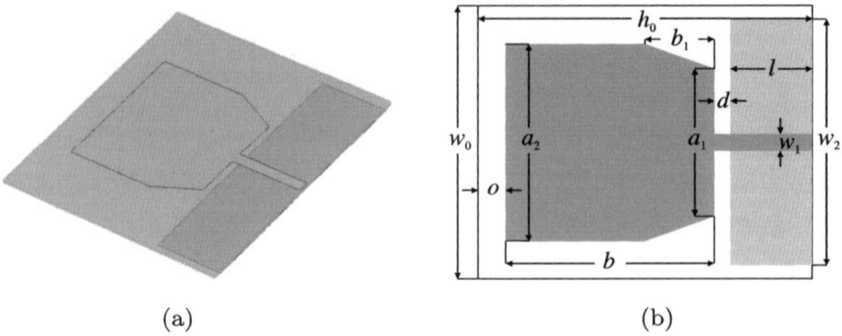

Fig. 9.7. Planar UWB monopole (Koziel *et al.*, 2014b): (a) visualization; (b) geometrical details.

The design objectives are: F_1 — minimization of $|S_{11}|$ within 3.1–10.6 GHz frequency range of interest, and F_2 — reduction the antenna footprint $A(\boldsymbol{x}) = w_0 \times h_0$, where $w_0 = o + w_2$ and $h_0 = l + d + b + o$.

9.2.2 *Electromagnetic Models Setup*

The high-fidelity model $\boldsymbol{f}$ of the antenna consists of about ~2,220,000 mesh cells and its average simulation time is 9 min. The low-fidelity model $\boldsymbol{c}$ contains ~42,000 mesh cells and its typical simulation time is 20 s. Both antenna models are implemented in CST Microwave Studio and simulated using its time domain solver (CST, 2013). The initial Kriging interpolation model $\boldsymbol{s}$ has been constructed using 1,000 $\boldsymbol{c}$ model samples allocated using Latin Hypercube Sampling (LHS) (Beachkofski and Grandhi, 2002).

9.2.3 *Numerical Results*

The surrogate model $\boldsymbol{s}$ created as described in Sec. 9.2.2 has been optimized using the MOEA running for 50 iterations and with a population of the size of 500. Subsequently, the obtained initial Pareto set has been refined using co-Kriging (Forrester *et al.*, 2008). Two refinement iterations were sufficient for convergence to the final Pareto front. Consequently, the high-fidelity representation of the Pareto front has been obtained after only 20 simulations of the $\boldsymbol{f}$ model. The initial and the refined sets obtained with the demonstrated

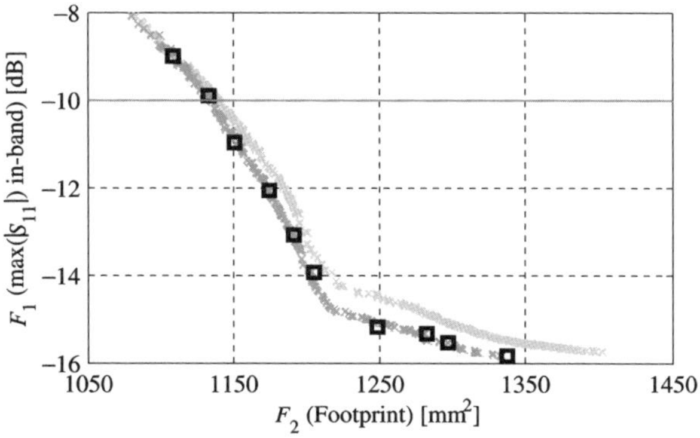

Fig. 9.8. UWB monopole antenna. Light- and dark-shade gray crosses represent the initial and the final approximation of the Pareto set, whereas rectangles denote selected high-fidelity model designs evaluated for the sake of verification (Koziel *et al.*, 2014b).

multi-objective design approach are shown in Fig. 9.8. An excellent agreement between the optimized co-Kriging model and the sampled high-fidelity model data can be observed.

The largest Pareto-optimal antenna design with a footprint of 1,315 mm^2 features the in-band reflection of −15.9 dB. The structure that maintains $|S_{11}|$ around −10 dB is characterized by a footprint of 1,133 mm^2. Thus, the variability of objectives F_1 and F_2 along the relevant fraction of the Pareto front is 5.9 dB and 182 mm^2 (almost 14 percent), respectively. Dimensions of the selected antenna designs are gathered in Table 9.2, whereas their corresponding frequency responses are shown in Fig. 9.9.

The cost of multi-objective antenna design corresponds to about 57 evaluations of the high-fidelity model $\boldsymbol{f}$ ($\sim$8.5 h of CPU-time) and it includes: 1,000 simulations of the low-fidelity model $\boldsymbol{c}$ and 20 evaluations of $\boldsymbol{f}$ required for a construction of the initial Kriging, and co-Kriging surrogates (two iterations of the refinement procedure).

9.3 Optimization of Dielectric Resonator Antenna Using Design Space Reduction and Multi-Objective Evolutionary Algorithm

In this section, we demonstrated multi-objective optimization of a quasi-isotropic dielectric resonator antenna (DRA) using a

Table 9.2 High-fidelity designs of the optimized UWB monopole

			Design Variables [mm]				
Design	F_1 [dB]	F_2 [mm^2]	a_1	a_2	b_1	l	w_2
$\boldsymbol{x}_f^{(1)}$	–9.0	1108	10.45	23.36	5.69	11.52	28.00
$\boldsymbol{x}_f^{(2)}$	–9.9	1133	10.61	23.31	5.93	11.54	28.01
$\boldsymbol{x}_f^{(3)}$	–11.0	1151	10.42	23.33	6.11	11.53	28.00
$\boldsymbol{x}_f^{(4)}$	–12.1	1174	10.59	23.25	6.31	11.63	28.01
$\boldsymbol{x}_f^{(5)}$	–13.1	1192	10.65	23.25	6.49	11.63	28.01
$\boldsymbol{x}_f^{(6)}$	–13.9	1205	10.61	23.23	6.62	11.63	28.03
$\boldsymbol{x}_f^{(7)}$	–15.2	1249	10.51	23.27	6.78	11.66	28.74
$\boldsymbol{x}_f^{(8)}$	–15.3	1283	10.74	23.88	6.87	11.54	29.53
$\boldsymbol{x}_f^{(9)}$	–15.5	1297	10.80	23.81	6.89	11.60	29.79
$\boldsymbol{x}_f^{(10)}$	–15.8	1338	10.78	23.78	6.89	11.61	30.89

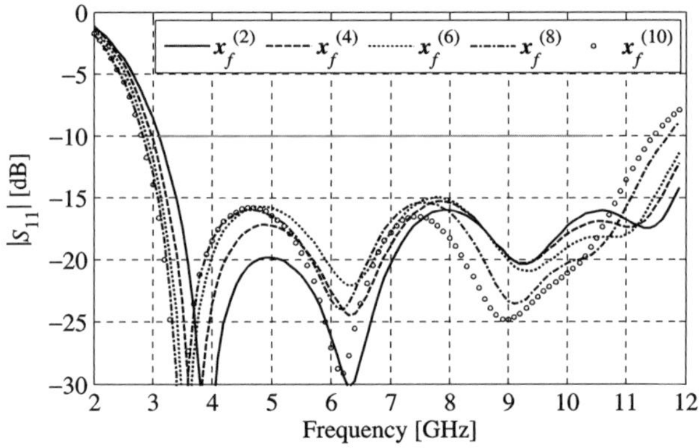

Fig. 9.9. Frequency characteristics of the selected designs from Table 9.2 (Koziel *et al.*, 2014b).

methodology of Sec. 7.1 supplemented by design space reduction. Three objectives are considered, including minimization of the antenna size, improving the radiation pattern shape (here, minimization of the difference between the maximum and minimum E-field strength), and also minimization of the in-band reflection level. Only the designs with $|S_{11}| \leq -10$ dB are considered acceptable, which makes the design process more complicated because only a small part of the Pareto front becomes of interest. In order to identify and further explore this part, design confinement method of Sec. 8.4 is utilized (Bekasiewicz *et al.*, 2016b). For the sake of comparison, the DRA is also optimized without space confinement and the result differences between the two approaches are discussed.

9.3.1 *Antenna Geometry*

The antenna structure is shown in Fig. 9.10 (Koziel and Bekasiewicz, 2015b). The design consists of a cuboid shape Taconic CER-10 dielectric resonator ($\varepsilon_r = 10$, $\tan\delta = 0.0035$) and a driven element in the form of a coaxial probe located within the material. The probe is fed from the bottom through a coaxial transmission line. The input impedance is 50 Ohm. The antenna introduces additional degrees of freedom for the probe location with respect to the original design of (Pan *et al.*, 2014). Consequently, a better control of its properties is possible during the optimization process. The vector of antenna design parameters is $\boldsymbol{x} = [a\ b\ c\ o_{1r}\ o_{2r}\ l_r]^T$. The relative variables are $o_1 = o_{1r} \cdot a$, $o_2 = o_{1r} \cdot b$ and $l = l_r \cdot c$, whereas dimensions $d = 1.26$ and $g = 0.82$ are fixed to ensure 50 Ohm input impedance.

Moreover, the ground plane thickness is $t = 0.05$ (see Fig. 9.11(b)). All non-relative dimensions are in mm. Note that utilization

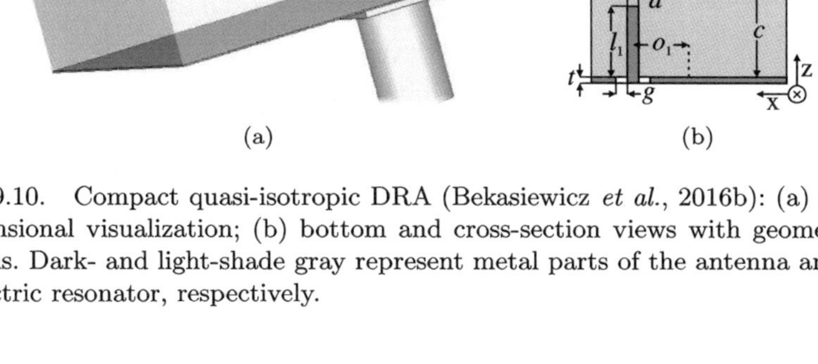

(a) (b)

Fig. 9.10. Compact quasi-isotropic DRA (Bekasiewicz *et al.*, 2016b): (a) three-dimensional visualization; (b) bottom and cross-section views with geometrical details. Dark- and light-shade gray represent metal parts of the antenna and the dielectric resonator, respectively.

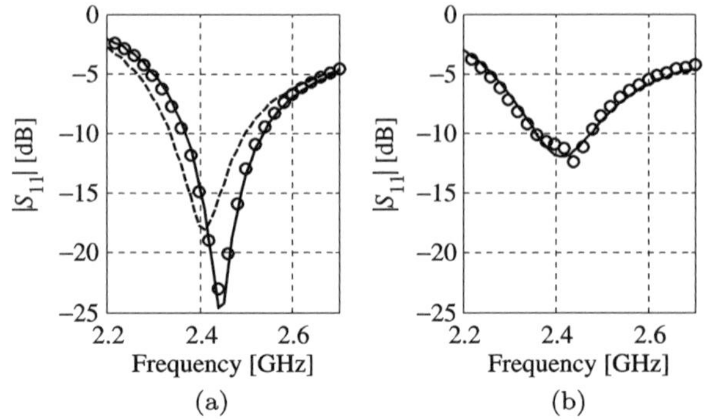

Fig. 9.11. Low-fidelity model correction of the reflection response (Koziel and Bekasiewicz, 2015b): (a) low- (– –) and high-fidelity model (—) responses at one of the reference designs, as well as corrected low-fidelity model response (◯) at the same design; (b) corrected low-fidelity (◯) and high-fidelity (—) model response at another design.

of relative variables is necessary to ensure geometrical consistency of the design. The initial lower and upper bounds of the design variables are $\boldsymbol{l} = [3\ 3\ 3\ {-0.45}\ {-0.45}\ 0]^T$ and $\boldsymbol{u} = [30\ 30\ 30\ 0.45\ 0.45\ 0.9]^T$ and they are determined based on engineering experience.

9.3.2 *Design Objectives and Antenna Models*

The following design objectives are considered: F_1 — minimization of the maximum reflection within 2.4–2.5 GHz frequency band of interest, F_2 — minimization of antenna volume $V(\boldsymbol{x}) = a \times b \times c$ and F_3 — reduction of difference between minimal and maximal E-field strength in x–z plane (see Fig. 9.10). The objective F_3 is defined as

$$F_3(\boldsymbol{x}) = \max\{0 \le \varphi \le 360\text{: } E(\boldsymbol{x}, \varphi)\} - \min\{0 \le \varphi \le 360\text{: } E(\boldsymbol{x}, \varphi)\}, \tag{9.2}$$

where $E(\boldsymbol{x}, \varphi)$ represents E-field strength for the given angle φ in x–z plane. The objective is calculated at the center frequency of 2.45 GHz. It should be noted that, in general, the objectives F_2 and F_3 are only weakly conflicting. The reason is that the radiation pattern in the x–z plane is the Fourier transform of the field in the aperture determined by the antenna dimension corresponding to this plane (in this case, the dimension b) (Balanis, 2005). As a consequence, the omnidirectional properties of the DRA in x–z plane will deteriorate when increasing b. However, other parameters may also contribute to the radiation pattern, to some extent. From this perspective, multi-objective optimization with respect to F_2 and F_3 is still considered interesting.

The high- (∼1,000,000 mesh cells, average evaluation time: 21 min) and low-fidelity (∼55,000 cells; evaluation time: 35 s) antenna models are both implemented in CST Microwave Studio (CST, 2013).

9.3.3 *Design Space Reduction and Surrogate Model Construction*

The first step of the design process is to carry out the initial design space reduction using the procedure of Sec. 8.2. The new lower and upper bounds are $\boldsymbol{l}^* = [3.3\ 24.5\ 14.5\ {-0.03}\ 0.37\ 0.63]^T$ and $\boldsymbol{u}^* = [29.8\ 30\ 15.8\ 0.16\ 0.45\ 0.98]^T$. The reduced design space is about four orders of magnitude smaller (volume-wise) than the initial one. In the next stage, the $\boldsymbol{f}$ model has been evaluated at the extreme Pareto samples utilized to define the initially reduced space. The obtained responses have been used to perform space mapping correction of the low-fidelity model $\boldsymbol{c}$. Because we deal with a narrow-band antenna, two types of correction are the most suitable, specifically, frequency scaling and multi-point output space mapping (OSM). The details of the correction procedure are discussed below (Koziel and Bekasiewicz, 2015b).

The frequency-scaled low-fidelity model $\boldsymbol{c}$ (denoted as $\boldsymbol{c}_F$) is defined as

$$\boldsymbol{c}_F(\boldsymbol{x}) = \boldsymbol{c}(\boldsymbol{x}, \boldsymbol{\alpha}F), \tag{9.3}$$

where $\boldsymbol{c}(\boldsymbol{x}, F)$ denotes explicit dependency of $\boldsymbol{c}_F$ on frequency F (in particular, it can be evaluation of the antenna $|S_{11}|$ for a range of frequencies). Further, we have

$$\boldsymbol{\alpha}F = \bar{\alpha}_0 + \bar{\alpha}_1 F, \tag{9.4}$$

where (9.4) describes the affine frequency scaling (Bandler *et al.*, 1995). The frequency scaling parameters are obtained as

$$[\bar{\alpha}_0\ \bar{\alpha}_1] = \arg \min_{[\alpha_0\ \alpha_1]} \sum_{k=1}^{N_r} ||\boldsymbol{f}(\boldsymbol{x}^{ref.k}) - \boldsymbol{c}(\boldsymbol{x}^{ref.k}, \alpha_0 + \alpha_1 F)||, \tag{9.5}$$

where $\boldsymbol{x}^{ref.k}$, $k = 1, \ldots, N_r$, are the reference designs (as mentioned before, here, these are the extreme points of the Pareto set). The remaining (vertical) misalignment can be reduced using multi-point response correction of the form

$$\boldsymbol{c}_{SM}(\boldsymbol{x}) = \bar{A} \cdot \boldsymbol{c}_F(\boldsymbol{x}), \tag{9.6}$$

where $\bar{A} = \text{diag}([a_1 \ \ldots a_m])$ is a diagonal correction matrix obtained as

$$\bar{A} = \arg\min_{A]} \sum_{k=1}^{N_r} ||\boldsymbol{f}(\boldsymbol{x}^{ref.k}) - A \cdot \boldsymbol{c}(\boldsymbol{x}^{ref.k}, \bar{\alpha}_0 + \bar{\alpha}_1 F)||^2. \tag{9.7}$$

The problem (9.7) can be reformulated as a linear regression problem which has an analytical least-square solution. Figure 9.11 shows the reflection response of the low- and high-fidelity model of the antenna considered in Sec. 9.3.1, as well as corrected low-fidelity model $\boldsymbol{c}_{SM}$ at one of the reference designs and some other (verification) design.

In the next step, the data acquisition process has been performed and the initial Kriging interpolation model $\boldsymbol{s}$ has been constructed using 576 samples of the corrected low-fidelity model $\boldsymbol{c}_{SM}$. The training set consists of 500 LHS-allocated samples supplemented with 64 designs obtained at the corners of the initially reduced space and additional 12 samples obtained during the low-fidelity model correction procedure. The cross-validation estimated root-mean square (RMS) error of the Kriging model is 3 percent.

The initial Pareto front has been obtained using MOEA (population size 2,000, maximum number of iterations 50). Subsequently, its feasible fraction (i.e., designs with maximum in-band reflection below –10 dB) has been utilized for confinement of the design space (cf. Sec. 8.4). It should be noted that the confined space is over 10 times smaller (volume-wise) than initially reduced one. The overall reduction rate of the confined space in comparison to the arbitrary defined one is 10^5. The surrogate model has been constructed again within the confined region using only 170 OSM-corrected LHS-allocated $\boldsymbol{c}$ samples (RMS error 1.5 percent).

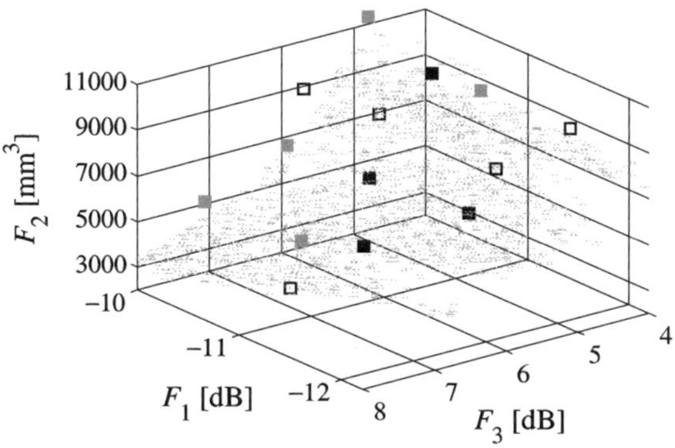

Fig. 9.12. Pareto front representations of the DRA obtained in confined space (Bekasiewicz *et al.*, 2016b). The $\boldsymbol{f}$ and $\boldsymbol{s}$ Pareto designs are denoted by squares and light-shade gray crosses, respectively. Detailed dimensions of the designs represented by non-empty squares are collected in Table 9.3. Responses of designs represented by gray squares are shown in Figs. 9.13 and 9.14.

9.3.4 *Numerical Results*

The final Pareto set has been found by optimizing the surrogate model (constructed within the confined space) using MOEA. No further refinement was necessary: due to space mapping correction of the low-fidelity model as described in Sec. 9.3.3, the surrogate can be considered sufficiently accurate. For the sake of verification, a set of Pareto-optimal designs has been evaluated using the high-fidelity model $\boldsymbol{f}$ (see Fig. 9.12). Although the results are in good agreement, a slight misalignment of reflection between $\boldsymbol{s}$ and $\boldsymbol{f}$ model responses can be observed. The reason is a residual inaccuracy of the corrected low-fidelity model. Notwithstanding, the discrepancies are below 0.5 dB. At the same time, the responses of F_2 and F_3 remain accurate.

Detailed data on selected high-fidelity Pareto-optimal designs is gathered in Table 9.3, whereas their reflection responses and radiation pattern characteristics are shown in Figs. 9.13 and 9.14, respectively. Among the evaluated high-fidelity designs, the antenna with the smallest volume features the largest E-field discrepancy of 7.9 dB and reflection of −10.6 dB. The design with the lowest E-field variations of 4.81 dB simultaneously exhibits the largest volume (over 11,000 mm^3) and barely acceptable reflection (−10 dB). Finally, the

Table 9.3 DRA: High-fidelity designs obtained in the confined space

		Objectives			Antenna parameters					
		F_1 [dB]	F_2 [mm^3]	F_3 [dB]	a	b	c	o_{1r}	o_{2r}	l_r
Antenna designs	$\boldsymbol{x}_f^{(1)}$	–10.0	11337	4.81	29.24	25.22	15.37	0.085	0.424	0.648
	$\boldsymbol{x}_f^{(2)}$	–10.7	10337	4.92	27.04	25.30	15.11	0.089	0.413	0.655
	$\boldsymbol{x}_f^{(3)}$	–11.3	10812	5.05	28.86	25.42	14.74	0.071	0.403	0.635
	$\boldsymbol{x}_f^{(4)}$	–10.8	6760	5.93	16.47	27.17	15.11	0.052	0.423	0.641
	$\boldsymbol{x}_f^{(5)}$	–9.9	4691	6.87	10.98	28.47	15.01	0.052	0.418	0.687
	$\boldsymbol{x}_f^{(6)}$	–10.8	4898	6.91	11.38	28.59	15.06	0.079	0.406	0.665
	$\boldsymbol{x}_f^{(7)}$	–11.4	5699	6.84	13.28	28.60	15.00	0.127	0.390	0.649
	$\boldsymbol{x}_f^{(8)}$	–10.6	3514	7.90	8.00	29.62	14.83	0.094	0.398	0.715
	$\boldsymbol{x}_f^{(9)}$	–11.2	4168	7.61	9.61	29.46	14.72	0.070	0.403	0.698
	$\boldsymbol{x}_f^{(10)}$	–10.5	8522	5.37	21.41	26.24	15.17	0.058	0.417	0.643

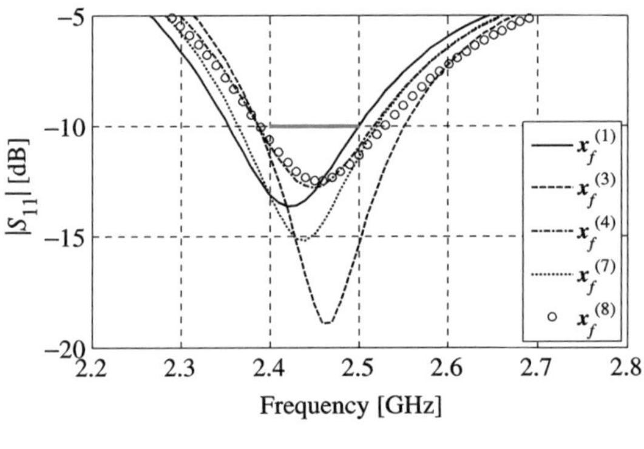

Fig. 9.13. Reflection responses of selected Pareto-optimal designs obtained in confined search space region (Bekasiewicz *et al.*, 2016b).

lowest in-band reflection of –11.4 dB has been obtained for the design with volume of almost 5,700 mm^3 and E-field discrepancy of 6.84 dB. The ranges of variability of objectives F_1, F_2 and F_3 along the Pareto front are 2.3 dB, 7,823 mm^3 and 3.1 dB, respectively. Also, it can be observed that omnidirectional antenna properties (objective F_3) are inversely proportional to variable b. However, for designs $\boldsymbol{x}_f^{(1)}$ and

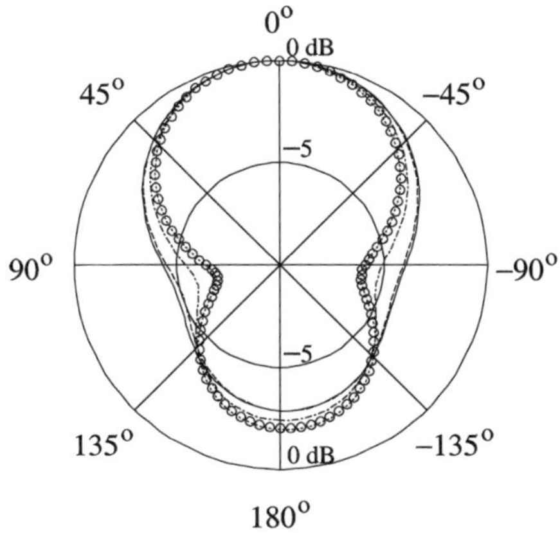

Fig. 9.14. E-field radiation patterns of the DRA at 2.45 GHz frequency for selected Pareto-optimal designs obtained within confined space (see Fig. 9.13 for the legend).

$\boldsymbol{x}_f^{(2)}$ a slight deterioration of the radiation pattern with the increase of b can be observed which suggests that other parameters also contribute to F_3.

The numerical cost of multi-objective antenna optimization corresponds to about 42 simulations of $\boldsymbol{f}$ (~14.5 h of CPU-time). The cost includes: 610 evaluations of $\boldsymbol{c}$ required to determine initially reduced search space, four simulations of $\boldsymbol{f}$ for pre-refinement of the $\boldsymbol{c}$ model, 567 evaluations of $\boldsymbol{c}$ for construction of the initial RSA, and $170 \times \boldsymbol{c}$ to establish the Kriging model $\boldsymbol{s}$ in the confined space. The estimated cost of direct MOEA-based optimization (100,000 simulations) of $\boldsymbol{f}$ and $\boldsymbol{c}$ model is about four years and over 40 days, respectively. The detailed cost breakdown of design optimization procedure can be found in Table 9.4.

9.3.5 *Multi-Objective Optimization in Initially Reduced Space*

For the sake of comparison, the DRA was also optimized in the initially reduced space (with no further space confinement). The RSA model utilized for MOEA optimization has been constructed using

Table 9.4 DRA optimization: cost breakdown

Algorithm step	Number of model evaluations		CPU time Absolute [h]	CPU time Relative to $\boldsymbol{f}$
Initial space reduction	$\boldsymbol{c}$	610	5.93	16.94
Multi-point OSM	$\boldsymbol{f}$	4	1.4	4
Data acquisition and RSA construction	$\boldsymbol{c}$	567	5.51	15.75
MOEA optimization	$\boldsymbol{s}$	100,000	N/A	N/A
Space confinement and RSA reset	$\boldsymbol{c}$	170	1.65	4.72
MOEA optimization	$\boldsymbol{s}$	100,000	N/A	N/A
Pareto set refinement	$\boldsymbol{f}$	0	0	0
Total cost	N/A	N/A	14.5	41.41
Optimization in initially reduced space (total cost)	N/A	N/A	26.14	74.69
Direct search (low-fidelity model)[#]	$\boldsymbol{c}$	100,000	972.2	2,778.3
Direct search (high-fidelity model)[#]	$\boldsymbol{f}$	100,000	35,000	100,000

[#]Estimated based on the number of $\boldsymbol{s}$ evaluations during MOEA optimization.

the same set of 576 $\boldsymbol{c}$ samples. However, multi-point OSM refinement of the training data has not been executed. The initial Pareto set has been obtained using MOEA with the same setup as in Sec. 9.3.3. In the next step, a set of 14 designs selected along the initial Pareto front has been refined using the response correction technique. A comparison of the low- and high-fidelity Pareto fronts is shown in Fig. 9.15.

The reflection characteristics and radiation patterns of the high-fidelity antenna designs selected from Table 9.5 are shown in Figs. 9.16 and 9.17. The reader can notice that the design with the lowest E-field discrepancy of 4.97 dB features the volume of 9,781 mm^3 and reflection of –10 dB. Antenna with the smallest volume also features the highest E-field strength discrepancy (8.1 dB). Reflection of the design is –10.8 dB. Finally, the lowest in-band reflection of –13.2 dB has been obtained for the antenna structure with volume

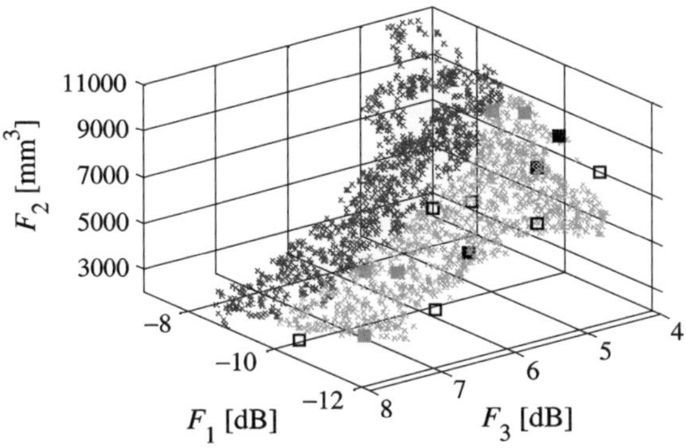

Fig. 9.15. Pareto front representations of the DRA obtained in the initially reduced space (Bekasiewicz *et al.*, 2016b). The high- and surrogate Pareto-optimal designs are denoted by squares and light-shade gray crosses, respectively. The dark-shade gray crosses represent designs that are outside the region of interest ($|S_{11}| > -10$ dB). Dimensions of the designs represented by non-empty squares are collected in Table 9.5.

Table 9.5 DRA: High-fidelity designs obtained in the initially reduced space

		Objectives			Antenna parameters					
		F_1 [dB]	F_2 [mm^3]	F_3 [dB]	a	b	c	o_{1r}	o_{2r}	l_r
Antenna designs	$\boldsymbol{x}_f^{(1)}$	−10.0	9781	4.97	25.49	25.47	15.07	0.039	0.432	0.659
	$\boldsymbol{x}_f^{(2)}$	−10.8	10424	5.02	27.10	25.64	15.00	0.035	0.439	0.659
	$\boldsymbol{x}_f^{(3)}$	−10.2	6564	5.94	16.09	27.17	15.01	0.061	0.435	0.657
	$\boldsymbol{x}_f^{(4)}$	−10.4	4793	6.95	11.33	28.54	14.82	0.074	0.406	0.686
	$\boldsymbol{x}_f^{(5)}$	−10.8	3116	8.11	6.74	29.91	15.45	0.117	0.430	0.671
	$\boldsymbol{x}_f^{(6)}$	−11.8	3963	7.84	9.18	29.67	14.55	0.110	0.408	0.690
	$\boldsymbol{x}_f^{(7)}$	−13.2	10164	5.43	26.62	26.28	14.53	0.048	0.422	0.633
	$\boldsymbol{x}_f^{(8)}$	−10.9	7342	5.82	18.33	27.01	14.83	0.070	0.448	0.633
	$\boldsymbol{x}_f^{(9)}$	−12.4	5057	7.22	11.82	29.05	14.73	0.003	0.402	0.633
	$\boldsymbol{x}_f^{(10)}$	−12.6	7870	5.95	19.91	27.26	14.50	0.082	0.449	0.633

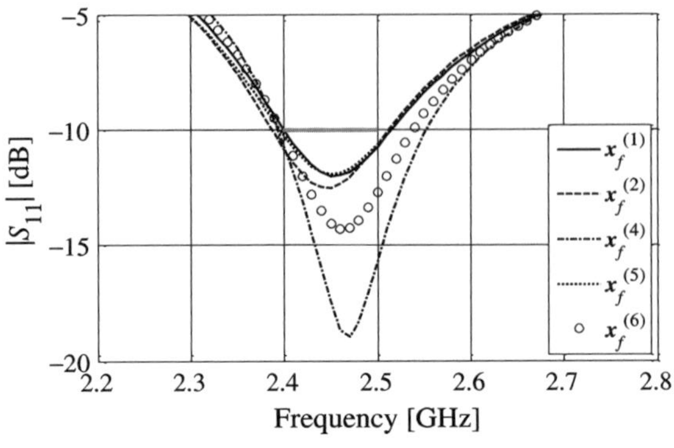

Fig. 9.16. Reflection responses of selected Pareto-optimal designs obtained in the initially reduced design space (Koziel and Bekasiewicz, 2016).

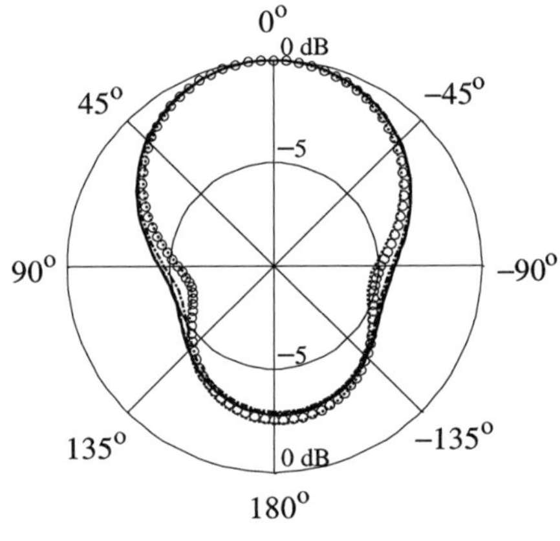

Fig. 9.17. E-field radiation patterns of the DRA at 2.45 GHz frequency for selected Pareto-optimal designs obtained within initially reduced space (see Fig. 9.16 for the legend).

of 10,164 mm^3 and E-field discrepancy of 5.43 dB. The ranges of variability of F_1, F_2 and F_3 along the Pareto front are 3.2 dB, 7,308 mm^3 and 3.1 dB, respectively.

The total cost of DRA optimization in the initially reduced space corresponds to ~75 simulations of $\boldsymbol{f}$ (~26.2 h of CPU-time). The cost includes: 610 simulations of $\boldsymbol{c}$ and additional 567 simulations of $\boldsymbol{c}$ required for determination of the initial space bounds and construction of the Kriging surrogate, as well as 42 evaluations of f required for the refinement of 14 designs selected along the initial Pareto front. It should be noted that the lack of space mapping significantly increased the CPU-time of the multi-objective design. A detailed cost breakdown of can be found in Table 9.4.

9.3.6 *Discussion*

In this section, the importance of design space reduction techniques for expedited multi-objective optimization of DRA with respect to three design objectives has been demonstrated. Two approaches have been exercised, specifically, MOEA-based optimization (Sec. 7.1) with rudimentary design space reduction (Sec. 8.2), and optimization with design space confinement (Sec. 8.4) aided by space mapping correction of the low-fidelity model. Both methods allow for obtaining the Pareto set at a reasonably low cost, however, utilization of confinement technique permits more precise allocation of the design space region containing the relevant part of the Pareto front. This, in turn, allows for local improvement of the low-fidelity model (within this very region) so that the Pareto set generated by MOEA does not need to be further refined. This considerably reduced the overall computational cost of the design process. On the other hand, more conventional approach, although more expensive, results in a larger Pareto set, in the sense of wider range of variation of the design objectives.

9.4 Design of a 12-Variable Yagi Antenna Using Design Space Reduction and Multi-Objective Evolutionary Algorithm

In this section, we demonstrate the MOEA-based multi-objective optimization of a planar Yagi–Uda antenna in multi-dimensional design space defined by twelve independent geometry parameters (Bekasiewicz *et al.*, 2014b). For problems of this complexity, initial design space reduction is of primary importance because construction of reliable data-driven surrogate in the entire space is not possible.

9.4.1 *Antenna Description and Design Objectives*

The considered antenna is shown in Fig. 9.18. It is a modified version of structure discussed in (Qian *et al.*, 1998) and it comprises a driven element feed by a microstrip-to-coplanar strip transition and a microstrip balun. The input impedance is 50 Ohm. The geometrical change with respect to the original structure is utilization of three directors instead of one. The antenna is implemented on the Taconic RF-35 substrate ($\varepsilon_r = 3.5$, $\tan\delta = 0.0018$, $h = 0.762$ mm). The vector of design variables is $\boldsymbol{x} = [s_1\ s_2\ s_3\ s_4\ v_1\ v_2\ v_3\ v_4\ u_1\ u_2\ u_3\ u_4]^T$. Parameters $w_1 = 1.7$, $w_2 = 3$, $w_3 = 0.85$ and $w_4 = 0.85$ remain fixed. All dimensions in are mm. The initial lower/upper bounds $\boldsymbol{l} = [2\ 2\ 2\ 2\ 18\ 7\ 7\ 7\ 3\ 7\ 2\ 1]^T$ and $\boldsymbol{u} = [10\ 10\ 10\ 10\ 30\ 15\ 15\ 15\ 12\ 16\ 6\ 3]^T$ are determined based on engineering experience.

The following design requirements are considered: F_1 — minimization of voltage standing wave ratio (VSWR), and F_2 — maximization

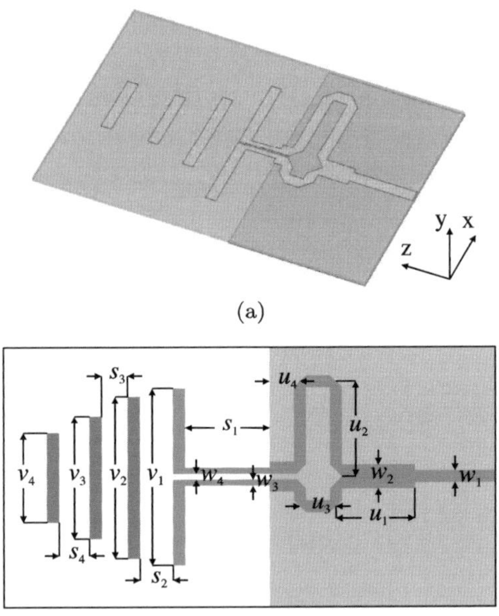

Fig. 9.18. A planar, 12-variable Yagi antenna with three directors (Bekasiewicz *et al.*, 2014b): (a) visualization; (b) geometrical details.

of average end-fire gain. Both objectives are considered within 5.2–5.8 GHz frequency range of interest. The maximum allowed VSWR is equal to for the entire bandwidth.

9.4.2 *Antenna Models and Design Space Reduction*

The high-fidelity model $\boldsymbol{f}$ of the Yagi antenna consists of about 600,000 mesh cells and its average evaluation time is 25 min. The low-fidelity model $\boldsymbol{c}$ contains ~110,000 mesh cells and its typical simulation time is 150 s. Both antenna models are implemented in CST Microwave Studio and evaluated using its transient solver (CST, 2013).

Here, we have utilized the design space reduction technique of Sec. 8.2 to identify region that contains Pareto-optimal designs. The refined lower and upper design bounds are $\boldsymbol{l}^* = [4.05\ 3.75\ 2.93\ 2\ 22.89\ 13\ 14.6\ 8\ 4.93\ 12.34\ 4.2\ 1.96]^T$, and $\boldsymbol{u}^* = [7.39\ 9.75\ 8.93\ 10\ 24.22\ 15\ 14.6\ 15\ 8.93\ 13.01\ 4.2\ 2.62]^T$. The refined design space is five orders of magnitude (volume-wise) smaller compared to the initial one. It should be reiterated that design space reduction is critical to permit construction of the accurate data-driven surrogate model using reasonable amount of training data. Later in the chapter we will demonstrate that for multi-dimensional problems, construction of the approximation model within initial design space is normally computationally infeasible.

9.4.3 *Numerical Results*

The response surface approximation model was constructed using 1,300 low-fidelity training samples allocated by means of LHS (Beachkofski and Grandhi, 2002). The generalization error of the model estimated using cross-validation technique (Queipo *et al.*, 2005) is only 1 percent for VSVR and 0.1 percent for gain.

The initial Pareto optimal set has been identified using a MOEA applied to the surrogate model $\boldsymbol{s}$ (setup: population size 500; maximum number of iterations 50). At the last stage, a set of 10 designs has been selected from the initial Pareto set and refined using the procedure described in Sec. 7.1.3. The results are shown in Table 9.6

Table 9.6 Yagi–Uda antenna: dimensions of selected high-fidelity Pareto-optimal designs

		Antenna designs									
		$x_f^{(1)}$	$x_f^{(2)}$	$x_f^{(3)}$	$x_f^{(4)}$	$x_f^{(5)}$	$x_f^{(6)}$	$x_f^{(7)}$	$x_f^{(8)}$	$x_f^{(9)}$	$x_f^{(10)}$
F_1 (VSWR)		2.00	1.86	1.80	1.62	1.57	1.47	1.24	1.23	1.21	1.18
F_2 (Gain [dB])		−8.04	−7.98	−7.96	−7.93	−7.77	−7.64	−7.42	−7.41	−6.59	−6.47
Antenna parameters	s_1	4.30	4.59	4.87	5.45	6.03	5.04	6.19	6.19	7.28	7.38
	s_2	9.75	9.40	9.46	8.99	7.57	6.56	3.82	3.82	3.75	3.79
	s_3	8.44	7.43	8.10	8.28	6.94	8.05	7.24	7.24	3.73	3.08
	s_4	9.92	9.99	9.98	9.75	9.75	9.49	9.61	9.64	9.77	9.41
	v_1	22.96	23.06	23.03	23.10	23.00	23.09	23.96	23.96	23.92	24.03
	v_2	14.99	14.96	14.92	14.93	14.97	14.89	14.95	14.95	14.71	14.65
	v_3	14.60	14.60	14.60	14.60	14.60	14.60	14.60	14.60	14.60	14.60
	v_4	15.00	15.00	15.00	15.00	14.78	15.00	14.81	14.75	11.70	11.52
	u_1	5.11	5.38	5.44	5.41	5.49	8.60	7.36	7.36	7.90	8.08
	u_2	12.34	12.39	12.35	12.40	12.40	12.79	12.39	12.39	12.36	12.36
	u_3	4.20	4.20	4.20	4.20	4.20	4.20	4.20	4.20	4.20	4.20
	u_4	2.35	2.23	2.01	2.02	1.98	1.98	2.59	2.59	2.39	2.24

(detailed antenna dimensions for the selected designs) and Fig. 9.19 (initial and refined Pareto sets).

It can be observed that the minimum antenna VSVR is 1.177 (with the corresponding average gain of 6.47 dB). The maximum average gain possible for this antenna is 8 dB while still maintaining the VSVR level of 2 within the entire frequency band of interest. Figures 9.20 and 9.21 show reflection and gain responses of the antenna for a few designs selected along the high-fidelity Pareto-optimal set (see Table 9.6 for their detailed dimensions).

It is interesting to analyze the cost of the multi-objective optimization process. At the first stage (design space reduction), 334 evaluations of the low-fidelity model were used to execute single-objective optimizations (220 and 114 evaluations to minimize F_1 and maximize F_2, respectively). The data-driven surrogate model was constructed using additional 1,300 low-fidelity model samples. Multi-objective optimization of the RSA model is a very fast process, the cost of which corresponds to less than one high-fidelity model evaluation. Finally, the refinement step requires 30 evaluations of the high-fidelity model. The total aggregated cost of Yagi–Uda antenna

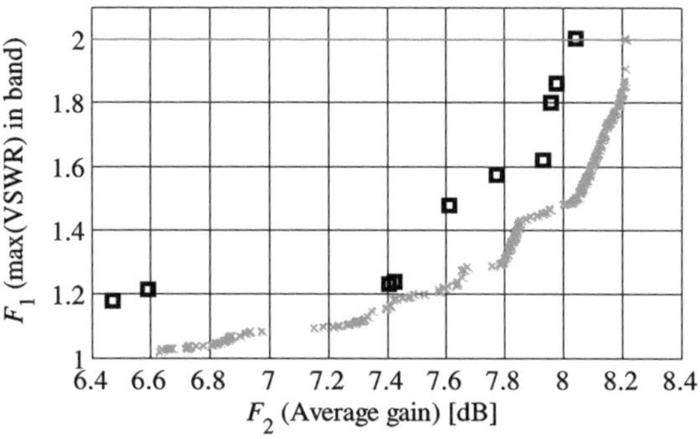

Fig. 9.19. Comparison of the low- (×) and high-fidelity (□) representations of the Pareto front (Bekasiewicz *et al.*, 2014b).

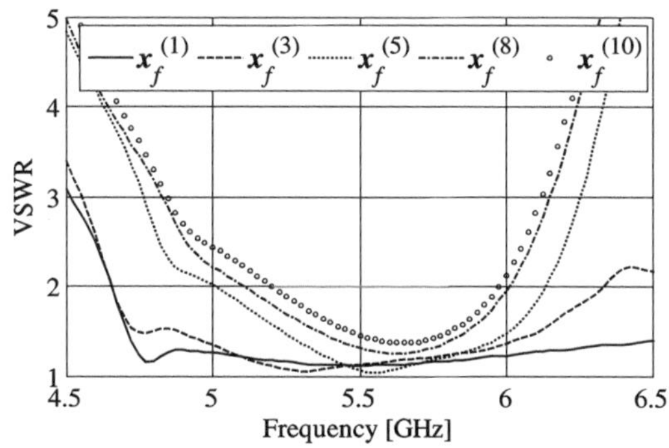

Fig. 9.20. VSWR responses of the Yagi antenna for designs selected from Table 9.6 (Bekasiewicz *et al.*, 2014b).

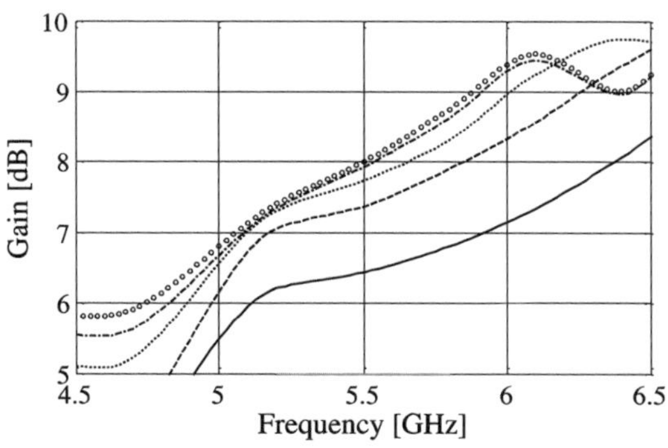

Fig. 9.21. Gain responses of the Yagi antenna for designs selected from Table 9.6 (see Fig. 9.20 for the legend; Bekasiewicz *et al.*, 2014b).

optimization is about 194 evaluations of $\boldsymbol{f}$ (~81 h of the CPU time), which is very low compared to direct multi-objective optimization using population-based metaheuristic, the latter requiring at least a few thousands of high-fidelity model evaluations (estimated using the number of evaluations of the surrogate model during MOEA optimization).

We would like to reiterate that multi-objective optimization is essential to obtain comprehensive information about the structure under design, here, the considered Yagi–Uda antenna. Also, the knowledge about possible trade-offs between conflicting objectives is fundamental for making design decisions, particularly in selecting the antenna structure for a particular application. The surrogate-assisted technique of Sec. 7.1 was utilized here, which allows us to gather this information at a very low computational cost. More importantly, high-fidelity Pareto set can be obtained without a supercomputer in a timeframe of hours rather than weeks (the latter is typical for metaheuristic-based optimization, see, e.g., Chamaani *et al.*, 2011).

9.4.4 *Measurements*

For the sake of experimental verification, selected antenna designs ($\boldsymbol{x}_f^{(1)}$, $\boldsymbol{x}_f^{(5)}$, $\boldsymbol{x}_f^{(10)}$ — see Table 9.6 for dimensions) have been fabricated and measured. Photographs of the manufactured designs can be found in Fig. 9.22, whereas a comparison of simulations and

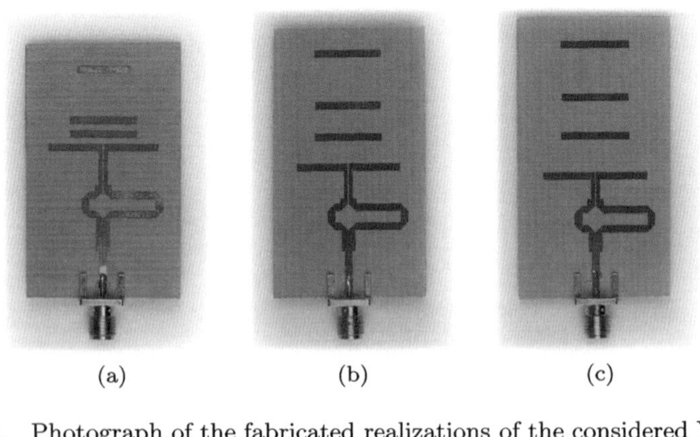

Fig. 9.22. Photograph of the fabricated realizations of the considered Yagi–Uda antenna: (a) $\boldsymbol{x}_f^{(1)}$, (b) $\boldsymbol{x}_f^{(5)}$, (c) $\boldsymbol{x}_f^{(10)}$.

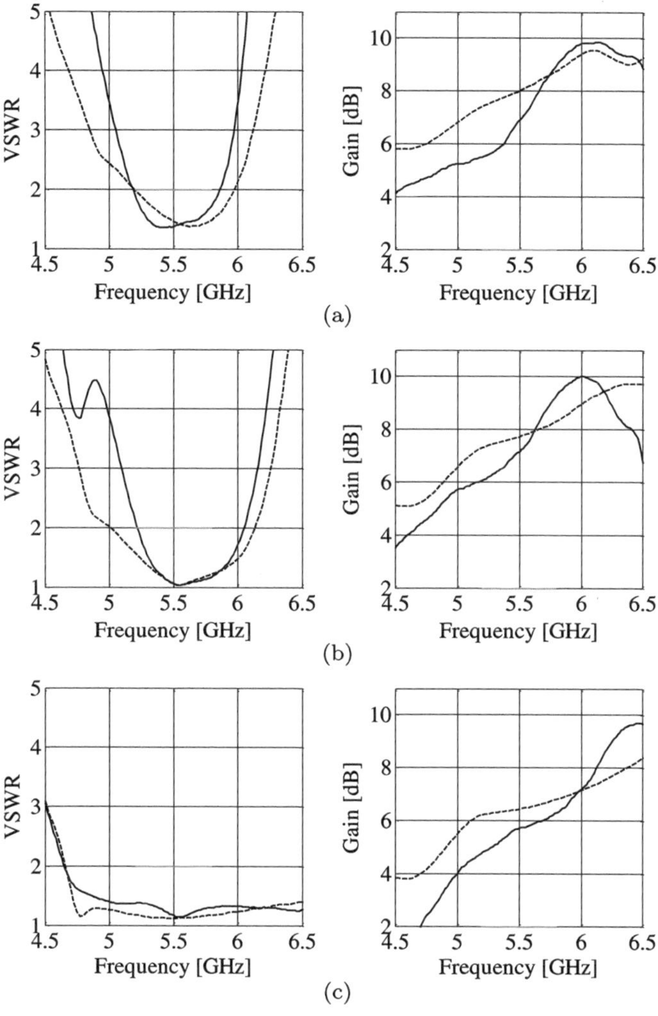

Fig. 9.23. Simulation (– –) and measurement (—) results of optimized Yagi–Uda antennas (Bekasiewicz *et al.*, 2014b) for designs: (a) $\boldsymbol{x}_f^{(1)}$; (b) $\boldsymbol{x}_f^{(5)}$; (c) $\boldsymbol{x}_f^{(10)}$. Left- and right-hand plots represent VSWR and gain (z direction; see Fig. 9.18).

measurements is shown in Fig. 9.23. It should be noted that the gain characteristics have been obtained indirectly using the three antenna method which is based on Friis transmission equation. The gain patterns have been measured in a direction parallel to the antenna surface (z direction in x–z plane; see Fig. 9.18(a)). More detailed explanation of the utilized gain measurement technique can be found in Balanis (2005).

Although the measurements coincide with the simulation results, some discrepancies can be observed. The in-band difference for considered antenna designs ($\boldsymbol{x}_f^{(1)}$, $\boldsymbol{x}_f^{(5)}$ and $\boldsymbol{x}_f^{(10)}$) are: 0.03, 0.56 and 0.2 for VSWR, as well as 0.71 dB, 0.61 dB, 2.17 dB for average gain. The discrepancies between the results are mostly due to utilization of the simplified high-fidelity antenna model that lacks the SMA connector. This resulted not only in the frequency shifts between the responses and also slightly different shapes of the characteristics. Misalignments between the simulated and the measured gain characteristics are mostly due to the lack of calibration of the measurement system.

9.5 Design of a Monopole Antenna Using Sequential Domain Patching

Our next example is multi-objective optimization of a monopole antenna using a sequential domain patching (SDP) technique of Sec. 7.3 (Koziel and Bekasiewicz, 2016c). It is compared here to MOEA-based optimization of Sec. 7.1 as well as direct (non-surrogate-based) optimization of the EM antenna model.

9.5.1 *Antenna Description and Design Objectives*

Geometry of the antenna is shown in Fig. 9.24 (Li *et al.*, 2012). It consists of a circular radiator fed through a tapered microstrip line. The fairly compact size of the structure has been obtained by means of an L-shaped ground plane stub and slot below the feed line which allow enhancing the current path within the antenna and improving its wideband impedance matching.

The considered structure is implemented on the Taconic RF-35 substrate of 0.762 mm thickness. The design parameters are $\boldsymbol{x} = [w_0\ l_1\ l_{2r}\ l_3\ l_4\ w_1\ w_2\ w_3\ s_1\ s_2\ o]^T$, whereas variable $l_2 = l_f + l_1 - 0.05r + l_{2r}$ is relative and dimensions $l_f = 10$, and $w_f = 1.7$ remain fixed. The unit for all geometrical parameters is mm. The initial

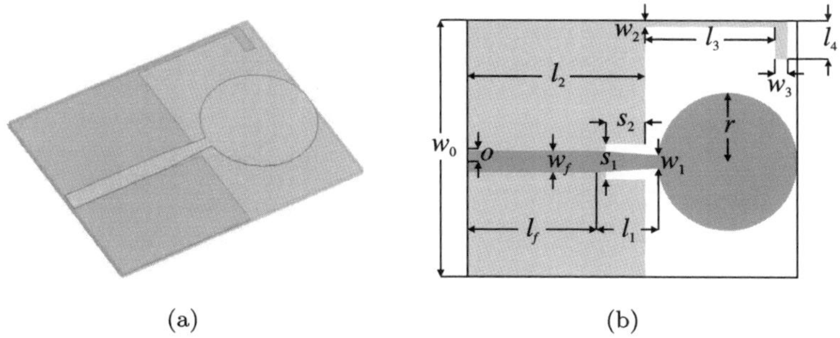

Fig. 9.24. UWB monopole with circular radiator (Li *et al.*, 2012): (a) visualization; (b) geometry with highlighted dimensions.

design space bounds $\boldsymbol{l}$ = [8.0 2.0 –2.0 5.0 1.0 0.5 0.2 0.5 0.2 1.0 4.0 –0.2]^T and $\boldsymbol{u}$ = [25.0 8.0 2.0 15.0 8.0 2.0 1.0 1.5 2.0 5.0 7.0 0.2]^T have been selected based on engineering experience.

The following design objectives are considered: F_1 — minimization of the antenna reflection in 3.1–10.6 GHz frequency band of interest and F_2 — minimization of antenna size defined as $w_0 \times l$ rectangle where $l = \max\{b_1, b_2\}$, $b_1 = l_f + l_1 - 0.05r + l_3 + l_4 + w_3$ and $b_2 = l_f + l_1 + 1.95r$.

9.5.2 *Antenna Models and Determination of Extreme Pareto Designs*

Both models of the considered antenna are implemented in CST Microwave Studio and simulated using its time domain solver (CST, 2013). The high-fidelity model $\boldsymbol{f}$ consists of about 2,200,000 hexahedral mesh cells and its average evaluation time is 18 min. The low-fidelity model $\boldsymbol{c}$ contains 170,000 cells and its typical simulation time is 63 s (it is over 17 times faster than $\boldsymbol{f}$).

Design space reduction has been performed using single-objective optimizations with respect to one design requirement at a time (see Sec. 8.2). The extreme Pareto optimal design featuring the smallest size is $\boldsymbol{x}_1^*$ = [10 2 –0.26 12.98 4.75 0.63 0.2 0.72 2 3.47 4.8 –0.05]^T, whereas $\boldsymbol{x}_2^*$ = [25 6.53 –0.5 14 3.01 1 0.5 1 0.93 3.17 6.3 –0.05]^T represents the structure with the lowest maximum in-band reflection.

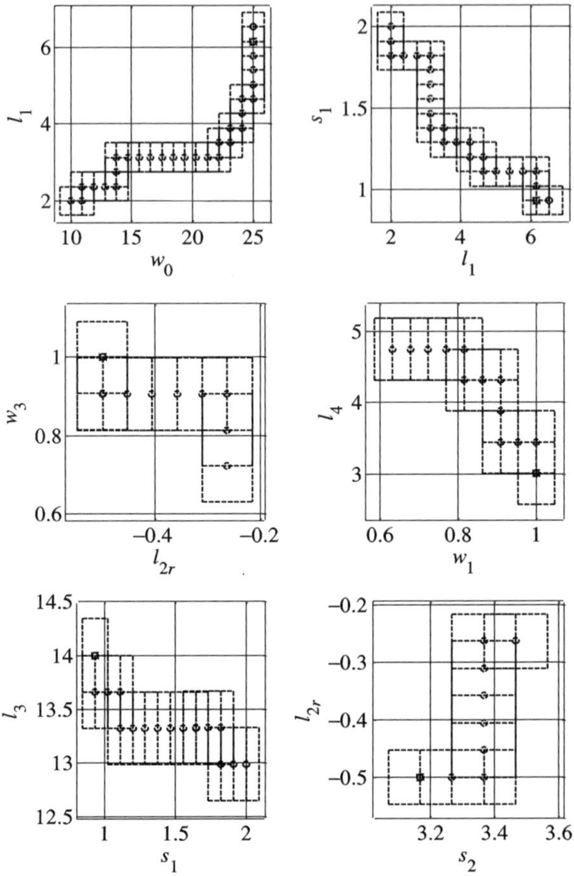

Fig. 9.25. Selected two-dimensional projections of the path created by the sequential patching algorithm for the monopole antenna of Fig. 9.24.

The total cost of finding these extreme solutions is 1,380 simulations of the low-fidelity model $\boldsymbol{c}$ (about 24 h of the CPU time).

9.5.3 *Multi-Objective Optimization Using Sequential Domain Patching Algorithm*

The optimization process carried out to produce the initial approximation of the Pareto set has been performed using the low-fidelity model. The sequential patching algorithm has been executed using

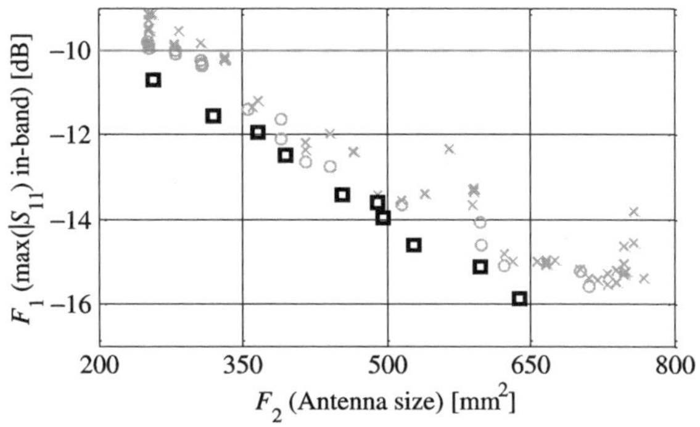

Fig. 9.26. The low- (○) and high-fidelity (□) Pareto sets obtained using sequential domain patching. Crosses denote dominated designs.

$\boldsymbol{m} = [m_1 \ldots m_n] = [16\ 12\ 5\ 3\ 4\ 8\ 4\ 3\ 12\ 3\ 8\ 2]$. The number of intervals along each dimension m_n has been selected manually, taking into account sensitivity of the antenna responses to particular geometry parameters. Figure 9.25 shows the selected two-dimensional projections of the path created by the SDP algorithm. Subsequently, 10 low-fidelity designs have been selected along the initial Pareto front and refined using local approximation models. Here, each surrogate $\boldsymbol{s}$ is constructed as a simplified second-order polynomial approximation of the $\boldsymbol{c}$ model at the selected design and $2n$ perturbations around it obtained using a star-distribution design of experiments (Koziel *et al.*, 2011a). The perturbation size in each direction corresponds to respective patch dimension.

The obtained high- and low-fidelity Pareto sets are shown in Fig. 9.26. The optimization results indicate that the change of objectives F_1 and F_2 along the Pareto front is 5.7 dB and 544 mm^2 (over 36 percent), respectively. Detailed dimensions of the refined antenna designs are gathered in Table 9.7, whereas the frequency responses for the selected antenna realizations are shown in Fig. 9.27.

The overall cost of multi-objective antenna optimization corresponds to about 158 simulations of the high-fidelity model $\boldsymbol{f}$ (~47 h of CPU-time) and it includes 1,380 simulations of $\boldsymbol{c}$ for identification of the extreme Pareto-optimal designs, 555 evaluations of $\boldsymbol{c}$ for SDP-based approximation of the Pareto front, 250 evaluations of

Table 9.7 UWB monopole: dimensions of selected high-fidelity Pareto designs

		Antenna designs								
		$x_f^{(1)}$	$x_f^{(2)}$	$x_f^{(3)}$	$x_f^{(4)}$	$x_f^{(5)}$	$x_f^{(6)}$	$x_f^{(7)}$	$x_f^{(8)}$	$x_f^{(9)}$	$x_f^{(10)}$
	F_1 [dB]	−10.7	−11.5	−11.9	−12.5	−13.4	−13.6	−14.0	−14.6	−15.1	−15.9
	F_2 [mm^2]	258	320	366	395	454	490	496	528	597	638
Antenna parameters	w_0	10.30	12.21	13.90	14.69	17.12	18.44	18.67	19.79	21.61	23.12
	l_1	2.33	2.71	2.71	2.87	3.06	3.13	3.36	3.23	3.77	3.44
	l_{2r}	−0.27	−0.38	−0.52	−0.45	−0.48	−0.50	−0.50	−0.50	−0.50	−0.53
	l_3	13.01	13.33	13.40	13.65	13.33	13.33	13.33	13.33	13.08	13.18
	l_4	4.34	4.15	3.82	3.75	3.60	3.44	3.44	3.27	3.44	3.37
	w_1	0.68	0.89	0.95	0.91	0.91	0.91	0.93	0.91	0.88	0.89
	w_2	0.20	0.27	0.27	0.27	0.27	0.27	0.27	0.27	0.27	0.27
	w_3	0.91	0.91	0.91	0.88	0.91	0.91	0.91	0.91	0.91	0.86
	s_1	2.02	1.91	1.89	1.81	1.73	1.64	1.72	1.60	1.29	1.35
	s_2	3.41	3.44	3.35	3.34	3.41	3.37	3.47	3.40	3.46	3.47
	r	4.76	4.85	5.06	5.23	5.49	5.55	5.55	5.74	5.87	6.06
	o	−0.05	−0.05	−0.05	−0.05	−0.05	−0.05	−0.05	−0.05	−0.05	−0.05

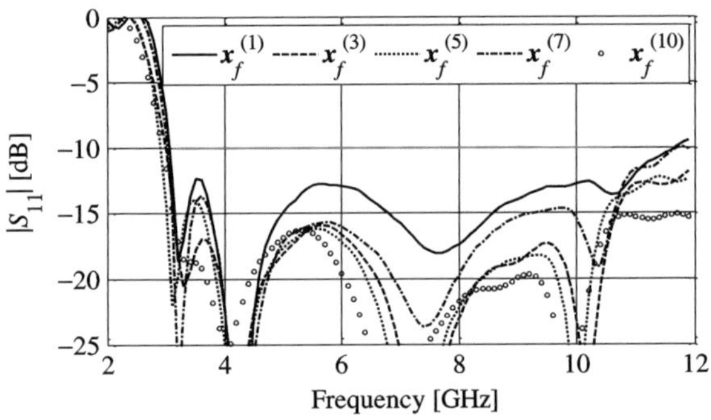

Fig. 9.27. Frequency characteristics of the high-fidelity Pareto-optimal designs selected from Table 9.7.

$\boldsymbol{c}$ for construction of 10 local approximation models, and 30 evaluations of $\boldsymbol{f}$ required by the refinement procedure. A detailed cost breakdown of the SDP-based multi-objective optimization (excluding the cost of identification of the extreme Pareto designs) is provided in Table 9.8.

9.5.4 *Comparison with Benchmark Techniques*

The sequential domain patching method has been compared with two other techniques considered in this book, both based on MOEA. The first method is direct multi-objective optimization of the low-fidelity antenna model $\boldsymbol{c}$ (algorithm setup: population size 100, maximum number of iterations 100). The second method is multi-objective optimization of the Kriging interpolation model constructed using data samples of the low-fidelity model $\boldsymbol{c}$ (setup: population size 500, maximum number of iterations 50). For fair comparison, both MOEA-based optimizations have been carried out within the search space region defined based on the dimensions of the extreme Pareto designs of Sec. 9.5.2 (cf. Sec. 8.2).

The comparison is executed for the low-fidelity model because direct high-fidelity model optimization would be prohibitively expensive (the estimated cost for the above evolutionary algorithm setup is over 100 days of the CPU time; 10,000 evaluations). The comparison results are shown in Fig. 9.28. The shapes of the obtained Pareto sets are noticeably different. Although variations of F_1 between MOEA-based and SDP optimization results are below 1 dB, the ranges of Pareto fronts along F_2 considerably vary between the techniques. At the same time, SDP outperforms MOEA because the extreme Pareto solutions are allocated at the corners of the 12-dimensional search

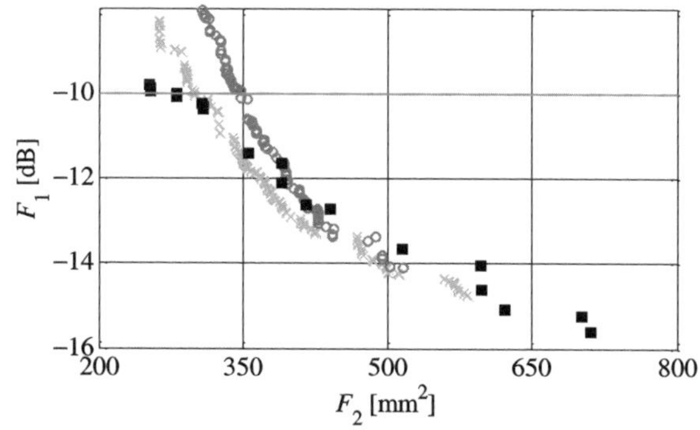

Fig. 9.28. Comparison of the Pareto sets obtained using SDP (■) and MOEA optimization of the Kriging interpolation model (×) and coarse-mesh EM model (○) of the considered UWB monopole antenna.

Table 9.8 Optimization of the compact UWB monopole: cost breakdown

Algorithm	Algorithm step	Number of model evaluations		CPU-time Absolute [h]	CPU-time Relative to $\boldsymbol{f}$
SDP	SDP-based optimization	$\boldsymbol{c}$	555	9.71	32.38
MOEA	Data acquisition	$\boldsymbol{c}$	1,802	31.54	105.12
	Optimization of $\boldsymbol{s}$ model	$\boldsymbol{s}$	25,000	N/A	N/A
	Total cost	N/A	N/A	31.54	105.12
	Optimization of $\boldsymbol{c}$ model	$\boldsymbol{c}$	10,000	175	583.33

space where exploration capabilities of evolutionary algorithm are limited. Consequently, the results of direct optimization performed using MOEA cover only about a half of the Pareto front where F_1 is below −10 dB. The Pareto front obtained by optimization of the Kriging model is larger, which is due to the increased population size with respect to direct search approach. The optimization costs of the considered methods are 10,000 simulations of $\boldsymbol{c}$ for direct search (~176 h), 1,802 evaluations of $\boldsymbol{c}$ model for MOEA optimization of Kriging model (~32 h), and 555 simulations of $\boldsymbol{s}$ for SDP algorithm (~10 h). A detailed cost breakdown of considered methods is provided in Table 9.8.

9.6 Optimization of Compact Monopole Antenna by Means of Pareto Front Exploration

Here, we consider another example of a UWB monopole antenna. It is optimized using Pareto front exploration, specifically, a technique of Sec. 7.2 (Koziel *et al.*, 2015b). Similarly as for the previous case, a comparison with MOEA-based methods is provided.

9.6.1 *Antenna Description and Design Objectives*

The antenna geometry is shown in Fig. 9.29. It has been extracted from the MIMO structure proposed in (Liu *et al.*, 2013). The antenna consists of a rectangular radiator fed through a 50 Ohm microstrip

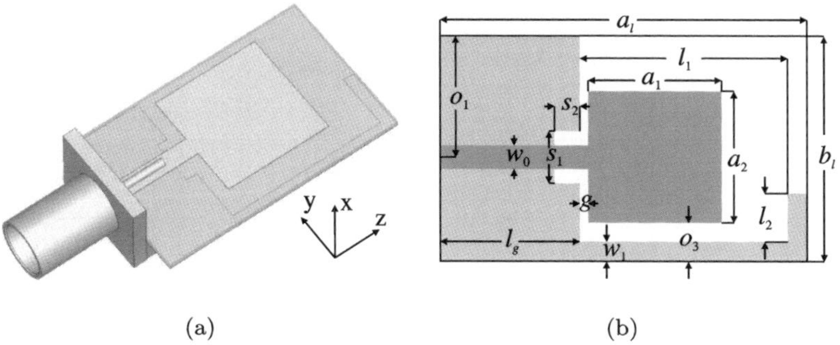

Fig. 9.29. Compact UWB monopole antenna: (a) three-dimensional visualization of the EM model with connector; and (b) geometry with highlighted design variables. Light-shade gray represent the ground plane.

line. Compact dimensions of the structure are ensured by introducing the L-shaped strip on the ground plane side which allows for current path enhancement. Moreover, the ground plane slit below the fed line allows for improving wideband impedance bandwidth. The antenna is implemented on a 0.762 mm thick Taconic RF-35 substrate. The vector of design variables is $\boldsymbol{x} = [l_g\ g\ a_1\ a_2\ l_1\ l_2\ w_1\ s_1\ s_2\ o_1\ o_3]^T$. Parameter $w_0 = 1.7$ ensures 50 Ohm input impedance, whereas $o_2 = 0.5 \cdot a_2$ provides central location of feed. The unit for all dimensions is mm. The design space bounds: $\boldsymbol{l} = [3.0\ 0.2\ 5.0\ 5.0\ 10.0\ 1.0\ 1.0\ 0.2\ 0.2\ 5.0\ 0.2]^T$ and $\boldsymbol{u} = [10.0\ 2.0\ 12.0\ 12.0\ 16.0\ 7.0\ 2.0\ 5.0\ 2.0\ 10.0\ 10.0]^T$ have been defined based on engineering experience.

The following design objectives are considered: F_1 — minimization of the antenna reflection within UWB frequency range (i.e., from 3.1 GHz to 10.6 GHz) and F_2 — minimization of size $S(\boldsymbol{x})$ defined as $a_l \times b_l$ rectangle, where $a_l = 0.5a_2 + o_1 + o_3$ and $b_l = l_g + l_1 + w_1$.

9.6.2 *Antenna Models and Initial Design*

The computational model of the antenna (referred to as high-fidelity model $\boldsymbol{f}$) is implemented in CST Microwave Studio and simulated using its transient solver (CST, 2013). It consists of about 8,000,000 hexahedral mesh cells and its average simulation time is 55 min. The low-fidelity representation of the structure, $\boldsymbol{c}$, contains ~100,000 mesh cells and its typical simulation time is about 60 s. For the sake of

Table 9.9 Compact UWB monopole: optimization results

		Antenna designs									
		$x_f^{(1)}$	$x_f^{(2)}$	$x_f^{(3)}$	$x_f^{(4)}$	$x_f^{(5)}$	$x_f^{(6)}$	$x_f^{(7)}$	$x_f^{(8)}$	$x_f^{(9)}$	$x_f^{(10)}$
F_1 [dB]		−15.9	−14.8	−13.6	−12.8	−11.9	−11.5	−10.7	−10.4	−10.3	−10.2
F_2[mm^2]		299	286	266	246	226	210	200	190	185	180
Antenna parameters	l_g	2.74	2.49	2.18	1.89	1.60	1.28	1.31	0.92	0.89	0.73
	g	6.44	6.48	6.35	6.25	6.11	6.18	5.94	5.93	5.90	5.85
	a_1	1.69	1.46	1.55	1.68	1.81	2.00	2.00	2.00	2.00	2.00
	a_2	3.06	3.21	3.25	2.29	3.31	3.43	3.25	3.37	3.37	3.49
	l_1	1.49	1.45	1.31	1.19	1.13	1.50	1.32	1.48	1.45	1.48
	l_2	5.71	5.71	5.87	5.92	6.07	6.25	6.36	6.57	6.56	6.64
	w_1	13.50	13.30	13.10	12.90	12.60	12.40	12.20	12.00	12.00	12.10
	s_1	9.07	8.93	8.47	7.91	7.47	6.93	6.62	6.34	6.31	6.27
	s_2	9.84	9.43	9.25	8.99	8.79	8.77	8.51	8.52	8.49	8.56
	o_1	0.44	0.56	0.63	0.69	0.73	0.87	0.82	0.81	0.83	0.85
	o_3	6.79	6.51	6.46	6.28	5.98	5.30	5.39	5.16	5.18	4.98

reliability, the SMA connector is included into the simulation models. This is important as the structure is electrically small.

The initial design for exploring the Pareto front is obtained as a solution to the single-objective optimization task of the form

$$x^{(1)} = \arg\min_{x} F_1\left(f(x)\right), \tag{9.8}$$

which gives the best possible value of the first objective. Solution to (9.8) is obtained using SBO with the coarse-discretization EM model c as an underlying low-fidelity model and OSM as a model correction method (Koziel and Ogurtsov, 2014a).

The starting point for solving (9.8) is $x^{(0)} = [8.0\ 1.0\ 10.0\ 10.0\ 17.0\ 5.0\ 1.0\ 4.0\ 1.0\ 8.0\ 2.0]^T$, and it has been extracted from the dimensions of the reference structure (Liu *et al.*, 2013). The detailed dimensions of optimized initial high-fidelity design $x^{(1)}$ are provided in Table 9.9. Comparison of the antenna response at the initial and the optimized designs is shown in Fig. 9.30.

9.6.3 *Numerical Results*

The point-by-point optimization procedure has been executed starting from the design obtained found by solving (9.8). Description of

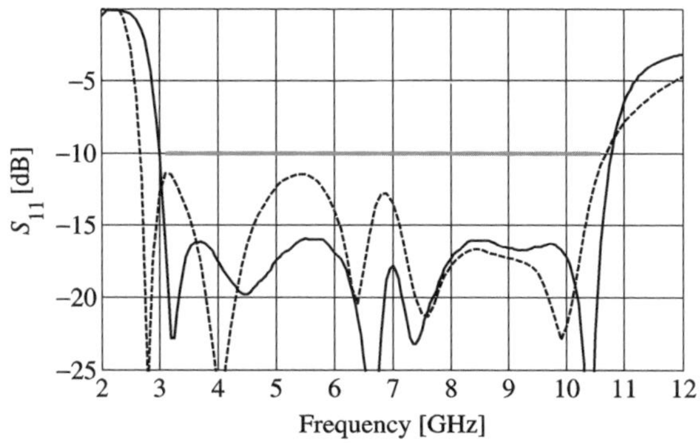

Fig. 9.30. Responses of the UWB monopole antenna at designs $\boldsymbol{x}^{(0)}$ (– –) and $\boldsymbol{x}^{(1)}$ (—). Detailed dimensions of the latter can be found in Table 9.9.

the method can be found in Sec. 7.2. Here, we only recall that subsequent Pareto-optimal designs are found by solving,

$$\boldsymbol{x}^{(k)} = \arg \min_{\boldsymbol{x},\ F_2(\boldsymbol{f}(\boldsymbol{x})) \leq F_2^{(k)}} F_1(\boldsymbol{f}(\boldsymbol{x})), \tag{9.9}$$

where $\boldsymbol{x}^{(k)}$ is the kth element of the Pareto set, and $F_2^{(k)}$ is the acceptance threshold for the second objective (cf. Sec. 7.3 for more details). The optimization process was terminated upon finding the 10th design that is close to the acceptable reflection level, i.e., $|S_{11}| \leq -10$ dB within the UWB frequency range.

The obtained Pareto set representation is shown in Fig. 9.31. Figure 9.32 shows the reflection responses for the selected designs, whereas the antenna dimensions for all the designs are provided in Table 9.9. The reader may find interesting that the "density" of the Pareto front representation has been increased close to its left-hand side edge in order to find a possibly small antenna design. The variability of the antenna size along the front is from 300 mm^2 to only 180 mm^2. The maximum in-band reflection of the extreme Pareto-optimal design with the largest occupied area is about –16 dB. At the same time, the smallest design features $|S_{11}|$ at the level of –10.2 dB. It should be noted that the smallest antenna realizations exhibit competitive miniaturization levels in comparison with some compact structures from the literature (e.g., Chahat *et al.*, 2011; Chu *et al.*, 2013).

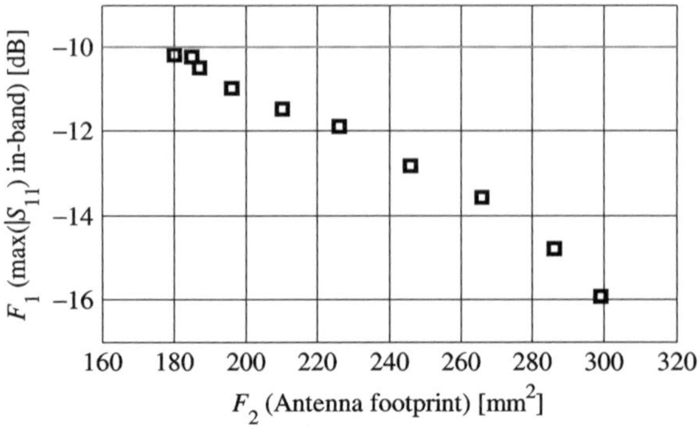

Fig. 9.31. Compact UWB monopole antenna: Pareto-optimal solutions obtained using point-by-point multi-objective optimization technique.

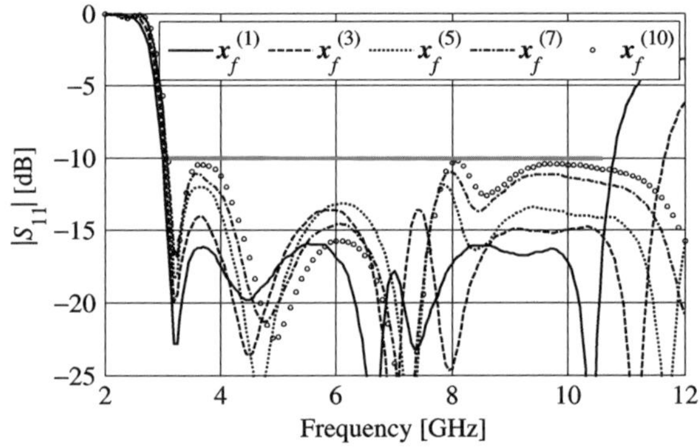

Fig. 9.32. Compact UWB monopole antenna: reflection characteristics corresponding to the selected Pareto-optimal designs of Table 9.9.

We would like to point out that the reference radiator of (Liu *et al.*, 2013) is characterized by an area of $15 \times 26 = 390\,\text{mm}^2$ and its in-band reflection is –11.5 dB (see Fig. 9.30). At the same time, Pareto design from Table 9.8 with similar $|S_{11}|$ of –10.4 dB features size of only $10.9 \times 19.2 = 210\,\text{mm}^2$ (over 46 percent smaller with respect to reference structure). Moreover, the design with the maximum in-band reflection of –15.9 dB (cf. Table 9.9) is characterized by a footprint of only $299\,\text{mm}^2$ (over 23 percent smaller than the reference design). In other words, the results indicate that the structure of (Liu *et al.*, 2013) represents a certain dominated solution within

Table 9.10 Optimization of UWB monopole antenna: cost breakdown and comparisons with the benchmark techniques

Algorithm	Algorithm step	Number of model evaluations		CPU-time Absolute [h]	CPU-time Relative to $\boldsymbol{f}$
PSE	Point-by-point optimization	$\boldsymbol{c}$	460	7.67	8.36
MOEA	Data acquisition	$\boldsymbol{c}$	802	13.37	14.58
	Optimization of $\boldsymbol{s}$ model	$\boldsymbol{s}$	25,000	N/A	N/A
	Total cost	N/A	N/A	13.37	14.58
	Optimization of $\boldsymbol{c}$ model	$\boldsymbol{c}$	10,000	166.67	181.82

the design space. It is important that — although it is not explicitly said in (Liu *et al.*, 2013) — the dimensions of the reference radiator have been obtained by means of parameter sweeps.

The cost of obtaining the initial design was 350 evaluations of the low-fidelity model $\boldsymbol{c}$ and four evaluations of $\boldsymbol{f}$. The average number of iterations of (9.9) required for the determination of the next design along the Pareto front was three. Therefore, the numerical cost of (9.9) corresponds to 46 evaluations of $\boldsymbol{c}$ and 3 evaluations of $\boldsymbol{f}$, respectively. A total cost of the multi-objective antenna optimization corresponds to only about 48 evaluations of the high-fidelity antenna model. A detailed cost breakdown of the design process driven by point-by-point algorithm is provided in Table 9.10.

9.6.4 *Comparison with Benchmark Techniques*

The discussed Pareto set exploration (PSE) technique has been compared with two benchmark methods exploiting MOEA. The first method directly optimizes the low-fidelity antenna model $\boldsymbol{c}$ using MOEA (algorithm setup: population size 100, number of iterations 100). In the second method, evolutionary algorithm is executed on the Kriging interpolation model constructed from the low-fidelity model data (MOEA setup: population size 500, number of iterations 50).

Similarly as in Sec. 9.5.4, the numerical results have been obtained at the low-fidelity model level because the computational cost of

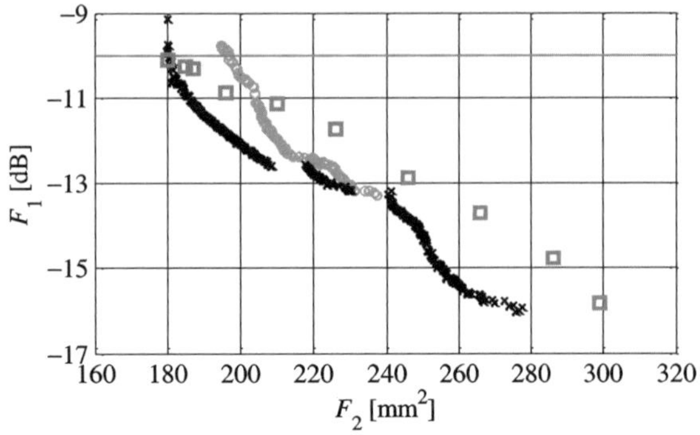

Fig. 9.33. Comparison of the low-fidelity Pareto-optimal sets obtained using point-by-point multi-objective optimization (□), MOEA optimization of Kriging interpolation model (×) and MOEA optimization of the low-fidelity model (○).

direct MOEA optimization of the high-fidelity model is prohibitive (estimated cost of 10,000 evaluations of $\boldsymbol{f}$ is almost 382 days of CPU-time). Kriging model has been set up within a region of the search space limited by designs $\boldsymbol{x}_f^{(1)}$ and $\boldsymbol{x}_f^{(10)}$ (cf. Table 9.9).

The results shown in Fig. 9.33 indicate that the shapes of the Pareto sets are noticeably different. Although variations between fronts along F_1 are below 1.5 dB, they significantly vary along F_2. The reason is that the extreme Pareto-optimal solutions are allocated at the corners of 11-dimensional search space which limits exploration capability of the MOEA. Due to insufficient number of low-fidelity data points, the first benchmark method captured only half of the Pareto front. The cost of antenna optimization (excluding the cost of obtaining extreme Pareto-optimal designs for MOEA, and the initial design for the point-by-point algorithm) is 10,000 evaluations of the low-fidelity model, 802 simulations of $\boldsymbol{c}$ and 460 evaluations of $\boldsymbol{c}$ for the two benchmark techniques, and PSE method, respectively. A detailed cost breakdown of multi-objective design optimization using considered techniques is provided in Table 9.10. We would like to emphasize that, for the considered design problem, point-by-point optimization allows considerable reduction of the computational cost of multi-objective optimization compared to MOEA-based techniques.

9.6.5 *Measurements*

The selected Pareto-optimal designs, i.e., $\boldsymbol{x}_f^{(1)}$, $\boldsymbol{x}_f^{(5)}$ and $\boldsymbol{x}_f^{(10)}$ have been fabricated (see Table 9.9 for detailed dimensions) and measured. Photographs of the manufactured prototypes are shown in Fig. 9.34. Reflection characteristics of the chosen structures are compared in Fig. 9.35. The obtained results indicate that all the measured designs fulfill the requirement upon maximum acceptable in-band $|S_{11}|$ (expected to be lower than –10 dB). The obtained values are -11 dB, –10.6 dB and –10 dB for the design $\boldsymbol{x}_f^{(1)}$, $\boldsymbol{x}_f^{(5)}$ and $\boldsymbol{x}_f^{(10)}$, respectively. Therefore, the maximum differences between simulations and measurements for the considered antennas are 4.9 dB, 1.3 dB and 0 dB. It should be noted that the visual discrepancy between the simulation and measurement results for the design $\boldsymbol{x}_f^{(1)}$ is actually small; the large difference in the maximum in-band $|S_{11}|$ is just a consequence of a slightly lower slope of the measured reflection around 3.1 GHz. In 3.2–10.6 GHz range the maximum difference between the responses of this antenna is below 1.5 dB.

Although the field characteristics of the antenna were not subjected to the optimization process, we have performed measurements of antenna gain patterns in a direction perpendicular to the antenna surface (x direction in the x–y plane; cf. Fig. 9.29(a)). The characteristics of the discussed antenna designs are compared in Fig. 9.35. The obtained results indicate that the average measured in-band gain levels of the designs $\boldsymbol{x}_f^{(1)}$, $\boldsymbol{x}_f^{(5)}$ and $\boldsymbol{x}_f^{(10)}$ are 0.05 dB, –0.41 dB and –0.85 dB, respectively. The corresponding simulated values are

(a) (b) (c)

Fig. 9.34. Photographs of the fabricated designs (front on the left-, back on the right-hand side): (a) $\boldsymbol{x}^{(1)}$, (b) $\boldsymbol{x}^{(5)}$ and (c) $\boldsymbol{x}^{(10)}$.

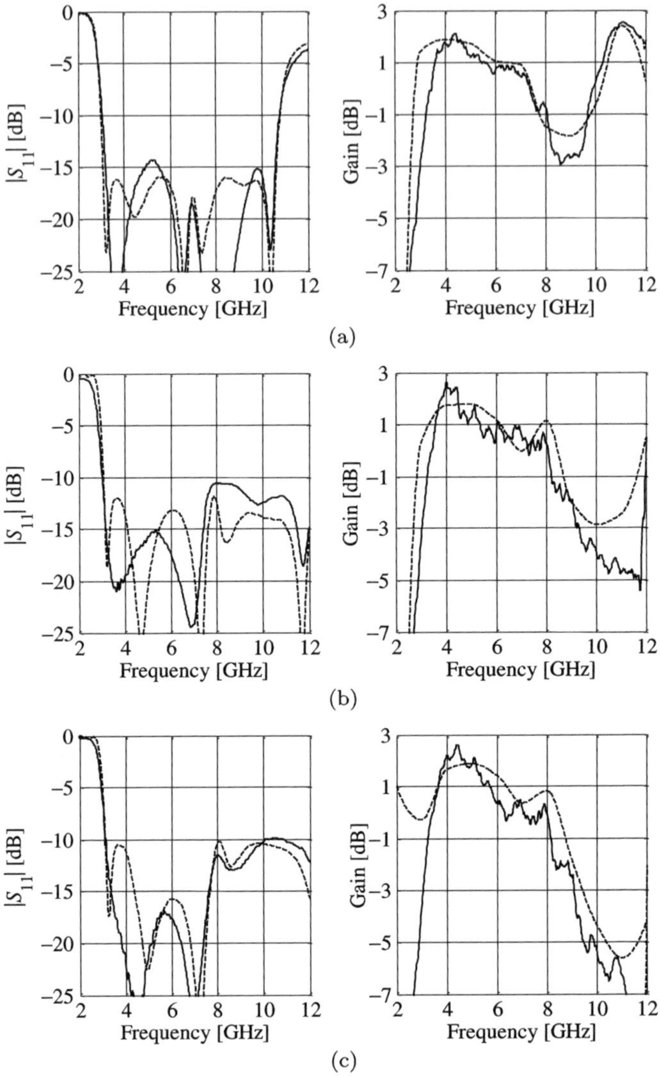

Fig. 9.35. A comparison of the simulated (– –) and measured (—) reflection (left-hand side) and gain (right-hand side) characteristics of the selected designs of the compact monopole antenna of Fig. 9.29: (a) $\boldsymbol{x}^{(1)}$, (b) $\boldsymbol{x}^{(5)}$ and (c) $\boldsymbol{x}^{(10)}$.

0.36 dB, 0.14 dB and –0.12 dB, so that the differences between the obtained results are minor (only 0.31 dB, 0.55 dB and 0.26 dB, respectively). The characteristics indicate that reduction of structure size noticeably influences the field properties of the antenna.

It should be mentioned that visible discrepancies between the simulation and measurement results are partially due to electrically large equipment utilized during measurements. Also, we would like to emphasize that the considered antennas are electrically small (i.e., dimensions of $\boldsymbol{x}_f^{(1)}$ and $\boldsymbol{x}_f^{(10)}$ are 21.8 mm × 13.7 mm and 18.6 mm × 9.7 mm, respectively). In particular, the size of the SMA connector is significant compared to the antenna footprints. Consequently, the precision of the prototype assembly (i.e., trimming of the design or allocation of the connector with respect to the center of the feed line) is important, since it may noticeably influence the measurement results.

9.7 Design of a Ultra-Wideband Monopole Antenna Using Sequential Domain Patching Algorithm with Automated Patch Size Selection

In this section, we consider another antenna optimized using a sequential domain patching algorithm of Sec. 7.3. The structure is more complex than the structures discussed before (13 geometry parameters) yet the optimization process is concluded with the computational cost corresponding to less than 80 evaluations of the high-fidelity EM model of the antenna.

9.7.1 *Antenna Geometry and Design Objectives*

Geometry of the considered monopole antenna is shown in Fig. 9.36 (Yang *et al.*, 2008). The structure consists of a driven element in the form of three trapezoids stacked together into a radiator which is fed through a 50 Ohm microstrip line. The antenna is designed on a 0.762 mm thick Taconic RF-35 substrate. Geometry is represented by 13-variable vector $\boldsymbol{x} = [a_1\ a_2\ a_3\ a_4\ a_5\ a_6\ b_1\ b_2\ b_3\ w_2\ l\ d\ o]^T$. Parameter $w_1 = 1.7$ remain fixed to ensure 50 Ohm input impedance

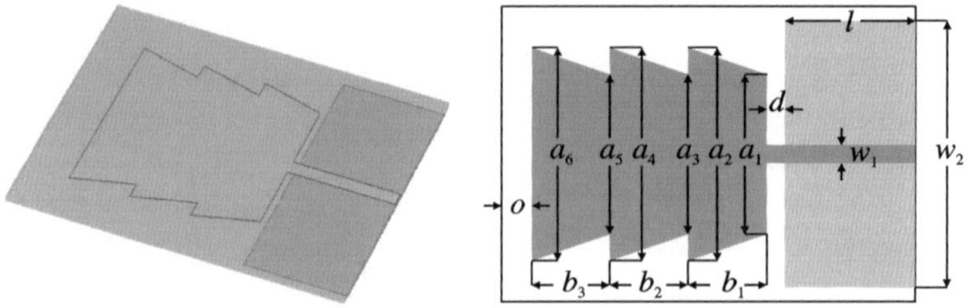

(a) (b)

Fig. 9.36. Planar UWB monopole antenna (Yang *et al.*, 2008): (a) visualization; (b) geometrical details of the structure.

(all dimensions are in mm). The initial (arbitrarily defined) design bounds are as follows $\boldsymbol{l}$ = [5.0 5.0 5.0 5.0 5.0 5.0 1.0 1.0 1.0 0.2 8.0 20.0 5.0]^T and $\boldsymbol{u}$ = [25.0 25.0 25.0 25.0 25.0 25.0 15.0 15.0 15.0 2.0 15.0 40.0 10.0]^T.

Design objectives concern minimization of the antenna reflection within 3.1–10.6 GHz frequency given by $F_1(\boldsymbol{x}) = \max(|S_{11}|_{3.1-10.6\,\text{GHz}})$ and reduction of the structure size defined as $F_2(\boldsymbol{x}) = l_0 \times w_0$, where $l_0 = l + d + b_1 + b_2 + b_3 + o$ and $w_0 = w_2 + o$.

9.7.2 *Antenna Models and Extreme Pareto Designs*

The computational models of the structure are implemented in CST Microwave Studio and simulated using its transient solver (CST, 2013). The high-fidelity model $\boldsymbol{f}$ contains ~2,500,000 hexahedral cells and its average simulation time is 10 min. The low-fidelity model $\boldsymbol{c}$ contains about 33,600 mesh cells and its average simulation time is 22 s.

The extreme Pareto designs $\boldsymbol{x}_1^*$ = [10.07 21.63 22.2 21 20.8 22.7 3.9 3.8 12.32 0.6 11.15 28.34 5]^T and $\boldsymbol{x}_2^*$ = [10.97 21.76 22.2 22.79 21.01 23.7 3.9 3.92 13.02 0.67 10.6 35.35 5]^T have been found through single-objective optimizations (cf. Sec. 8.2). The numerical cost of this step is 800 simulations of the low-fidelity model (about 4.9 h of CPU-time).

9.7.3 *Numerical Results*

The antenna has been optimized using the SDP algorithm of Sec. 7.3. The vector of intervals determining the patch sizes has been obtained

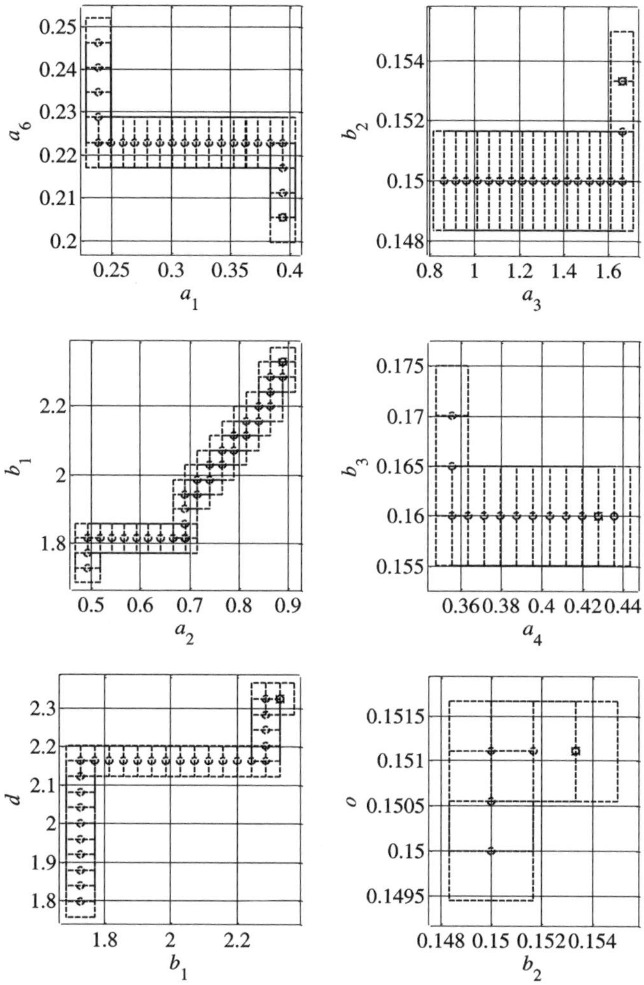

Fig. 9.37. Selected two-dimensional projections of the path created by the sequential patching algorithm.

automatically as described in Sec. 7.3.3. The maximum number of intervals along a single dimension has been set to $m_{\max} = 15$ and the resulting vector of interval numbers is $M = [4\ 3\ 2\ 4\ 3\ 3\ 3\ 2\ 4\ 5\ 3\ 15\ 3]$. Figure 9.37 shows the selected two-dimensional projections of the path created by the algorithm. The initial representation of

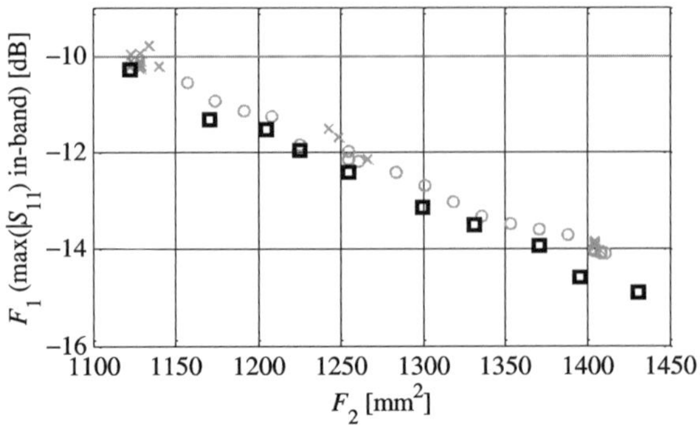

Fig. 9.38. The low- (○) and high-fidelity (□) Pareto sets obtained using SDP. Crosses represent dominated designs on the path generated by the SDP algorithm.

the Pareto front obtained using the algorithm and the set of 10 high-fidelity designs refined using local approximation models are shown in Fig. 9.38.

The obtained simulation results indicate that the variability of the objectives F_1 and F_2 along the feasible fraction of the Pareto front is 4.6 dB and 308 mm^2 (almost 22 percent), respectively. Detailed dimensions of the high-fidelity antenna designs are gathered in Table 9.11, whereas the frequency responses for the selected geometries are provided in Fig. 9.39.

The total cost of multi-objective antenna optimization using the domain patching algorithm corresponds to about 78.8 simulations of $\boldsymbol{f}$ (~13 h of CPU-time) and it includes 800 simulations of the low-fidelity model $\boldsymbol{c}$ for identification of extreme Pareto-optimal designs, 13 evaluations of the low-fidelity model for automated determination of the intervals vector, and 248 simulations of $\boldsymbol{c}$ for SDP optimization. The refinement of the selected antenna designs required 270 simulations of the low-fidelity model and 30 evaluations of the high-fidelity model, respectively.

9.7.4 *Comparison with Benchmark Algorithms*

Similarly as for the two antenna structures discussed previously, the SDP algorithm has been compared (at the low-fidelity model level) with the two MOEA-based benchmark techniques. The first method

Table 9.11 Optimization results of the trapezoid-based monopole

	F_1 [dB]	F_2 [mm^2]	Design variables												
			a_1	a_2	a_3	a_4	a_5	a_6	b_1	b_2	b_3	w_2	l	D	o
$\boldsymbol{x}_f^{(1)}$	–10.3	1123	10.7	21.63	22.20	21.0	20.94	23.70	3.9	3.8	12.32	0.60	10.79	28.34	5.00
$\boldsymbol{x}_f^{(4)}$	–11.5	1205	11.2	21.60	22.20	21.5	21.01	23.81	3.9	3.8	12.42	0.60	10.61	30.68	5.00
$\boldsymbol{x}_f^{(5)}$	–12.4	1254	11.4	21.63	22.20	21.9	20.94	23.70	3.9	3.8	12.67	0.62	10.79	31.61	5.00
$\boldsymbol{x}_f^{(7)}$	–13.5	1331	10.8	21.64	22.20	21.7	20.97	23.77	3.9	3.8	13.12	0.61	10.70	33.34	5.00
$\boldsymbol{x}_f^{(10)}$	–14.6	1395	11.0	21.59	22.20	21.5	21.01	24.00	3.9	3.9	13.19	0.60	10.81	34.85	5.00

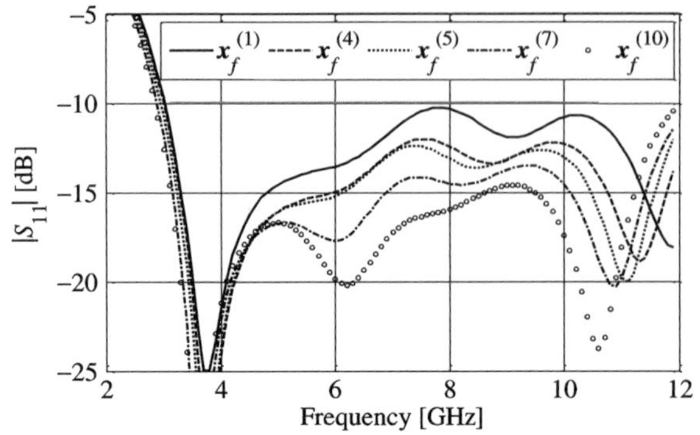

Fig. 9.39. Frequency characteristics of the antenna designs listed in Table 9.11.

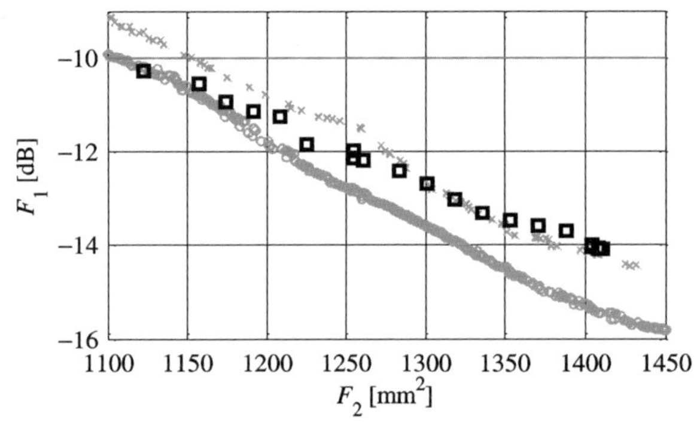

Fig. 9.40. Comparison of low-fidelity Pareto sets obtained using SDP (□), as well as first (○) and second (×) benchmark method.

exploits the approximation surrogate model constructed using the low-fidelity data acquired within the part of the search space defined by the extreme designs (MOEA setup: 500 individuals and 50 iterations). The second benchmark algorithm is MOEA directly optimizing the low-fidelity model (setup: 100 individuals, 100 iterations). The results shown in Fig. 9.40 indicate that the SDP-based front is similar to the MOEA-based ones. Small vertical differences between the obtained Pareto sets are negligible from the practical point of view. Detailed cost breakdown (excluding identification of extreme Pareto designs) of the antenna optimization using discussed techniques is provided in Table 9.12.

Table 9.12 Optimization of UWB monopole antenna of Fig. 9.36: cost breakdown and comparisons with the benchmark techniques

Algorithm	Algorithm step	Number of model evaluations		CPU-time Absolute [h]	CPU-time Relative to $\boldsymbol{f}$
SDP	Patch size determination	$\boldsymbol{c}$	13	0.08	0.48
	SDP-based optimization	$\boldsymbol{c}$	248	1.52	9.09
MOEA	Data acquisition	$\boldsymbol{c}$	1,500	9.17	55
	Optimization of $\boldsymbol{s}$ model	$\boldsymbol{s}$	25,000	N/A	N/A
	Total cost	N/A	N/A	9.17	55
	Optimization of $\boldsymbol{c}$ model	$\boldsymbol{c}$	25,000	152.8	916.7

The reason for using low-fidelity model in our numerical comparisons is that direct optimization of the high-fidelity model is prohibitively expensive (its estimated cost is about 70 days of the CPU time). Also, we would like to emphasize that it is essential to perform identification of the Kriging interpolation model for expedited MOEA optimization within the reduced search space. The Kriging model constructed using 1,500 low-fidelity data samples acquired within the initial design space exhibits the RMS error of 22 percent which makes it unsuitable for reliable optimization. More detailed discussion on the importance of space reduction for MOEA optimization of considered antenna can be found in (Koziel *et al.*, 2014c).

9.8 Design of a 14-Variable Multi-Input Multi-Output Antenna Using Design Space Reduction and Co-Kriging

In this section, we demonstrate the use of MOEA-based surrogate-assisted optimization of a UWB MIMO antenna (Koziel and Bekasiewicz, 2016g). The problem at hand is challenging not only because of a highly-dimensional design space (14 geometry parameters) and also because of the necessity of controlling reflection response of two antennas. The obtained numerical results are supplemented by physical measurements of the selected antenna designs.

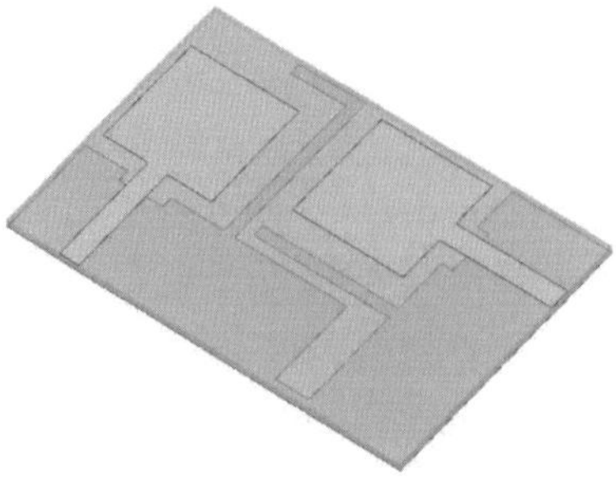

(a)

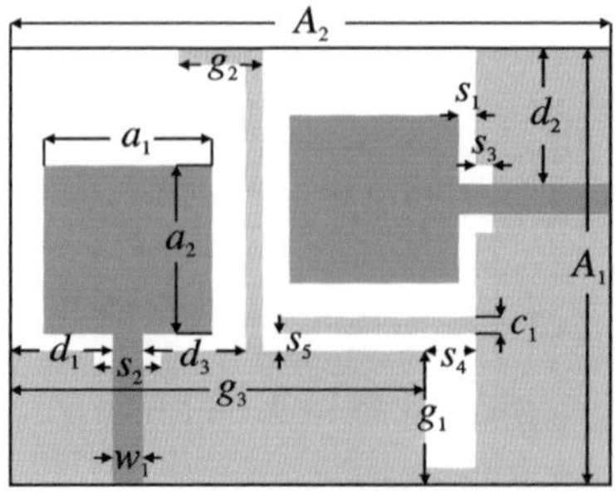

(b)

Fig. 9.41. A compact UWB MIMO antenna: (a) structure visualization; (b) geometrical details (Liu *et al.*, 2013).

9.8.1 *Antenna Description and Design Objectives*

A UWB MIMO antenna considered here is shown in Fig. 9.41 (Liu *et al.*, 2013). The structure is implemented on a 0.762 mm thick Taconic RF-35 substrate. The antenna consists of two rectangular monopole radiators fed through separate 50 Ohm microstrip lines. The slots within the ground plane allow for obtaining wideband impedance matching. Moreover, compact geometry of the structure is achieved by using the I- and L-shaped ground plane stubs both of which increase the current path. The driven elements are located perpendicularly to each other in order to achieve high isolation. There are fourteen geometry parameters to be adjusted $\boldsymbol{x} = [a_1\ a_2\ c_1\ d_1\ d_2\ d_3\ \ g_1\ g_2\ g_3\ s_1\ s_2\ s_3\ s_4\ s_5]^T$. The parameter $w_1 = 1.7$ is fixed to ensure 50 Ohm input impedance. All dimensions are in mm.

The design objective F_1 is to minimize of the maximum reflection coefficients $|S_{11}|$ and $|S_{22}|$ of both radiators within the UWB (3.1–10.6 GHz). In other words only the designs satisfying $\max\{|S_{11}|_{3.1-10.6\,\text{GHz}}\} \leq -10$ dB and max $\{|S_{22}|_{3.1-10.6\,\text{GHz}}\} \leq -10$ dB are of interest from the optimization standpoint. The objective F_2 is to minimize the antenna footprint $V(\boldsymbol{x}) = A_1 \times A_2$, where $A_1 = g_1 + d_2 + w_1 + d_3 + s_5$ and $A_2 = g_3 + s_4 + g_1$.

9.8.2 *Antenna Models and Design Space Reduction*

The high-fidelity model $\boldsymbol{f}$ consists of ~1,600,000 mesh cells and its average evaluation time is 20 min. The low-fidelity model $\boldsymbol{c}$ contains ~140,000 mesh cells and its typical simulation time is 71 s. Both are implemented in CST Microwave studio and simulated using its time domain solver (CST, 2013). The initial search space is defined by the following bounds $\boldsymbol{l} = [5\ 5\ 0.2\ 2\ 2\ 2\ 2\ 1\ 20\ 0\ 1\ 0.2\ 0.2\ 0.2]^T$, and $\boldsymbol{u} = [15\ 15\ 2\ 10\ 10\ 10\ 15\ 10\ 30\ 2\ 8\ 2\ 6\ 5]^T$.

Here, design space reduction has been performed using the enhanced sequential method that utilizes both the high- and low-fidelity extreme Pareto designs to determine the refined region of the solution space (cf. Sec. 8.2). The reduced bounds are $\bar{\boldsymbol{l}}^* = [9.8\ 9.5\ 1\ 6.14\ 6.7\ 5.83\ 6.33\ 5\ 27\ 0.72\ 3.94\ 1\ 3.24\ 0.44]^T$, $\bar{\boldsymbol{l}}^* = [10\ 10\ 1.3\ 6.82\ 7.21\ 6.1\ 10.89\ 5\ 29.17\ 1\ 4.2\ 1.3\ 3.64\ 0.98]^T$. It should be noted that effective dimensionality of the problem has been reduced to 13 parameters because $g_2 = 5$ for all (the $\boldsymbol{c}$ and $\boldsymbol{f}$) extreme designs. The reduced space is 10^{12} times smaller (volume-wise) than the initial one.

9.8.3 *Numerical Results*

The Kriging interpolation model $\boldsymbol{s}$ has been constructed using 1,000 LHS-allocated low-fidelity training samples and four low-fidelity extreme designs obtained during the space reduction step. The cross-validation-estimated RMS error of the surrogate is below 2 percent. The initial Pareto set has been obtained by MOEA optimization (setup: 500 individuals, 50 iterations) of the surrogate model $\boldsymbol{s}$. Subsequently, the Pareto front has been refined using co-Kriging (cf. Sec. 7.1.4). The number of MOEA iterations has been reduced to 25 because the initial population for co-Kriging-based correction is already a good representation of the Pareto front. The convergence of the process has been obtained after three iterations, i.e., a total of 30 high-fidelity model evaluations.

For the sake of numerical validation, 10 designs have been selected along the obtained Pareto front and evaluated using the high-fidelity antenna model (see Fig. 9.42). The agreement between the optimized

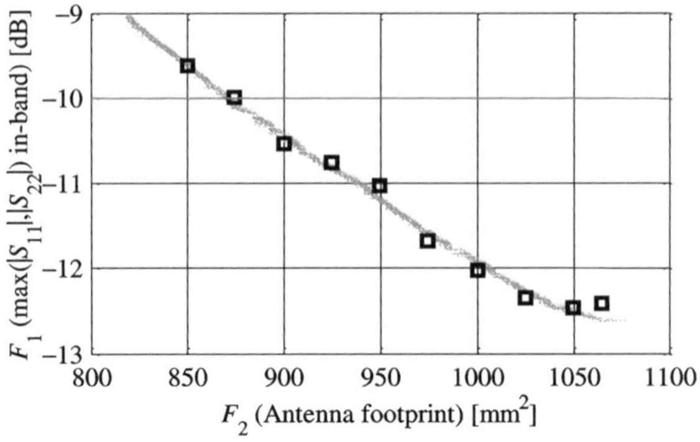

Fig. 9.42. The surrogate (×) and high-fidelity (□) Pareto sets obtained using co-Kriging (Koziel and Bekasiewicz, 2016g).

co-Kriging surrogate and the high-fidelity test samples is very good. The largest design with footprint of 1,050 mm^2 features the lowest in-band reflection of –12.5 dB. The area of the smallest antenna design that still fulfills design specifications is 875 mm^2. The variability of the objectives F_1 and F_2 along the Pareto front is 2.5 dB and 175 mm^2 (almost 17 percent), respectively. The details concerning the selected high-fidelity Pareto set are gathered in Table 9.13, whereas the frequency responses of selected optimal designs are shown in Fig. 9.43.

The cost of antenna optimization corresponds to about 204 evaluations of the high-fidelity model (~493 h of CPU-time) and it includes: 1,869 evaluations of the low-fidelity model $\boldsymbol{c}$ (1,025 and 844 evaluations for optimization with respect to F_1 and F_2, respectively), four high-fidelity model evaluations to identify the reduced space, 1,000 evaluations of $\boldsymbol{c}$ for construction of the surrogate model, and a total of 30 evaluations of the high-fidelity model for co-Kriging-based refinement of the initial Pareto set. At the same time, the estimated cost of direct MOEA optimization (25,000 evaluations) within the initial design space is 493 h (8,333 h) for low- (high-) fidelity model, respectively. A detailed cost breakdown of the design optimization procedure is provided in Table 9.14.

Table 9.13 Compact MIMO antenna: selected Pareto-optimal designs

	Antenna designs									
	$\boldsymbol{x}_f^{(1)}$	$\boldsymbol{x}_f^{(2)}$	$\boldsymbol{x}_f^{(3)}$	$\boldsymbol{x}_f^{(4)}$	$\boldsymbol{x}_f^{(5)}$	$\boldsymbol{x}_f^{(6)}$	$\boldsymbol{x}_f^{(7)}$	$\boldsymbol{x}_f^{(8)}$	$\boldsymbol{x}_f^{(9)}$	$\boldsymbol{x}_f^{(10)}$
F_1 [dB]	−9.6	−10.0	−10.5	−10.8	−11.0	−11.7	−12.0	−12.3	−12.5	−12.4
F_2 [mm^2]	850	874	900	925	950	975	1,000	1,025	1,050	1,064
Antenna parameters[#] a_1	9.90	9.91	9.89	9.89	9.85	9.86	9.87	9.88	9.87	9.89
a_2	9.64	9.66	9.63	9.67	9.63	9.61	9.61	9.62	9.61	9.67
c_1	1.21	1.21	1.22	1.24	1.24	1.24	1.24	1.24	1.24	1.25
d_1	6.50	6.50	6.47	6.46	6.44	6.43	6.40	6.42	6.42	6.42
d_2	6.97	6.97	6.95	6.90	6.86	6.86	6.83	6.83	6.86	6.86
d_3	5.93	5.93	5.95	5.94	5.98	5.94	5.91	5.91	5.91	5.90
g_1	7.25	7.62	7.98	8.38	8.72	9.12	9.44	9.78	10.09	10.30
g_3	27.22	27.23	27.25	27.39	27.50	27.47	27.75	27.79	27.86	27.92
s_1	0.79	0.76	0.80	0.78	0.80	0.79	0.84	0.79	0.81	0.81
s_2	4.05	4.04	4.05	4.04	4.04	4.05	4.03	4.01	4.05	4.01
s_3	1.10	1.07	1.06	1.05	1.07	1.05	1.06	1.08	1.08	1.09
s_4	3.38	3.40	3.42	3.40	3.41	3.44	3.41	3.42	3.40	3.42
s_5	0.61	0.64	0.71	0.70	0.71	0.71	0.75	0.79	0.83	0.80

[#]Parameter $g_2 = 5$ obtained in the course of design space reduction is excluded from table.

9.8.4 *Measurements*

In this section, we present experimental verification of the designed MIMO antenna structures. The three designs, denoted as $\boldsymbol{x}_f^{(k)}$, $k = 1$, 2, 3, have been selected for fabrication. Their detailed dimensions, as well as the objective function values are gathered in Table 9.13. Figure 9.44 shows the comparison of simulations and measurements in terms of reflection characteristics of both radiators, as well as isolation between them. Overall, the measured and simulated responses are in good resemblance, however, some discrepancies can be observed.

The difference between maximum in-band values with respect to S_{11} and S_{22} is 0.8 dB, 0.5 dB and 0.6 dB for $\boldsymbol{x}_f^{(1)}$, $\boldsymbol{x}_f^{(2)}$ and $\boldsymbol{x}_f^{(3)}$, respectively. At the same time, the difference between the maximum values of simulated and measured isolation is 1.3 dB, 0.6 dB and 0.2 dB for $\boldsymbol{x}_f^{(1)}$, $\boldsymbol{x}_f^{(2)}$ and $\boldsymbol{x}_f^{(3)}$ respectively. It should be noted that the obtained vertical misalignments are considered acceptable for compact antenna structures. The discrepancies between the simulated

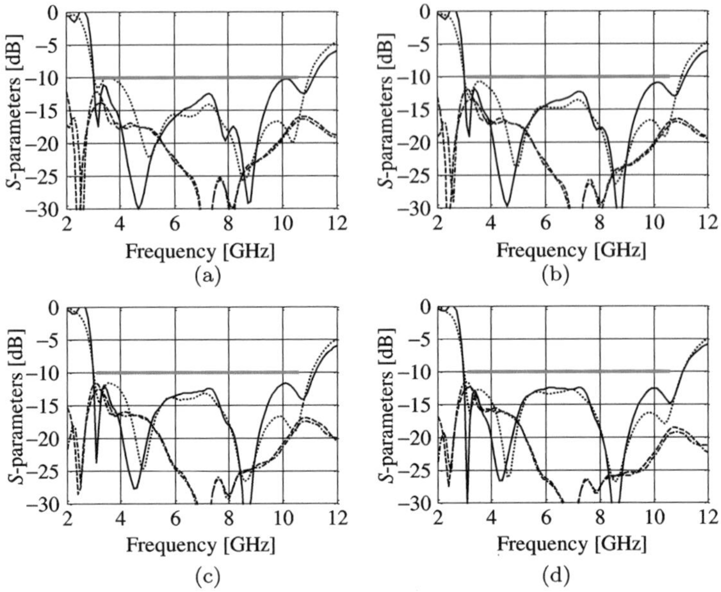

Fig. 9.43. Frequency characteristics of the UWB MIMO antenna: $|S_{11}|$ (—), $|S_{21}|$ (– –), $|S_{31}|$ (– · –) and $|S_{41}|$ (· · ·). Selected designs are: (a) $\boldsymbol{x}_f^{(2)}$; (b) $\boldsymbol{x}_f^{(4)}$; (c) $\boldsymbol{x}_f^{(6)}$; and (d) $\boldsymbol{x}_f^{(9)}$. Note that reduction of antenna footprint has limited influence on the isolation (denoted as $|S_{21}|$ and $|S_{12}|$) between radiators.

and the measured results are mostly due to the miniaturized design. In other words, geometrical limitations for operation of the compact monopole structures come from reduction of the current path within the circuit (Valderas *et al.*, 2011).

This has serious consequences, since electrical components of the antenna such as connectors contribute to its electrical length and thus allows obtaining smaller designs. At the same time, connectors influence on the structure operation and performance. Here, the differences between the simulated and the measured responses are mostly introduced by the simplified model of antenna lacking connector, as well as rough structure assembly (i.e., the manual cutting of the substrate and connector trimming). All of these factors influence the current path within the MIMO antenna, and therefore distort the measurement results.

Table 9.14 Compact MIMO optimization: cost breakdown

Algorithm step		Number of model evaluations	CPU-time Absolute [h]	CPU-time Relative to f
Design space reduction	c	1,869	36.86	110.58
	f	4	1.33	4
Data acquisition and RSA construction	c	1,000	19.72	59.17
MOEA optimization	s	25,000	N/A	N/A
Pareto set refinement	f	30	10	30
	s	37,500	N/A	N/A
Total cost	N/A	N/A	67.91	203.75
Direct search (low-fidelity model)#	c	25,000	493.1	1,479.3
Direct search (high-fidelity model)#	f	25,000	8333.3	25,000

#Estimated based on the number of s evaluations during MOEA optimization.

It should be mentioned, that modifications of the antenna geometry aimed at reduction of its size also influences its field characteristics such as gain or radiation pattern. This should be accounted for to provide real-world antenna designs for practical applications. The optimization problem concerning field antenna characteristics has been described and solved in Secs. 9.5, 9.6 and 9.9.

9.9 Optimization of Broadband Quasi-Yagi Antenna Using Multi-Objective Evolutionary Algorithm and Rotational Space Reduction

The last antenna structure considered in this chapter demonstrates the use of rotational design space reduction for multi-objective optimization in highly-dimensional parameter space (Koziel and Bekasiewicz, 2016d). A quasi-Yagi antenna considered here is described by 24 independent parameters so that even upon initial space reduction, it is still difficult (and expensive) to construct a reliable approximation surrogate. Application of the rotational reduction scheme limits this cost by a factor of four, which results in almost 30 percent reduction of the overall optimization expense.

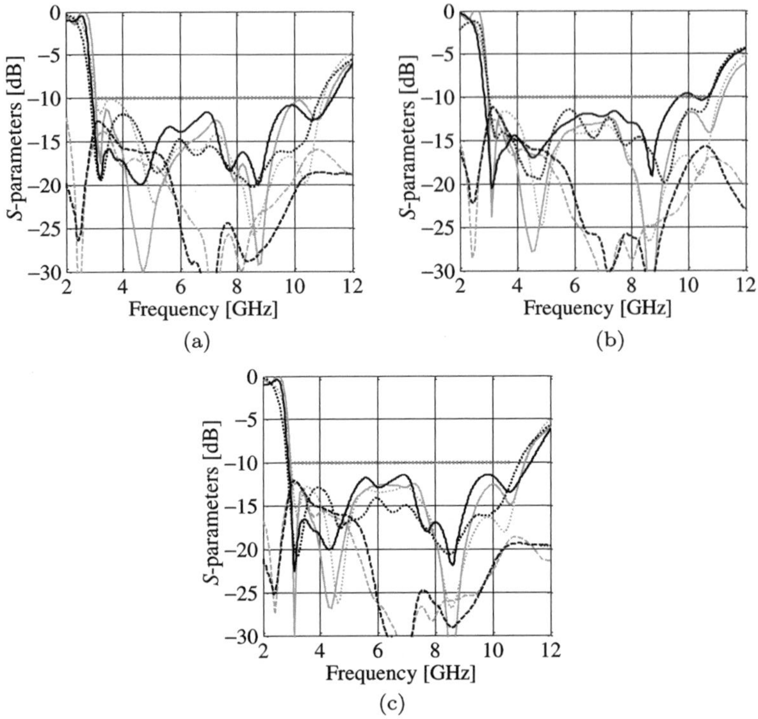

Fig. 9.44. Pareto-optimal designs of the MIMO antenna: (a) $\boldsymbol{x}_f^{(2)}$, (b) $\boldsymbol{x}_f^{(6)}$ and (c) $\boldsymbol{x}_f^{(9)}$. Comparison of simulation (gray lines) and measurement (black lines) results in terms of $|S_{11}|$ (—), $|S_{21}|$ (– –), and $|S_{22}|$ (···).

9.9.1 *Antenna Description and Design Objectives*

The considered quasi-Yagi antenna is shown in Fig. 9.45 (Wu *et al.*, 2014). Similarly as for the previous example, the structure is implemented on the Taconic RF-35 substrate.

The antenna consists of a dipole-driven element excited by a coplanar stripline (CPS) and two directors. The dipole is fed through a microstrip-to-CPS transition. An impedance transformer allows for obtaining wideband operation of the antenna. Moreover, compact geometry is ensured by two symmetrical ground plane stubs which increase current path. The design is represented by a 24-variable vector: $\boldsymbol{x} = [w_0\ d_1\ d_2\ g_1\ l_1\ l_2\ l_3\ l_4\ l_6\ l_7\ l_8\ l_f\ o_1\ o_2\ s_1\ s_2\ s_3\ w_1\ w_2\ w_3\ w_4\ w_5\ w_7\ w_8]^T$. The relative parameter is $l_5 = 3l_3$, whereas

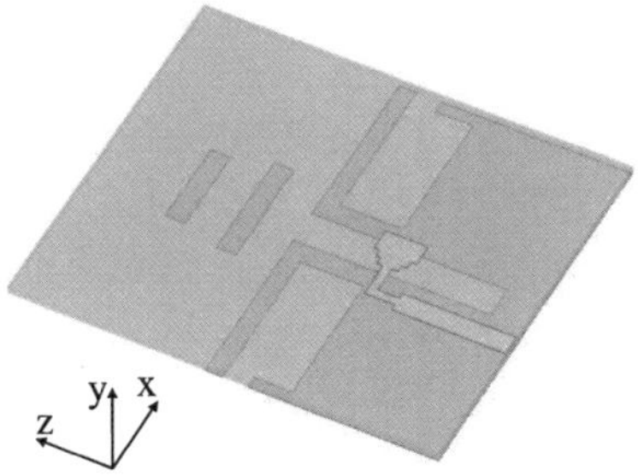

(a)

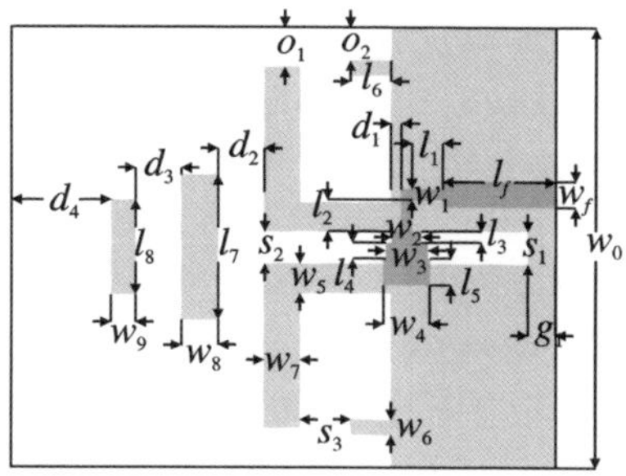

(b)

Fig. 9.45. Planar quasi-Yagi antenna (Koziel and Bekasiewicz, 2016d): (a) visualization; and (b) geometrical details of the structure.

$d_3 = 3$, $d_4 = 6.5$, $w_f = 1.7$ and $w_6 = 1$ (all dimensions are in mm). The bounds of the initial search space are $\boldsymbol{l} = [20\ 0\ 1\ 1\ 0.1\ 1\ 0.2\ 0.2\ 1\ 5\ 5\ 5\ 0\ 0\ 1\ 1\ 1\ 0.2\ 1.2\ 2\ 2.9\ 0.5\ 0.5\ 0.5]^T$ and $\boldsymbol{u} = [40\ 1\ 5\ 5\ 4\ 4\ 1.2\ 1.2\ 5\ 13\ 13\ 13\ 2\ 2\ 4\ 4\ 4\ 1.2\ 2.2\ 3\ 3.9\ 2.5\ 2.5\ 2.5]^T$.

The high-fidelity model $\boldsymbol{f}$ (~2,100,000 mesh cells; average simulation time: 13 min) and the low-fidelity model $\boldsymbol{c}$ (~80,000 cells; simulation: 40 s) of the antenna are both implemented in CST Microwave Studio and simulated using its transient solver (CST, 2013).

Design requirements are as follows: F_1 — minimization of maximal in-band reflection and F_2 — maximization of antenna average gain. Both objectives are considered within 4–10 GHz frequency band of interest.

9.9.2 *Design Space Reduction and Kriging Model Construction*

In order to enable expedited MOEA-based optimization, the initial design space has been reduced using the rotational technique of Sec. 8.3. After the first step of the process (reduction by means of sequential single-objective optimizations), the dimensionality of the problem has been limited to 16 parameters because $l_f = 7.5$, $w_1 = 0.7$, $w_0 = 28.86$, $l_4 = 0.58$, $o_2 = 0.13$, $d_2 = 3$, $l_7 = 9.47$ and $l_8 = 6.21$ are the same for both extreme Pareto designs. Therefore, the initially reduced region of the search space is represented by the vector $\boldsymbol{x} = [d_1\ g_1\ l_1\ l_2\ l_3\ l_6\ o_1\ s_1\ s_2\ s_3\ w_2\ w_3\ w_4\ w_5\ w_7\ w_8]^T$. The refined lower and upper bounds of are $\boldsymbol{l}^* = [0.65\ 1.63\ 1.9\ 2.02\ 0.72\ 2.6\ 0.09\ 2\ 2\ 2\ 1.7\ 2.3\ 2.9\ 1.6\ 1.68\ 2.24]^T$ and $\boldsymbol{u}^* = [0.68\ 3.09\ 2\ 2.07\ 0.96\ 2.83\ 1.09\ 2.2\ 2.13\ 3.73\ 2.04\ 2.92\ 3.4\ 2.08\ 2.43\ 2.45]^T$. We would like to emphasize that the initially reduced space is 10^{18} times (volume-wise) smaller as compared to the initial one. In the next step, reduced space has been rotated. Also, all dimensions of the rotated box except the one corresponding to its main diagonal have been scaled down by a factor of 3. The rotated design space region is seven orders of magnitude smaller than the initially reduced one, thus the total reduction factor with respect to space region of Sec. 9.9.1 is 10^{25}.

In the next step, the Kriging interpolation model $\boldsymbol{s}$ has been identified within the rotated space. The acceptable accuracy of the surrogate has been obtained with only 502 LHS-allocated training samples. The cross-validation-estiamted RMS error of $\boldsymbol{s}$ is 2 percent. For the sake of comparison, the RSA has been also constructed within the initial and the initially reduced space regions. In both cases, the data acquisition process has been terminated after 2,000 $\boldsymbol{c}$ simulations. RMS error of the former model was over 40 percent which makes it useless for optimization. At the same time, the error of the model constructed within initially reduced space is 3 percent.

9.9.3 *Numerical Results*

Data-driven surrogate models constructed within the initially reduced and the rotated search space regions have been both utilized for MOEA optimization (algorithm setup: 50 iterations, 500 individuals). Their corresponding initial Pareto sets have been refined using response correction technique. A comparison between the high-fidelity and surrogate representations of the Pareto fronts from both search space regions is shown in Fig. 9.46. The results indicate that the slope of the front generated within the initially reduced space is noticeably larger than for the one from rotated region, which is due to limited exploration capabilities of the MOEA near corners of the 16-dimensional search space. This problem is mitigated in the

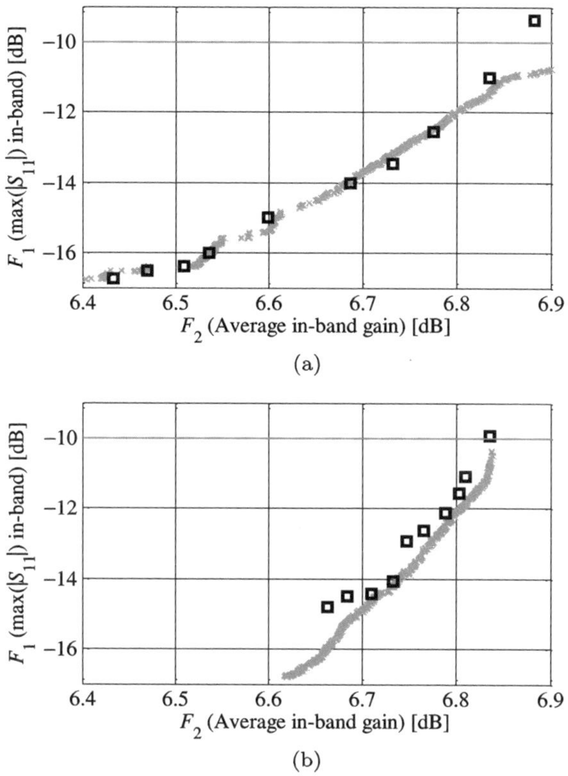

Fig. 9.46. The $\boldsymbol{R}_f$ (□) and $\boldsymbol{R}_s$ (×) Pareto sets of the quasi-Yagi antenna obtained in: (a) rotated; and (b) initially reduced search spaces (Koziel *et al.*, 2015c).

rotated space because the extreme designs are located on its walls. It should be noted that apart from limited accuracy, the cost of surrogate model construction within the initially reduced space is four times higher compared to data acquisition in the rotated space.

The high-fidelity model design from the rotated space with the highest average gain of 6.84 dB exhibits the worst acceptable in-band reflection of −11 dB. The design with the lowest reflection of −16.7 dB features the smallest gain of 6.43 dB. The variability of objectives F_1 and F_2 along the Pareto front is 5.7 dB and 0.41 dB, respectively. The detailed dimensions of the obtained Pareto-optimal designs are gathered in Table 9.15. The reflection and gain characteristics for the

Table 9.15 Optimization results of quasi Yagi–Uda antenna

	Antenna designs									
	$\boldsymbol{x}_f^{(1)}$	$\boldsymbol{x}_f^{(2)}$	$\boldsymbol{x}_f^{(3)}$	$\boldsymbol{x}_f^{(4)}$	$\boldsymbol{x}_f^{(5)}$	$\boldsymbol{x}_f^{(6)}$	$\boldsymbol{x}_f^{(7)}$	$\boldsymbol{x}_f^{(8)}$	$\boldsymbol{x}_f^{(9)}$	$\boldsymbol{x}_f^{(10)}$
F_1 [dB]	−9.4	−11.0	−12.6	−13.5	−14.0	−15.0	−16.0	−16.4	−16.5	−16.7
F_2 [dB]	6.88	6.83	6.78	6.73	6.69	6.60	6.54	6.51	6.47	6.43
Antenna parameters#										
d_1	0.68	0.67	0.67	0.67	0.67	0.66	0.66	0.66	0.65	0.65
g_1	1.48	1.84	2.15	2.31	2.32	2.62	2.75	2.76	2.91	2.95
l_1	1.99	1.99	1.96	1.96	1.95	1.93	1.93	1.91	1.92	1.91
l_2	2.08	2.07	2.05	2.06	2.05	2.03	2.04	2.04	2.03	2.03
l_3	0.75	0.73	0.80	0.81	0.82	0.88	0.92	0.91	0.93	0.93
l_6	2.65	2.66	2.66	2.69	2.70	2.78	2.79	2.79	2.84	2.85
o_1	0.90	1.01	0.72	0.64	0.58	0.31	0.19	0.21	0.16	0.29
s_1	2.19	2.14	2.12	2.10	2.09	2.05	2.08	2.03	2.04	2.02
s_2	2.14	2.09	2.09	2.07	2.06	2.03	2.04	2.02	2.03	2.02
s_3	3.59	3.35	3.15	3.05	2.90	2.60	2.46	2.32	2.24	2.08
w_2	1.97	1.95	1.95	1.90	1.86	1.83	1.77	1.79	1.80	1.76
w_3	2.88	2.76	2.68	2.64	2.61	2.49	2.55	2.50	2.50	2.43
w_4	3.04	3.05	3.14	3.13	3.15	3.21	3.31	3.28	3.36	3.42
w_5	2.13	1.92	1.87	1.82	1.78	1.74	1.70	1.71	1.65	1.59
w_7	2.49	2.33	2.25	2.16	2.07	1.99	1.91	1.87	1.83	1.84
w_8	2.44	2.43	2.38	2.37	2.36	2.33	2.28	2.30	2.30	2.28

#Parameters l_f, w_1, w_0, l_4, o_2, d_2, l_7 and l_8 are excluded from the table.

selected high-fidelity antenna geometries are shown in Figs. 9.47 and 9.48, respectively.

The detailed cost breakdown of the design optimization procedure is provided in Table 9.16. The numerical cost of the multi-objective antenna design corresponds to about 205 simulations of $\boldsymbol{f}$ (~45 h of CPU-time). The design space reduction process involved 2,917 simulations of the low-fidelity model $\boldsymbol{c}$ (1,076 evaluations for minimization of the maximal reflection and 1,841 simulations for maximization of average gain), 502 evaluations of $\boldsymbol{c}$ for the acquisition of training samples and 30 simulations of $\boldsymbol{f}$ for refinement of selected Pareto-optimal designs (three iterations).

It should be noted that the estimated cost of direct MOEA-based optimization (25,000 model evaluations) of the $\boldsymbol{f}$ model of the antenna is over 260 days of the CPU-time. The estimated cost of direct $\boldsymbol{c}$ model optimization corresponds to over 1,280 evaluations

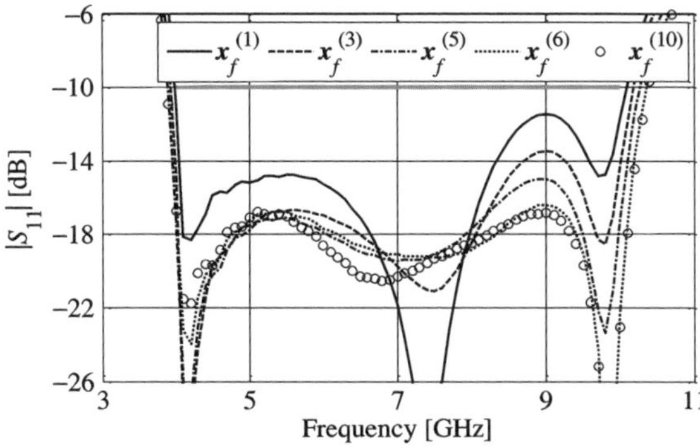

Fig. 9.47. Reflection characteristics of quasi-Yagi antenna for high-fidelity designs selected from Table 9.15 (Koziel and Bekasiewicz, 2016d).

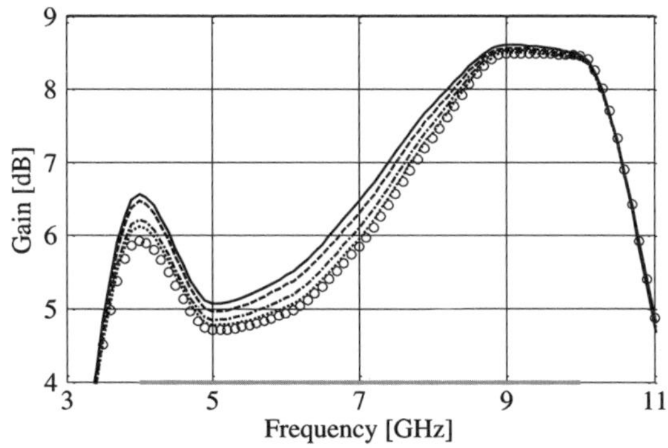

Fig. 9.48. Gain characteristics of quasi-Yagi antenna for designs selected from Table 9.15 (Koziel and Bekasiewicz, 2016d). For legend see Fig. 9.48.

of $\boldsymbol{f}$ (~278 h). Moreover, the cost of expedited MOEA optimization within the initially reduced space is over 61 h of the CPU-time.

9.9.4 *Experimental Validation*

The Pareto-optimal designs $\boldsymbol{x}_f^{(2)}$, $\boldsymbol{x}_f^{(5)}$ and $\boldsymbol{x}_f^{(10)}$ of the quasi-Yagi antenna have been fabricated and measured. Photographs of the manufactured structures are shown in Fig. 9.49, whereas their detailed dimensions are gathered in Table 9.15. Geometrical differences between the considered geometries that can be distinguished

Table 9.16 Quasi-Yagi antenna: design cost breakdown

Reduction method	Algorithm step	Number of model evaluations		CPU time Absolute [h]	CPU time Relative to $\boldsymbol{f}$
Rotational	Space reduction	$\boldsymbol{c}$	2,917	32.41	149.59
	Data acquisition and RSA construction	$\boldsymbol{c}$	502	5.58	25.74
	MOEA optimization	$\boldsymbol{s}$	25,000	N/A	N/A
	Pareto set refinement	$\boldsymbol{f}$	30	6.50	30.0
	Total cost	N/A	N/A	44.49	205.33
Sequential	Space reduction	$\boldsymbol{c}$	2,917	32.41	149.59
	Data acquisition and RSA construction[#]	$\boldsymbol{c}$	2,002	22.24	102.67
	MOEA optimization	$\boldsymbol{s}$	25,000	N/A	N/A
	Pareto set refinement	$\boldsymbol{f}$	30	6.50	30.0
	Total cost	N/A	N/A	61.15	282.26
N/A	Direct search (low-fidelity model)[!]	$\boldsymbol{c}$	25,000	277.78	1,282.1
	Direct search (high-fidelity model)[!]	$\boldsymbol{f}$	25,000	5416.7	25,000

[#]Data acquisition terminated after 2,002 $\boldsymbol{c}$ simulations.
[!]Estimated based on the number of $\boldsymbol{s}$ evaluations during MOEA optimization.

with bare eye include slightly different sizes (design $\boldsymbol{x}_f^{(2)}$ is the smallest whereas $\boldsymbol{x}_f^{(10)}$ is the largest; note that antenna miniaturization was not the purpose of the optimization process). Moreover, all antennas exhibit different allocations of ground plane stubs, as well as different widths and lengths of reflectors.

A comparison of antenna reflection characteristics is shown in Fig. 9.50. The obtained results indicate that design $\boldsymbol{x}_f^{(2)}$ slightly violates the assumed design specification ($|S_{11}| > -10$ dB). The maximum measured in-band reflection is –11.3 dB, –12 dB and –9.6 dB for the design $\boldsymbol{x}_f^{(10)}$, $\boldsymbol{x}_f^{(5)}$ and $\boldsymbol{x}_f^{(2)}$, respectively. The corresponding simulated values are –16.7 dB, –14 dB and –11 dB, so that the peak reflection discrepancies between simulations and measurements are 5.4 dB, 2 dB and 1.4 dB. It should be noted that considerable difference of

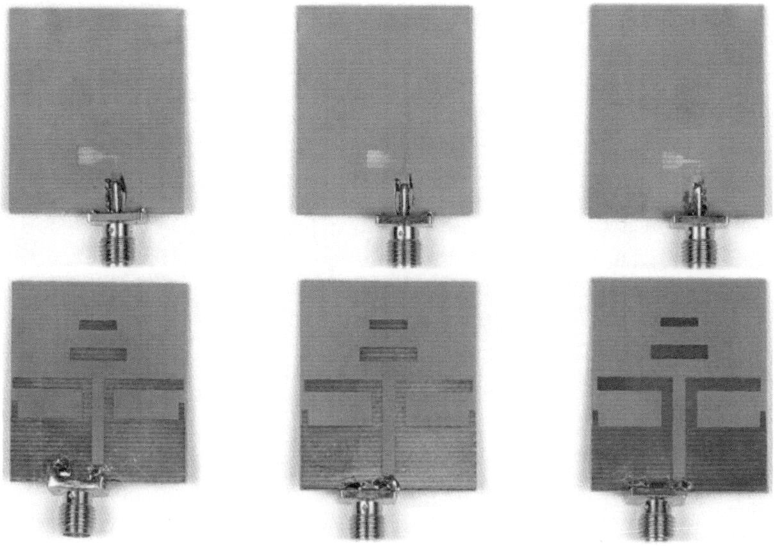

(a) (b) (c)

Fig. 9.49. Planar quasi-Yagi antenna — top- and bottom-view photographs of fabricated prototypes. The selected Pareto-optimal designs are: (a) $\boldsymbol{x}_f^{(2)}$; (b) $\boldsymbol{x}_f^{(5)}$; and (c) $\boldsymbol{x}_f^{(10)}$. For detailed dimensions of manufactured structures see Table 9.15 (Koziel and Bekasiewicz, 2016d).

the maximum in-band $|S_{11}|$ for the design $\boldsymbol{x}_f^{(10)}$ is a consequence of misalignment between simulated and measured characteristics. The measured bandwidth of the antenna at $\boldsymbol{x}_f^{(10)}$ is 350 MHz narrower than simulated one. At the same time, the reflection measured for the design $\boldsymbol{x}_f^{(2)}$ is 400 MHz broader than the one obtained from simulations. The discrepancies between responses of the EM model and fabricated prototypes also include different number of resonances.

The gain characteristics of considered antenna designs are compared in Fig. 9.50. The shapes of simulated and measured characteristics are similar. On the other hand, in a range from 4 GHz to 6 GHz gain levels of measured prototypes are noticeably lower than simulated characteristics. The measured average in-band gain for designs $\boldsymbol{x}_f^{(2)}$, $\boldsymbol{x}_f^{(5)}$ and $\boldsymbol{x}_f^{(10)}$ is 6.05 dB, 6.16 dB and 6.23 dB, respectively. The corresponding simulated values are 6.83 dB, 6.68 dB and 6.43 dB,

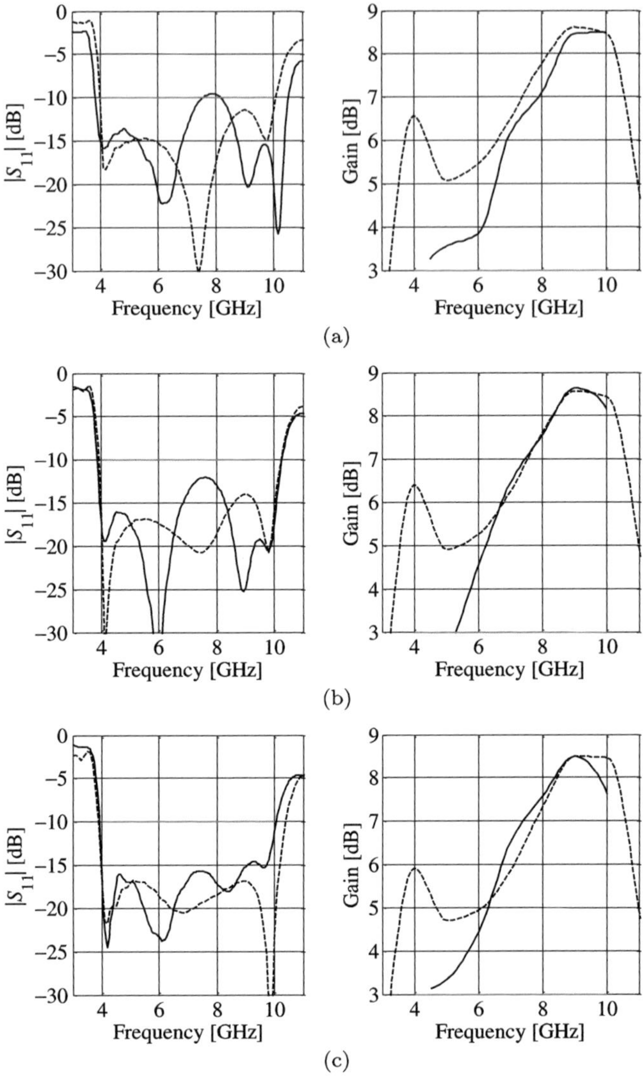

Fig. 9.50. Comparison of simulated (– –) and measured (—) characteristics of quasi-Yagi antenna: (a) design $\boldsymbol{x}_f^{(2)}$; (b) design $\boldsymbol{x}_f^{(5)}$; (c) design $\boldsymbol{x}_f^{(10)}$. Left- and right-hand side plots represent reflection and gain characteristics, respectively.

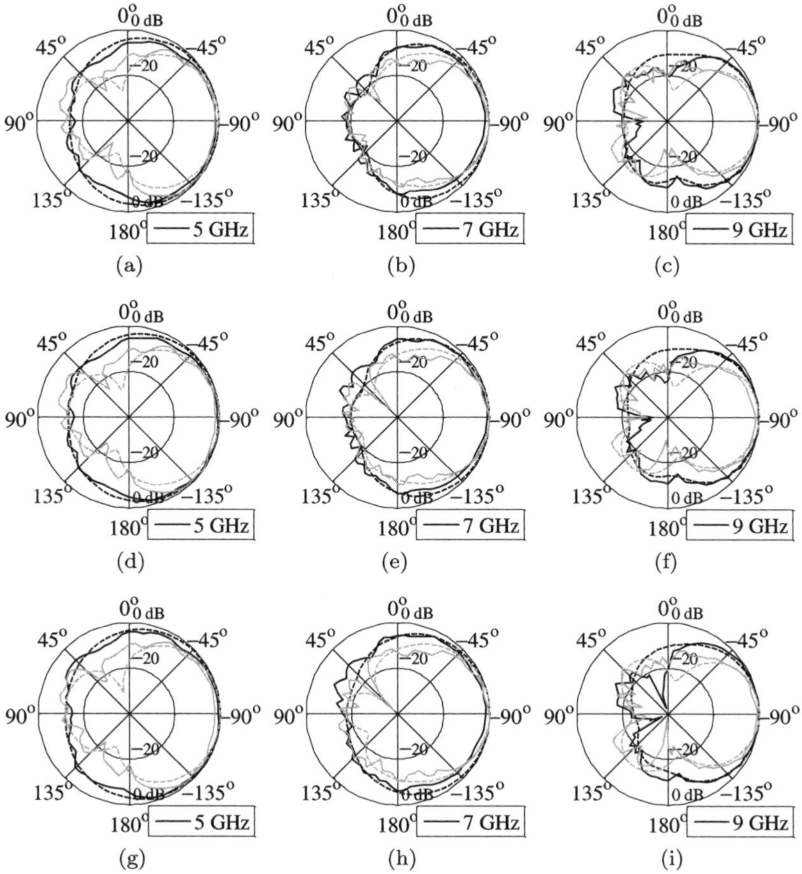

Fig. 9.51. Comparison of simulated (– –) and measured (—) E-field radiation patterns of the quasi-Yagi antenna in y–z (black lines) and x–z (gray lines) planes: (a)–(c) design $\boldsymbol{x}_f^{(2)}$; (d)–(f) design $\boldsymbol{x}_f^{(6)}$; (g)–(i) design $\boldsymbol{x}_f^{(10)}$.

so that the differences between obtained results are 0.78 dB, 0.52 dB and 0.2 dB.

The simulated and measured E-field radiation patterns of the antenna designs are shown in Fig. 9.51. The results are compared in y–z (elevation) and x–z (horizontal) planes (see Fig. 9.45 for visualization) for 5 GHz, 7 GHz and 9 GHz frequencies. The measured and simulated front-to-back (F/B) ratios for selected

Table 9.17 Quasi-Yagi antenna: comparison of F/B ratios

		Simulation			Measurement		
	Frequency	$\boldsymbol{x}_f^{(2)}$	$\boldsymbol{x}_f^{(5)}$	$\boldsymbol{x}_f^{(10)}$	$\boldsymbol{x}_f^{(2)}$	$\boldsymbol{x}_f^{(5)}$	$\boldsymbol{x}_f^{(10)}$
y–z plane	5 GHz	14.52	13.65	12.11	15.11	15.37	14.35
	7 GHz	21.85	20.93	17.84	17.91	17.48	16.99
	9 GHz	21.06	22.92	26.04	27.97	31.30	33.75
x–z plane	5 GHz	26.44	27.54	30.40	24.58	22.47	19.25
	7 GHz	23.07	25.50	30.71	24.94	29.63	27.82
	9 GHz	25.85	25.83	25.14	26.12	27.03	28.76

antenna designs are gathered in Table 9.17. The obtained results are similar; however, measured F/B rates are slightly larger.

The differences between simulations and measurements result from utilization of the EM antenna model that lacks the SMA connector. As shown in Bekasiewicz and Koziel (2015a), the SMA connector should be included into the EM models of compact antennas because it notably influences the structure characteristics (e.g., for considered antenna structure, measured $|S_{11}|$ feature additional in-band resonance which is most likely introduced by the SMA connector).

To some extent, discrepancies between the results are also influenced by the fabrication tolerances, electrically large measurement equipment, as well as imperfections of the measurement setup (e.g., manual positioning of the antennas for gain measurements based on the three antenna method). Moreover, the considered antenna is small and thus precision of the prototype assembly (board cutting, SMA positioning, etc.) is rather important.

9.10 Summary

In this chapter, comprehensive numerical studies undertaken to demonstrate the multi-objective optimization procedures of Chapters 7 and 8 have been presented. We illustrated operation and performance of the surrogate-assisted algorithms using several antenna structures of various levels of complexity and with various number of geometry parameters from a few to over 20. For some of the antenna structures considered here, a comparison of different approaches (including direct MOEA-based optimization of the EM-simulation model) has been provided. It can be observed that surrogate-based methods do allow for efficient multi-objective design of antenna structures. Depending on the problem complexity (which depends on design space dimensionality and also computational cost of the EM model), the optimization cost varies from a few dozen to a few (two, three) hundreds of high-fidelity EM-simulations of the respective antenna structure. Such cost is practically acceptable.

Further studies concerning the considered design optimization methods, including investigation of various practical issues and also applications in other engineering disciplines, have been discussed in the remaining chapters of this book.

Chapter 10

Selected Topics and Practical Issues

As demonstrated in Chapter 9, surrogate-assisted algorithms of Chapters 7 and 8 can be utilized for expedited multi-objective optimization of antenna structures. The same methods can also handle expensive computational models in other engineering disciplines (cf. Chapter 11). Here, we discuss several aspects and practical issues of these techniques. In particular, we look into their scalability properties. More specifically, we are interested in the relationship between dimensionality of the design space and the computational cost of the optimization algorithm (Sec. 10.1). Another issue is statistical analysis of surrogate-assisted multi-objective optimization algorithm of Sec. 7.1, where several components are of stochastic nature. In particular, the initial Pareto front approximation is obtained using multi-objective evolutionary algorithm (MOEA). In Sec. 10.2, we investigate the influence of MOEA operation on the quality of the final Pareto set representation found by the optimization procedure. Finally, in Sec. 10.3, we study the effect of various patch size setups on the cost and performance of sequential domain patching (SDP) algorithm of Sec. 7.3. The studies presented in this chapter lead to certain conclusions concerning the robustness of the considered optimization procedures, indicate potential applicability for solving more challenging problems (than those presented so far in the book), as well as give some guidelines and recommendations in terms of the algorithm setup.

10.1 Scalability of Surrogate-Assisted Multi-Objective Optimization Algorithm

Computational complexity as a function of the problem dimensionality is a very important property of optimization algorithms. If the computational cost of the optimization process grows too quickly with the number of design variables of the problem at hand, it poses serious limitations for applicability of a given method for solving more demanding problems.

In this section, a numerical study is performed concerning the scalability of the surrogate-based multi-objective antenna design algorithm of Sec. 7.1, based on a set of benchmark problems with the dimensionality of the design space ranging from six to 24 and the simulation time of the high-fidelity antenna model from 10 min to 20 min. The obtained results indicate that the computational cost of the optimization process increases in a close-to-quadratic fashion with respect to the problem dimension. This conjecture gives promising perspectives for multi-objective antenna design using surrogate-based optimization (SBO) in the context of even higher dimensional problems. A discussion of various factors contributing to the overall optimization cost and their dependence on the problem dimensionality is also provided. Furthermore, possible ways of further reduction of the cost are outlined.

10.1.1 *Test Cases*

For the sake of this numerical study, three antennas are considered as shown in Fig. 10.1. The first structure is a planar, single layer dipole antenna (Fig. 10.1(a)). It consists of the main radiator element and two parasitic strips (Spence and Werner, 2006; Bekasiewicz *et al.*, 2014a). The design variables are $\boldsymbol{x} = [l_0\ w_0\ a_0\ l_p\ w_p\ s_0]^T$, while $a_1 = 0.5\,\text{mm}$ and $w_1 = 0.5\,\text{mm}$ remain fixed (see Fig. 3). Substrate is a 1.58 mm thick Rogers RT5880. Both the low-fidelity model $\boldsymbol{c}$ (~167,900 mesh cells, 30 s evaluation time) and the high-fidelity model $\boldsymbol{f}$ (~12,510,000 mesh cells, 20 min evaluation time) are simulated in CST Studio (CST, 2013). The following objectives are considered: F_1 — minimization of antenna reflection within 3.1–10.6 GHz band, and F_2 — reduction of the structure footprint defined as a $w_s \times l_s$ rectangle, where $w_s = 2w_p + 2s_0 + w_0$ and $l_s = 2l_0 + 2a_0 + a_1$.

The second test case is a 13-variable ultra-wideband (UWB) monopole antenna (Fig. 10.1(b)). The radiator is composed of three stacked trapezoids fed by a 50 Ohm microstrip line (Yang *et al.*, 2008; Koziel *et al.*, 2014c). The antenna is implemented on a Taconic RF-35 dielectric substrate ($\varepsilon_r = 3.5, \tan\delta = 0.0018, h = 0.762\,\text{mm}$). Design

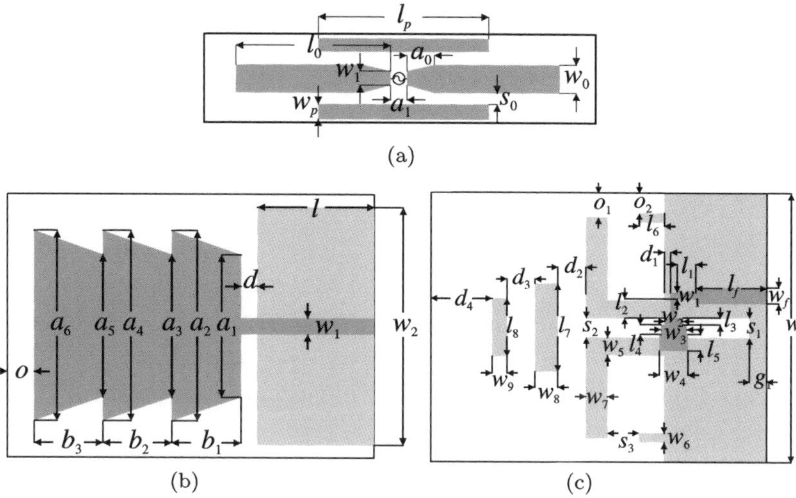

Fig. 10.1. Antenna structures considered in multi-objective optimization algorithm scalability study: (a) six-variable UWB dipole (Spence and Werner 2006), (b) 13-variable UWB monopole (Yang *et al.*, 2008), (c) 24-variable quasi-Yagi antenna (Wu *et al.*, 2014).

variables are $\boldsymbol{x} = [a_1\ a_2\ a_3\ a_4\ a_5\ a_6\ b_1\ b_2\ b_3\ w_2\ l\ d\ o]^T$, whereas $w_1 = 1.7\,\text{mm}$ to ensure the 50 Ohm input impedance. Both the high-fidelity model $\boldsymbol{f}$ and the low-fidelity model $\boldsymbol{c}$ are simulated using the transient solver of CST Microwave Studio (CST, 2013). Average evaluation time of the former is 10 min (~2,500,000 mesh cells), whereas the simulation time of the $\boldsymbol{c}$ model is 22 s (~33,600 mesh cells). The following design objectives are considered: F_1 — minimization of reflection $|S_{11}| \leq -10\,\text{dB}$ for 3.1–10.6 GHz band of interest, and F_2 — reduction of antenna size defined as $A \times B$ rectangle, where $A = l + d + b_1 + b_2 + b_3 + o$ and $B = w_2 + o$.

The last structure is a planar quasi-Yagi antenna (Fig. 10.1(c)) (Wu *et al.*, 2014; Koziel *et al.*, 2015c). As the previous structure, it is also implemented on a 0.762 mm thick Taconic RF-35 substrate. The antenna is constructed using a dipole driven element excited by a coplanar stripline (CPS) and two directors. The dipole is fed through a microstrip-to-CPS transition. The antenna models are implemented and simulated in CST Microwave Studio (CST, 2013). The average evaluation time of the high-fidelity model (~2,100,000 mesh cells) is 13 min. The low-fidelity model simulation time is 40 s (~80,000 mesh cells). The design objectives are: F_1 — minimization of reflection and F_2 — maximization of the antenna average gain. Both are considered within the 4–10 GHz frequency band (only solutions with $|S_{11}| \leq -10$ dB are acceptable).

10.1.2 *Numerical Results*

All antenna structures have been optimized using the methodology of Sec. 7.1 with design space reduction scheme of Sec. 8.2 (Koziel and Bekasiewicz, 2016k).

For the first antenna structure, the initial design space is defined by $\boldsymbol{l} = [10\ 5\ 0.5\ 2\ 1\ 0.1]^T$ and $\boldsymbol{u} = [30\ 20\ 5\ 20\ 15]^T$. The reduced lower/upper bounds are $\bar{\boldsymbol{l}}^* = [18\ 7.96\ 0.5\ 12.8\ 4.01\ 1.08]^T$ and $\bar{\boldsymbol{u}}^* = [18.7\ 12.98\ 0.53\ 13.72\ 8.45\ 1.54]^T$, which results in the reduction factor of 10^6 (volume-wise). The total cost of the corresponding two single-objective optimization runs is 350 evaluations of the coarse-mesh electromagnetic (EM) model $\boldsymbol{c}$ and four high-fidelity simulations for the extreme points refinement. A Kriging interpolation model $\boldsymbol{s}$ has been constructed within a reduced design space using a total of 500 $\boldsymbol{c}$ samples (423 obtained from Latin hypercube sampling (LHS) (Beachkofski and Grandhi, 2002), supplemented with 64 space corners and 13 points generated using a star-distribution scheme (Koziel *et al.*, 2011a)). The Pareto-optimal set obtained by optimizing the $\boldsymbol{s}$ model using MOEA is shown in Fig. 10.2. The high-fidelity representation of the Pareto front (Fig. 10.2) has been obtained using 10 designs selected along the initial Pareto-optimal set. The obtained minimum in-band $|S_{11}|$ level is –13.6 dB, whereas the smallest size that only slightly violates the requirements upon acceptable reflection is 792 mm^2. The variation of the antenna reflection and footprint along the Pareto front is 3.6 dB and 21 percent, respectively.

The aggregated cost of the optimization process corresponds to 57.4 evaluations (~19.1 h of the CPU-time) of the high-fidelity model (850$\times \boldsymbol{c}$, and 34 $\times \boldsymbol{f}$ required for correction of the extreme designs and

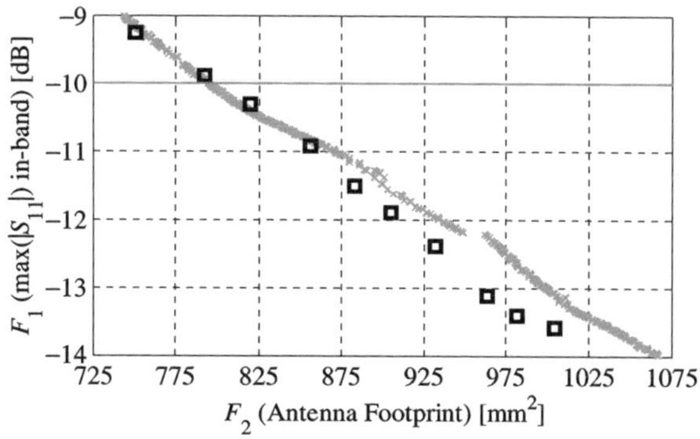

Fig. 10.2. Planar UWB dipole antenna: Pareto set obtained by optimization of the surrogate model (×) and the corresponding $\boldsymbol{f}$ representation (□) (Bekasiewicz *et al.*, 2014a).

the Pareto-optimal set refinement), which is negligible comparing to the estimated cost of direct multi-objective optimization of the high-fidelity model $\boldsymbol{f}$ (around 25,000 evaluations). The cost breakdown is shown in Table 10.1.

For the second antenna structure, the initial bounds for the design parameters are $\boldsymbol{l} = [5\ 5\ 5\ 5\ 5\ 5\ 1\ 1\ 1\ 0.2\ 8.0\ 20.0\ 5.0]^T$ and $\boldsymbol{u} = [25.0\ 25.0\ 25.0\ 25.0\ 25.0\ 25.0\ 15.0\ 15.0\ 15.0\ 2.0\ 15.0\ 40.0\ 10.0]^T$. The reduced design space is $\bar{\boldsymbol{l}}^* = [10.07\ 21.63\ 22.2\ 21.0\ 20.8\ 22.7\ 3.2\ 3.8\ 12.32\ 0.57\ 8.3\ 22.07\ 5.0]^T$ and $\bar{\boldsymbol{u}}^* = [11.3\ 21.96\ 24.30\ 24.15\ 21.27\ 24.6\ 3.9\ 4.0\ 13.08\ 0.74\ 11.2\ 39.35\ 5.75]^T$. It is 14 orders of magnitude smaller (volume-wise) than the initial one. Moreover, the ranges for most of the parameters are considerably reduced (Koziel *et al.*, 2014c). The Kriging interpolation model $\boldsymbol{s}$ is generated using a total of 1,500 low-fidelity model training samples allocated using LHS. The initial Pareto-optimal set has been generated using the MOEA algorithm (Fig. 10.3).

The high-fidelity Pareto front representation consisting of 10 designs upon refinement is shown in Fig. 10.3. The minimum antenna footprint that satisfies the requirements upon reflection is 1,134 mm^2, whereas the minimum in-band reflection corresponds to –14.9 dB.

Table 10.1 Optimization cost breakdown for the three antenna design cases

Design case	Design space dimension	Design stage	Number of model evaluations	CPU time Absolute [min]	CPU time Relative to f	CPU time Total [f]	CPU time Total [h]
1	6	Design Space Reduction	$350 \times c + 4 \times f$	255	12.8	57.4	19.1
		Kriging Model Construction	$500 \times c$	250	12.5		
		Finding Initial Pareto Set	$25{,}000 \times s$	42	2.1		
		Pareto Set Refinement	$30 \times f$	600	30		
2	13	Design Space Reduction	$800 \times c + 4 \times f$	333	33.3	122.1	20.4
		Kriging Model Construction	$1{,}500 \times c$	550	55		
		Finding Initial Pareto Set	$25{,}000 \times s$	38	3.8		
		Pareto Set Refinement	$30 \times f$	300	30		
3	24	Design Space Reduction	$2{,}840 \times c + 4 \times f$	1,893	145.6	322.9	69.9
		Kriging Model Construction	$2{,}800 \times c$	1,867	143.6		
		Finding Initial Pareto Set	$25{,}000 \times s$	48	3.7		
		Pareto Set Refinement	$30 \times f$	390	30		

Moreover, the minimum size of the antenna, which satisfies requirements upon reflection, is almost 20 percent smaller than the structure optimized with respect to $|S_{11}|$.

The total computational cost of multi-objective design of the considered UWB monopole antenna corresponds to 122 evaluations of the high-fidelity model (~20.4 h of the CPU time). The cost includes: $800\times \boldsymbol{c}$ and $4\times \boldsymbol{f}$ evaluations for the design space reduction, $1,500\times \boldsymbol{c}$ samples acquired to form the training set for identification of the response surface approximation (RSA) model, and 30 $\boldsymbol{f}$ evaluations for the refinement of the 10 selected antenna designs (see Table 10.1 for a detailed cost breakdown). The above cost is negligible compared

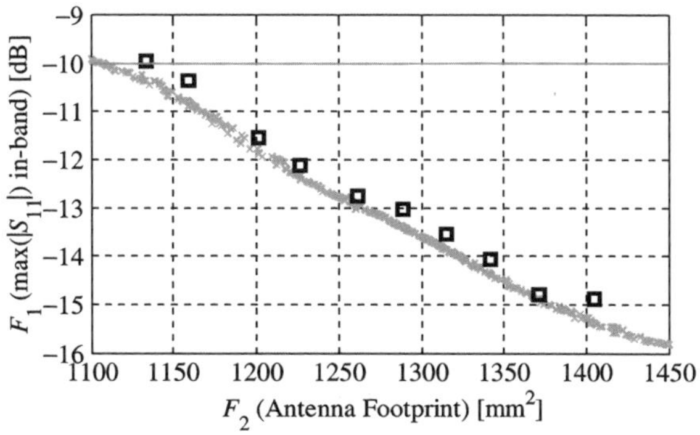

Fig. 10.3. UWB monopole antenna: Pareto set obtained by optimization of the surrogate model (×) and the corresponding $\boldsymbol{f}$ representation (□) (Koziel *et al.*, 2014c).

to the estimated CPU-time required by 25,000 evaluations during direct MOEA optimization of the high-fidelity model $\boldsymbol{f}$.

For the last test case, the initial parameter set is $\boldsymbol{x}^{(0)} = [w_0\ d_1\ d_2\ g_1\ l_1\ l_2\ l_3\ l_4\ l_6\ l_7\ l_8\ l_f\ o_1\ o_2\ s_1\ s_2\ s_3\ w_1\ w_2\ w_3\ w_4\ w_5\ w_7\ w_8]^T$. The parameter $l_5 = 3l_3$, whereas $d_3 = 3, d_4 = 6.5, w_f = 1.7, w_6 = 1$ are fixed (dimensions in mm). The initial design bounds $\boldsymbol{l} = [20\ 0\ 1\ 1\ 0.1\ 1\ 0.2\ 0.2\ 1\ 5\ 5\ 5\ 0\ 0\ 1\ 1\ 1\ 0.2\ 1.2\ 2\ 2.9\ 0.5\ 0.5\ 0.5]^T$ and $\boldsymbol{u} = [40\ 1\ 5\ 5\ 4\ 4\ 1.2\ 1.2\ 5\ 13\ 13\ 13\ 2\ 2\ 4\ 4\ 4\ 1.2\ 2.2\ 3\ 3.9\ 2.5\ 2.5\ 2.5]^T$ are considerably narrowed down after performing the design space reduction procedure.

In particular, some of the variables assume the same value for both extreme designs along the Pareto front, i.e., $l_f = 7.5, w_1 = 0.7, w_0 = 28.86, l_4 = 0.58, o_2 = 0.13, d_2 = 3, l_7 = 9.47$ and $l_8 = 6.21$ (Koziel *et al.*, 2015c). Consequently, these variables are fixed in the later steps of the design process. The reduced solution space is confined to 16 independent parameters: $\boldsymbol{x} = [d_1\ g_1\ l_1\ l_2\ l_3\ l_6\ o_1\ s_1\ s_2\ s_3\ w_2\ w_3\ w_4\ w_5\ w_7\ w_8]^T$. The design space reduction rate is 10^{18} (volume-vise). The Kriging interpolation model was constructed using 2,000 $\boldsymbol{c}$ samples allocated using LHS algorithm. The set of 10 selected Pareto-optimal designs have been refined using the algorithm of Sec. 7.1. Figure 10.4 shows the comparison between low- and high-fidelity model representations of the Pareto front. It

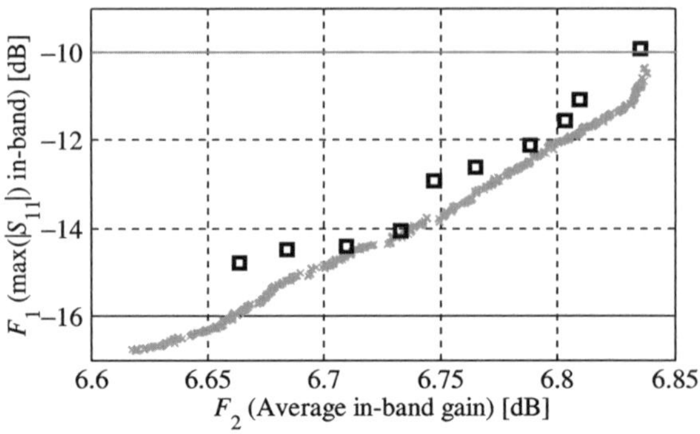

Fig. 10.4. Quasi-Yagi antenna: Pareto set obtained by means of the surrogate model optimization (×) and the corresponding $\boldsymbol{f}$ representation (□) (Koziel *et al.*, 2015c).

should be noted that the design with the highest average gain of 6.83 dB exhibits the worst acceptable in-band reflection (–10 dB). The average gain varies by 0.17 dB along the Pareto front. The solution that exhibits the lowest in-band reflection of –14.8 dB features the smallest gain (reflection variability is 4.8 dB).

The computational cost of the multi-objective antenna design corresponds to about 323 $\boldsymbol{f}$ simulations (~70 h of the CPU-time), cf. Table 10.1. The initial design space reduction required $2840 \times \boldsymbol{c}$ and $4 \times \boldsymbol{f}$ simulations for the optimization and correction of the extreme Pareto-optimal designs, respectively. The RSA model setup required 2,800 $\boldsymbol{c}$ evaluations, whereas the refinement of the 10 selected low-fidelity Pareto-optimal designs required 30 $\boldsymbol{f}$ simulations. The estimated cost of direct optimization of discussed quasi-Yagi antenna is 25,000 $\boldsymbol{f}$ model evaluations (over 260 days of the CPU-time).

10.1.3 *Analysis of the Algorithm Scalability*

The detailed breakdown of the multi-objective optimization cost for the three antenna design cases presented in Sec. 10.1.2 has been gathered in Table 10.1. Apart from the cost of each design stage (design space reduction, construction of the data-driven surrogate model, identification of the initial Pareto set and its refinement), the table contains information about the overall cost. The basis

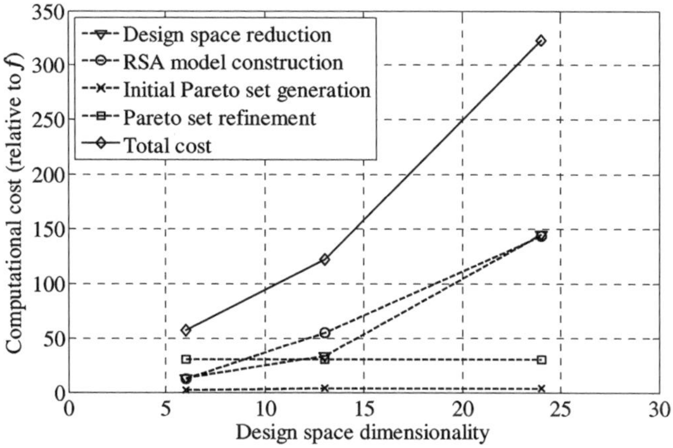

Fig. 10.5. Computational cost of multi-objective antenna optimization versus the problem dimensionality. The cost is expressed in terms of the corresponding number of high-fidelity model evaluations. The plots indicate the overall cost as well as the costs of various stages of the design process (Bekasiewicz and Koziel, 2016k).

for the subsequent discussion is the relative cost expressed in terms of the corresponding number of high-fidelity model evaluations for the respective antenna structure. Figure 10.5 shows a graphical representation of the relevant data, in particular, the dependence of the individual design stage cost as well as the total cost on the problem dimensionality.

The first and obvious observation is that both the cost of the initial Pareto set generation (i.e., MOEA-based optimization of the Kriging surrogate) and the Pareto set refinement are almost independent of the number of antenna geometry parameters. This is due to the way both of these steps are realized (the same MOEA setup for all the problems and the same number of the designs to be refined, as well as a very low cost of the surrogate model evaluation compared to that of the high-fidelity EM model).

The two most expensive steps are design space reduction and data acquisition for surrogate model construction. The reduction phase requires single-objective optimizations (using SBO techniques). At this stage, a pattern search algorithm is utilized (Kolda *et al.*, 2003), computational complexity of which is more or less $O(n^2)$

where n is the number of design parameters. The cost of data acquisition for RSA model construction grows very quickly with the problem dimensionality, despite the fact that the design space is reduced beforehand. Space reduction limits the ranges of the design variables (critical for feasible construction of the surrogate) but not the space dimensionality in general. Still, for the third test case, the effective number of variables has been also reduced (from 24 to 16) because the updated lower and upper bounds were found to be virtually the same for some of the parameters.

On the other hand, the aforementioned and fast increase of the cost of the space reduction and RSA model construction refer to low-fidelity model evaluations. Consequently, the overall cost of the multi-objective optimization increases more or less in a quadratic fashion as indicated in Fig. 10.5. Thus, the scalability of the method can be considered as very good from the practical standpoint, at least for the considered range of design problem dimensionalities.

Further reduction of the cost (and, consequently, better scalability properties) can be achieved by (i) applying more efficient algorithms for low-fidelity model optimization in the first stage of the design process, and (ii) better design space reduction schemes that aim at more precise location of the Pareto front in the feature space of the problem (cf. Secs. 8.3 and 8.4). The latter would lead to limiting the number of the training samples for RSA model construction. With respect to (i) it seems that the use of sequential approximation optimization methods (Koziel and Ogurtsov, 2013a; Leifsson and Koziel, 2015a) could result in improved efficiency of low-fidelity model optimization.

10.2 Statistical Analysis of Multi-Objective Evolutionary Algorithm-Based Optimization with Kriging Surrogates

The optimization procedure of Sec. 7.1 exploits — to generate initial approximation of the Pareto front — MOEA. In this section, we investigate the effect of limited repeatability of MOEA outcome on the quality of the produced Pareto front (Koziel and Bekasiewicz,

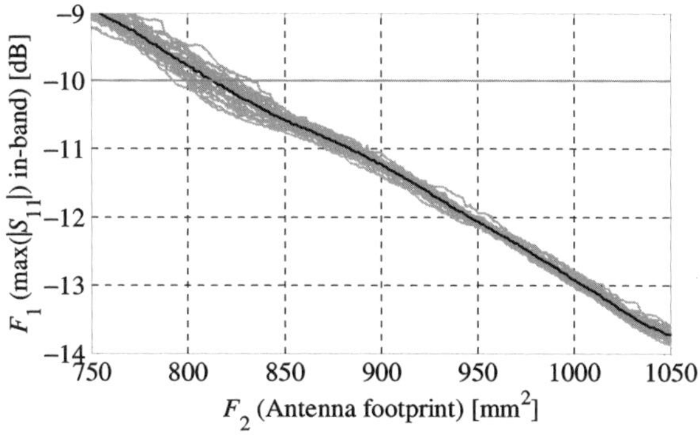

Fig. 10.6. UWB dipole antenna: black and gray lines represent Pareto-optimal sets obtained in 30 runs of MOEA and the mean value (Koziel and Bekasiewicz, 2016k).

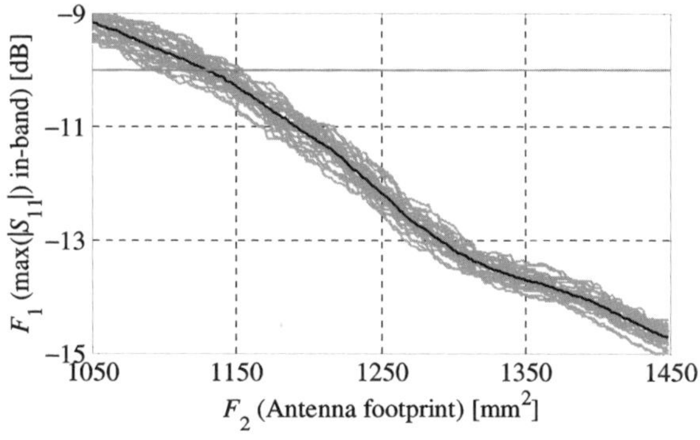

Fig. 10.7. UWB monopole antenna: black and gray lines represent Pareto-optimal sets obtained in 30 runs of MOEA and the mean value (Koziel and Bekasiewicz, 2016k).

2016k). We utilize the same three antennas as considered in Sec. 10.1. Before discussing the numerical results, it should be emphasized that the quality of the Pareto set at this stage is not critical, since the designs selected from it are merely utilized as starting points for further SBO-based refinement to the high-fidelity model level. Notwithstanding, the stochastic nature of the algorithm influences the optimization process.

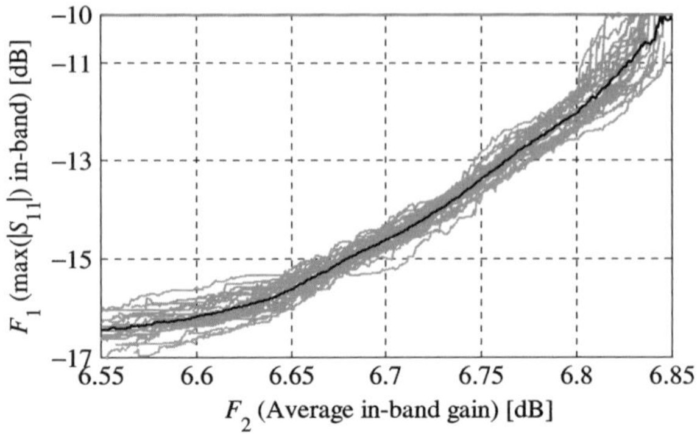

Fig. 10.8. Quasi-Yagi antenna: Pareto-optimal sets obtained in 30 runs of MOEA (gray lines) and the mean value (black line) (Koziel and Bekasiewicz, 2016k).

Table 10.2 MOEA statistics for the antenna design cases of Sec. 10.2.1

Design case	Average standard deviation	Average distance from the mean Pareto front	Worst case peak distance from the mean Pareto front
1	0.059	0.073	0.426
2	0.097	0.147	0.546
3	0.164	0.214	1.559

Figures 10.6–10.8 show the Pareto fronts obtained for 30 independent runs of MOEA for all three antennas. For the first test case (UWB dipole), the variability of the fronts is minor. The largest discrepancy between the individual Pareto sets and the mean value is around 0.4 dB which is considered negligible from the antenna operation standpoint. Moreover, the standard deviation of the results and average distance from the mean value are 0.06 dB and 0.07 dB, respectively. For that reason, the initial Pareto set has negligible influence on the final results obtained in the subsequent refinement stage. The detailed data statistics are gathered in Table 10.2.

For the second test case (UWB monopole), the largest discrepancy between the Pareto sets and the mean value is around 0.55 dB, whereas standard deviation and average distance from the mean value are 0.1 dB and 0.14 dB, respectively. Similarly as for the previous example, this level of variability is considered insignificant from the point of view of subsequent design refinement. Finally, in the last test case (quasi-Yagi antenna), the worst case peak distance from the mean value is 1.55 dB. Moreover, the standard deviation is only 0.16 dB and the average distance from the mean value is only 0.2 dB. Although the Pareto front variability is considerably larger than for the previous cases, its influence on the design refinement stage can still be neglected.

10.3 Patch Size Setup Trade-Offs for Sequential Domain Patching Algorithm

The SDP algorithm of Sec. 7.3 is fully automated but requires setup of the number of intervals along each dimension of the design space (a vector $\boldsymbol{m}$). This setup influences the cost of the optimization process but it may also affect the algorithm performance because the resolution of the search depends on the values of the vector $\boldsymbol{m}$ components. In Sec. 7.3.3, a sensitivity-based procedure was described for automated determination of patch sizes.

Here, we investigate how the patch size setup influences the SDP algorithm performance. We consider various scenarios including the sensitivity-based one (for various values of $m_{\max}$; cf. Sec. 7.3.3) and uniform (also for various number of intervals along the search space dimensions).

10.3.1 *Test Cases*

The first test case is a UWB monopole antenna shown in Fig. 10.9(a). The structure is designed with Taconic RF-35 substrate (h = 0.762 mm, $\varepsilon_r = 3.5, \tan\delta = 0.0018$). The antenna consists of a circular patch fed through a 50 Ohm microstrip line and modified ground plane with L-shaped strip (Li *et al.*, 2012). The design variables are: $\boldsymbol{x}_I = [w_0\ l_1\ l_2\ l_3\ l_4\ w_1\ w_2\ w_3\ r\ o_r]^T$, whereas parameters $w_f = 1.7, l_f = 10$ and $o = 0.5w_0 \cdot r \cdot o_r$ (all dimensions in mm). Antenna models are implemented in CST (CST, 2013) ($\boldsymbol{f}$: ~2,600,000 mesh cells, simulation time 24 min; $\boldsymbol{c}$: ~370,000 cells, 150

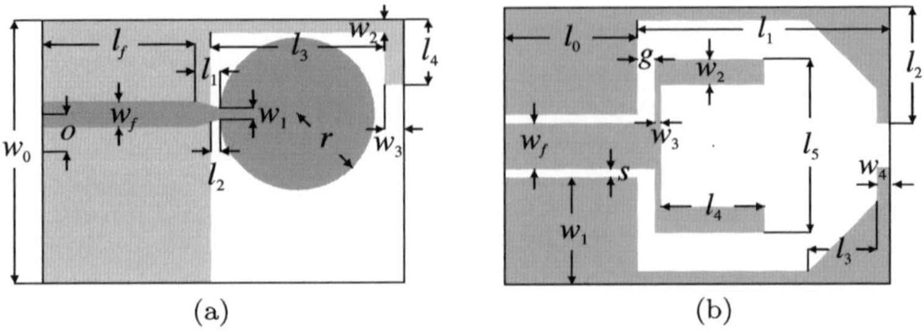

Fig. 10.9. Test cases used to analyze the SDP algorithm performance for various patch size setups: (a) UWB monopole (Li *et al.*, 2012), (b) uniplanar antenna (Qing and Chen, 2009).

s). The variable ranges are: $\boldsymbol{l} = [15\ 2\ {-2}\ 10\ 3\ 0.2\ 0.2\ 0.2\ 4.5\ {-1}]^T$ and $\boldsymbol{u} = [30\ 8\ 2\ 15\ 8\ 1.7\ 1.2\ 1.2\ 8.5\ 1]^T$. We consider two objectives: F_1 — minimization of reflection in 3.1–10.6 GHz band and F_2 — minimization of footprint. The latter is defined as $A(\boldsymbol{x}) = w_0 \cdot \max\{l_f + l_1 + 2r - 0.05r, l_f + l_1 - 0.05r + l_2 + l_3 + w_3\}$.

The second test case is a uniplanar antenna shown in Fig. 10.9(b). The structure consists of a driven element in the form of fork-shaped radiator fed through a coplanar waveguide and an open slot (Qing and Chen, 2009). Geometry ensures wideband impedance which covers UWB bandwidth. The design variables are $\boldsymbol{x} = [l_0\ l_1\ l_{2r}\ l_{3r}\ l_4\ l_5\ w_1\ w_2\ w_3\ w_4\ g]^T$. Parameters $w_f = 3.5$ and $s = 0.16$ remain fixed to ensure 50 Ohm input impedance. Relative variables are: $l_2 = (0.5w_f + s + w_1) \cdot \max\}l_{2r},\ l_{3r}\}$ and $l_3 = (0.5w_f + s + w_1) \cdot l_{3r}$. The unit for all non-relative dimensions is mm. Two antenna models — implemented in CST Microwave Studio and evaluated using its time domain solver (CST, 2013) — are considered. The high-fidelity model $\boldsymbol{f}$ contains ~1,600,000 cells (simulation time 10 min). The low-fidelity model $\boldsymbol{c}$ consists of ~215,000 cells and its typical simulation time is 67 s. The ranges of design parameters given by the following bounds: $\boldsymbol{l} = [5\ 15\ 0.2\ 0.2\ 4\ 8\ 7\ 0.5\ 0.2\ 0.2\ 0.2]^T$ and $\boldsymbol{u} = [15\ 25\ 1\ 0.8\ 11\ 16\ 15\ 3.5\ 2.5\ 2\ 2]^T$ are determined based on engineering experience. The initial design $\boldsymbol{x}^{\text{init}} = [10\ 19\ 0.8\ 0.6\ 7.7\ 12\ 11.3\ 2\ 0.5\ 1\ 0.8]^T$ for single objective optimizations aimed at determination of the extreme Pareto-optimal

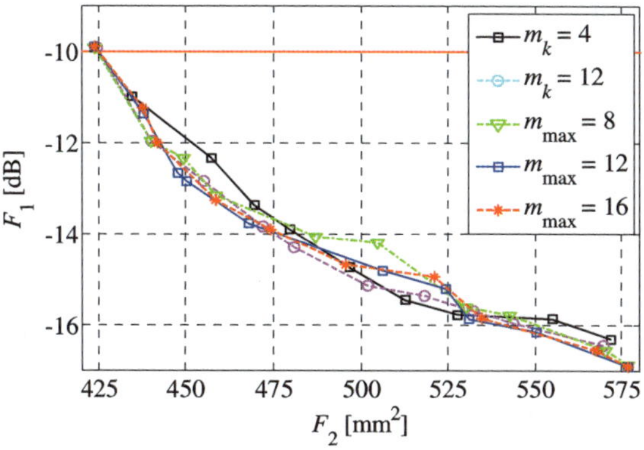

Fig. 10.10. UWB monopole antenna: Pareto sets obtained using SDP for five setups of the patch size vector $\boldsymbol{m}$. Note that m_k and $m_{\max}$ correspond to uniform and non-uniform setup of the patch size vector.

designs is based on (Qing and Chen, 2009). The following objectives are considered: F_1 — minimization of the reflection within 3.1–10.6 GHz band and F_2 — minimization of footprint defined as $A(\boldsymbol{x}) = (2(w_1 \cdot s) + w_f) \cdot l_0 \cdot l_1$.

10.3.2 *Numerical Results*

For the first test case, the SDP algorithm was used to generate initial Pareto sets using five different setups for $\boldsymbol{m} = [m_1 \ldots m_n]$: (i) $m_k = 5$, (ii) $m_k = 12, k = 1, \ldots, n$, (iii) $\boldsymbol{m} = [4\ 5\ 8\ 3\ 2\ 2\ 2\ 2\ 4\ 2]$, (iv) $\boldsymbol{m} = [6\ 7\ 12\ 5\ 3\ 2\ 3\ 2\ 5\ 3]$ and (v) $\boldsymbol{m} = [8\ 9\ 16\ 6\ 3\ 2\ 4\ 2\ 7\ 4]$ (the last three are sensitivity-based with $m_{\max} = 8, 12$ and 16; see also Sec. 7.3.3). The final Pareto sets found for all setups are shown in Fig. 10.10, see also Table 10.3.

For the second test case, the SDP algorithm has been executed using seven different setups for $\boldsymbol{m} = [m_1 \ldots m_n]$: (i) $m_k = 4$, (ii) $m_k = 8$, (iii) $m_k = 12, k = 1, \ldots, n$, (iv) $\boldsymbol{m} = [2\ 2\ 2\ 2\ 2\ 2\ 6\ 2\ 3\ 5\ 2]$, (v) $\boldsymbol{m} = [3\ 3\ 2\ 2\ 2\ 2\ 8\ 2\ 4\ 7\ 2]$, (vi) $\boldsymbol{m} = [4\ 4\ 2\ 2\ 2\ 2\ 12\ 2\ 6\ 10\ 2]$ and (vii) $\boldsymbol{m} = [5\ 6\ 2\ 2\ 2\ 2\ 16\ 2\ 8\ 13\ 2]$ (the last four are sensitivity-based with $m_{\max} = 6, 8, 12$ and 16, respectively). The final Pareto sets found for all setups are shown in Fig. 10.11; detailed data on their computational cost is gathered in Table 10.4.

Table 10.3 UWB monopole: cost of multi-objective optimization for various setups of the patch size vector $\boldsymbol{m}$

	Initial Pareto set		Design refinement			Total cost	
M setup	$\boldsymbol{c}$ evaluations	Time [h]	$\boldsymbol{c}$ evaluations	$\boldsymbol{f}$ evaluations	Time [h]	Absolute [h]	Relative to $\boldsymbol{f}$
(i)	319	22.1	100	30	16.1	38.2	95.5
(ii)	800	55.6	100	30	16.1	71.7	179.3
(iii)	198	13.7	100	30	16.1	29.8	74.5
(iv)	298	20.3	100	30	16.1	36.4	91.0
(v)	388	26.9	100	30	16.1	43.0	107.5

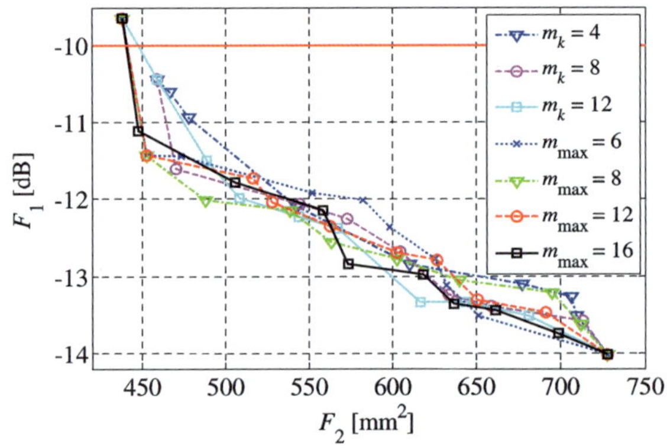

Fig. 10.11. Uniplanar antenna: Pareto sets obtained using SDP for seven setups of the patch size vector $\boldsymbol{m}$. Note that m_k and $m_{\max}$ correspond to uniform and non-uniform setup of the patch size vector.

10.3.3 *Discussion*

It can be observed that — for the first test case — all Pareto sets are of similar quality although those obtained for setups (ii), (iv) and (v) are more regular. At the same time, the optimization cost for the setup (ii) is considerably higher than for (iv) and (v) indicating that sensitivity-based patch size setup offers good quality at lower computational expenses.

For the second antenna, the results indicate that the patch size setup also has limited influence on the accuracy of the algorithm (the

Table 10.4 Uniplanar antenna: cost of multi-objective optimization for various setups of the patch size vector $\boldsymbol{m}$

	Initial Pareto set		Design refinement			Total cost	
M setup	$\boldsymbol{c}$ evaluations	Time [h]	$\boldsymbol{c}$ evaluations	$\boldsymbol{f}$ evaluations	Time [h]	Absolute [h]	Relative to $\boldsymbol{f}$
(i)	312	5.8	110	30	7.1	12.9	77.4
(ii)	618	11.5	110	30	7.1	18.6	111.6
(iii)	992	18.5	110	30	7.1	25.6	153.6
(iv)	167	3.1	110	30	7.1	10.2	61.2
(v)	201	3.7	110	30	7.1	10.8	64.8
(vi)	246	4.6	110	30	7.1	11.7	70.2
(vii)	284	5.3	110	30	7.1	12.4	74.4

discrepancy between Pareto sets along F_1 is below 1.5 dB) while it can noticeably contribute to the computational cost.

The Pareto sets obtained for setups (v) and (vii) seem to be the best with respect to the number of non-dominated $\boldsymbol{f}$ designs (6 of 10 in both cases). At the same time, the responses for setups (ii) and (vi) are the most regular among compared ones. The optimization cost for setups (v) and (vii) is 2.5 and 2.2 times lower than for (ii) which indicates that sensitivity-based patch size setup offers good quality at lower computational expenses. It should be also noted that patch size determines the ranges of perturbations for the local RSA model. For large $m_{\max}$, as in setup (iii), the patch size (which is a domain of the local RSA model constructed for design refinement purposes) is small, which limits the refinement range. The models built using perturbations obtained by means of sensitivity analysis may be better from this perspective because they cover larger space regions.

Overall conclusion is that the most suitable setup for SDP algorithm is a sensitivity-based one with a moderate value of $m_{\max}$. It should also be noted that the obtained results reconfirm the robustness of the SDP technique.

10.4 Summary

In this chapter, we investigated various properties of the surrogate-assisted multi-objective design optimization procedures considered in this book. The first aspect was scalability of the optimization algorithm exploiting RSA surrogates (Sec. 7.1). The presented results indicate that the computational cost of the method grows relatively slow (at most as n^2 with n being the design space dimensionality) and can potentially be reduced by exploiting more efficient methods for design space reduction and identification of the extreme Pareto-optimal designs.

The second issue investigated here was related to the effect of MOEA performance on the quality of the Pareto front found by the optimization algorithm (Sec. 7.2). The numerical results obtained for three test cases indicate that the influence of MOEA is minor (and can be further reduced by the refinement procedure).

Finally, sensitivity of the SDP algorithm (Sec. 7.3) to the patch size setup has been investigated using two antenna examples and multiple patch size setups, including both sensitivity-based and uniform arrangements. The results indicate that the quality of the Pareto front is weakly dependent on the number of patches between the extreme Pareto-optimal designs and, therefore, the algorithm setup that leads to lower computational cost is generally preferred.

All of the results presented in this chapter essentially confirm the robustness of the optimization methods considered in the book, which is encouraging, particularly from the point of view of applying these techniques for even more demanding antenna design problems. Obviously, the list of issues that needs further investigation is not limited to those discussed in this chapter. One of the important topics is dependence of the algorithm performance (both cost- and quality-wise) on the low-fidelity EM antenna model. The latter is the foundation of all the presented methods; its accuracy and speed is likely to affect both the computational cost and the quality of the Pareto front generated by the algorithm. This and other practical issues will be considered in the future work.

Chapter 11

Applications in Other Engineering Disciplines

In this chapter, we discuss applications of surrogate-assisted optimization methods considered in this book for solving multi-objective design problems in other engineering areas. A common issue is a high cost of computational models that prevents the straightforward applications of the off-the-shelf algorithms such as population-based metaheuristics. On the other hand, utilization of the methods described in Chapters 7 and 8 along with variable-fidelity simulations allows for yielding the Pareto-optimal designs in practically acceptable timeframes. Three specific case studies are presented, including multi-objective design of impedance matching transformer using sequential domain patching (SDP), optimization of compact microstrip coupler by means of Pareto front exploration, and two-objective design of transonic airfoils using response surface surrogates and co-Kriging.

11.1 Multi-Objective Design of Impedance Matching Transformers

In this section, application of SDP technique (Sec. 7.3) to multi-objective optimization of miniaturized impedance matching transformers has been discussed (Bekasiewicz *et al.*, 2016a). We begin by explaining the concepts behind microwave circuit miniaturization exploited here. Description of the transformer structure utilized as a working example is followed by presentation of the numerical results and comparison of SDP with other optimization techniques.

11.1.1 *Compact Microwave Circuits: Design Challenges*

Many classes of passive circuits such as filters, couplers, or impedance transformers are characterized by large dimensions which limit their usefulness for modern microwave systems. At the same time, due to their modular architecture, they can be decomposed into simpler sub-components, in particular, transmission line (TL) sections. This permits circuit miniaturization by replacing its conventional blocks by certain artificial sections, such as metamaterials (Buell *et al.*, 2006), defected ground structures (Kurgan and Kitlinski, 2009) and compact microstrip resonant cells (Kurgan *et al.*, 2012). The last option is especially interesting due to single layer implementation and considerable geometrical flexibility.

Example realizations of compact microstrip resonant cells (CMRCs) are shown in Fig. 11.1. CMRC offers shortening of the physical length of the section compared to TL at the expense of increased width. The cells are universal in the sense that they are applicable for a wide range of circuits. Also, they allow for obtaining size reduction rates of up to over 90 percent (Bekasiewicz and Kurgan, 2014; Tseng and Chen, 2008). Design miniaturization using CMRCs is illustrated in Fig. 11.2.

Circuit miniaturization by means of TL-to-CMRC replacement incurs certain design challenges. A typical CMRC has four or more variables (transmission and coupled lines have two and three parameters, respectively) which makes the design more complex in terms of finding the optimum geometry. Also, miniaturized designs exhibit complex topologies with significant electromagnetic (EM) cross-couplings. Accurate evaluation of such structures requires numerically expensive EM analysis (Koziel *et al.*, 2016a). Finally, CMRC-based design miniaturization involves seeking for a trade-off between

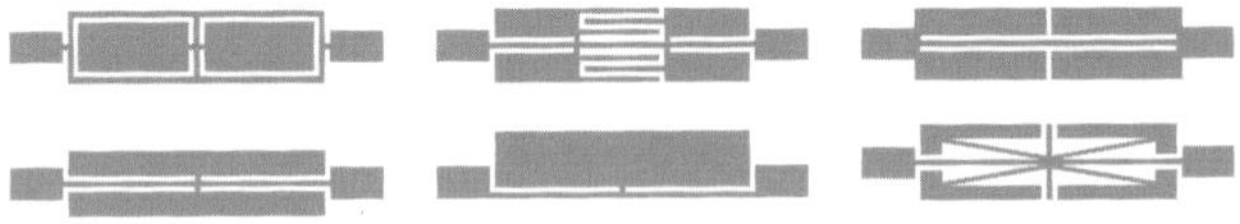

Fig. 11.1. CMRCs — typical realizations (Kurgan *et al.*, 2012).

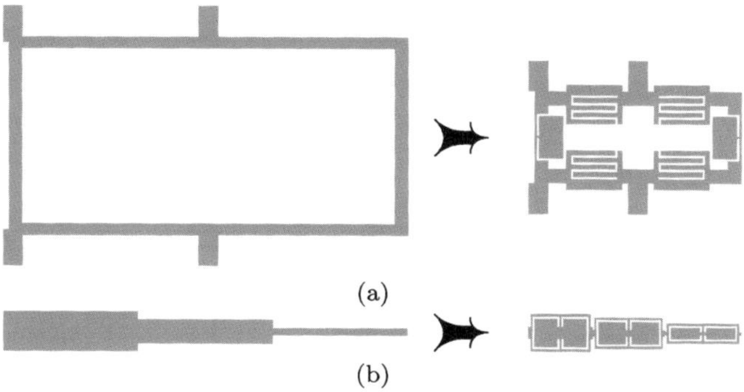

Fig. 11.2. Design miniaturization using CMRCs: (a) compact rat-race coupler (RRC) with two different cells; (b) impedance transformer obtained as a cascade connection of CMRCs.

the circuit size and its electrical performance. For circuits such as impedance transformers or couplers, reduction of geometry results in distortion of the frequency response, increase of the return loss or degradation of bandwidth (BW) (Bekasiewicz and Kurgan, 2014; Koziel *et al.*, 2016a). Finding acceptable compromises between the aforementioned figures of merit requires multi-objective optimization.

11.1.2 *Transformer Structure and Models*

Consider a compact broadband impedance transformer shown in Fig. 11.3 (Bekasiewicz *et al.*, 2016a). The structure contains three CMRCs and it is supposed to match the 50 Ohm source to the 130 Ohm load. The dielectric substrate is a 0.762 mm thick Taconic RF-35 ($\varepsilon_r = 3.5$; $\tan\delta = 0.0018$). Design is represented by a 15-variable vector $\boldsymbol{x} = [w_{1.1}\ w_{2.1}\ w_{3.1}\ l_{1.1}\ l_{2.1}\ w_{1.2}\ w_{2.2}\ w_{3.2}\ l_{1.2}\ l_{2.2}\ w_{1.3}\ w_{2.3}\ w_{3.3}\ l_{1.3}\ l_{2.3}]^T$. Parameters $w_{i1} = 1.7$ and $w_{i2} = 0.17$ ensure desired input and output impedances, whereas $l_0 = 1$. The unit for all dimensions is mm.

The transformer is implemented in CST Microwave Studio and simulated using its time domain solver (CST, 2013). The high-fidelity model $\boldsymbol{f}$ consists of ~1,200,000 mesh (simulation time 20 min). The low-fidelity model $\boldsymbol{c}$ contains ~120,000 cells (simulation time 92 s).

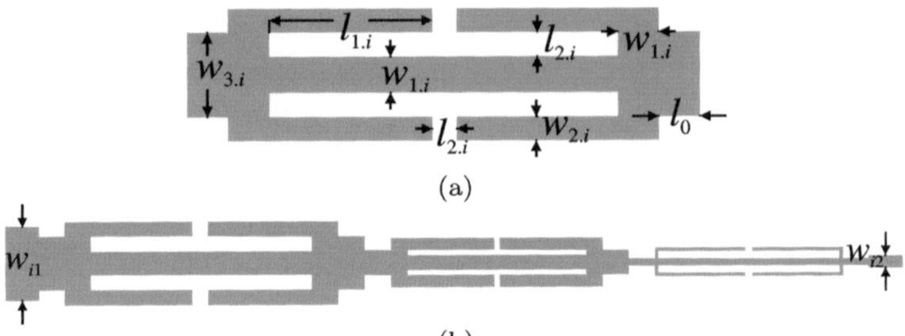

Fig. 11.3. Compact impedance transformer: (a) CMRC section with highlighted geometrical details (i denotes the index of the section in which the cell is implemented); and (b) transformer geometry (Bekasiewicz *et al.*, 2016a).

We consider two design objectives: minimization of the maximum in-band return loss defined as $F_1(\boldsymbol{x}) = \max(|S_{11}|_{2\text{–}5.5\,\text{GHz}})$ and miniaturization of the circuit area $F_2(\boldsymbol{x}) = A \times B$, where $A = \sum_k (2\cdot(l_{1.k}+w_{1.k}+l_0)+l_{2.k})$ and $B = \max(2\cdot(l_{2.k}+w_{2.k})+w_{1.k})$, where $k = 1, 2, 3$. Note that only the solutions for which $F_1(\boldsymbol{x}) \leq -10$ dB are of interest.

11.1.3 *Results and Comparisons*

The transformer has been optimized using the SDP algorithm of Sec. 7.3. The extreme Pareto designs $\boldsymbol{x}_1^* = [0.27\ 0.1\ 0.96\ 2.06\ 0.14\ 0.11\ 0.18\ 1\ 2.17\ 0.12\ 0.11\ 0.13\ 0.12\ 2.07\ 0.2]^T$ and $\boldsymbol{x}_2^* = [0.79\ 0.44\ 0.32\ 2.63\ 0.17\ 0.33\ 0.35\ 0.17\ 2.48\ 0.22\ 0.16\ 0.13\ 0.32\ 3.32\ 0.16]^T$ have been obtained using (7.13). The SDP algorithm was executed using the following vector of intervals: $M = [16\ 14\ 7\ 8\ 2\ 15\ 7\ 6\ 6\ 5\ 9\ 2\ 4\ 10\ 3]$, obtained using the procedure of Sec. 7.3.3. Subsequently, the selected Pareto designs have been refined using method of Sec. 7.1.3.

The results shown in Fig. 11.4 indicate that the F_1 and F_2 vary along the Pareto front by 11.2 dB and 24.6 mm^2 (over 50 percent). The geometrical details of selected high-fidelity Pareto-optimal designs are gathered in Table 11.1, whereas their corresponding frequency responses are shown in Fig. 11.5. For the sake of comparison, the Pareto front has been also obtained using procedure of Sec. 7.1. The results shown in Fig. 11.4 indicate that Pareto sets obtained

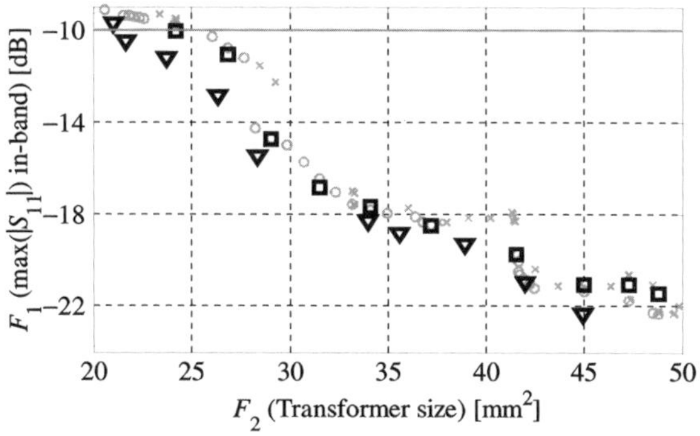

Fig. 11.4. Optimization results of the compact impedance transformer. Dominated and Pareto-optimal low-fidelity model solutions obtained using SDP the algorithm are denoted by (×) and (○), respectively. High-fidelity designs based on SDP algorithm and benchmark method of Sec. 7.1 are marked using (□) and (∇), respectively.

using both procedures are similar (vertical misalignment between them is below 1.5 dB). At the same time, the range of the front obtained using method of Sec. 7.1 along F_2 is slightly narrower due to limited exploration capabilities of the population-based algorithm in the vicinity of the search space corners.

The computational cost of multi-objective design optimization corresponds to about 173 high-fidelity model simulations (~58 h of the CPU-time) and it includes: 750 low-fidelity model simulations for single-objective optimizations, 30 evaluations of $\boldsymbol{c}$ for determination of patching intervals and 910 simulations of $\boldsymbol{c}$ for optimization using the SDP algorithm, respectively, as well as 310 and 20 evaluations of the low- and high-fidelity model for the refinement of the selected design objectives. The cost of the metaheuristic-based algorithm corresponds to ~241 evaluations of the high-fidelity model (~81 h of the CPU-time) which is almost 40 percent higher than for the SDP algorithm.

11.1.4 *Discussion*

The results presented in this section indicate that both deterministic (here, SDP) and metaheuristic-based algorithms enable very efficient optimization of miniaturized impedance transformer circuits. The

Table 11.1 Impedance transformer: selected Pareto-optimal designs

Design	Design parameters														
	$w_{1.1}$	$w_{2.1}$	$w_{3.1}$	$l_{1.1}$	$l_{2.1}$	$w_{1.2}$	$w_{2.2}$	$w_{3.2}$	$l_{1.2}$	$l_{2.2}$	$w_{1.3}$	$w_{2.3}$	$w_{3.3}$	$l_{1.3}$	$l_{2.3}$
$\boldsymbol{x}_f^{(1)}$	0.33	0.25	0.1	2.2	0.14	0.2	0.20	0.17	2.2	0.12	0.13	0.13	0.2	2.6	0.2
$\boldsymbol{x}_f^{(3)}$	0.53	0.25	0.4	2.2	0.14	0.2	0.35	0.17	2.2	0.16	0.13	0.13	0.2	2.6	0.2
$\boldsymbol{x}_f^{(5)}$	0.73	0.25	0.4	2.2	0.14	0.2	0.35	0.17	2.2	0.22	0.14	0.13	0.2	2.6	0.2
$\boldsymbol{x}_f^{(7)}$	0.76	0.27	0.4	2.2	0.14	0.2	0.35	0.17	2.2	0.22	0.14	0.13	0.2	2.9	0.2
$\boldsymbol{x}_f^{(9)}$	0.76	0.39	0.4	2.4	0.14	0.3	0.35	0.17	2.3	0.22	0.16	0.15	0.2	3.2	0.3

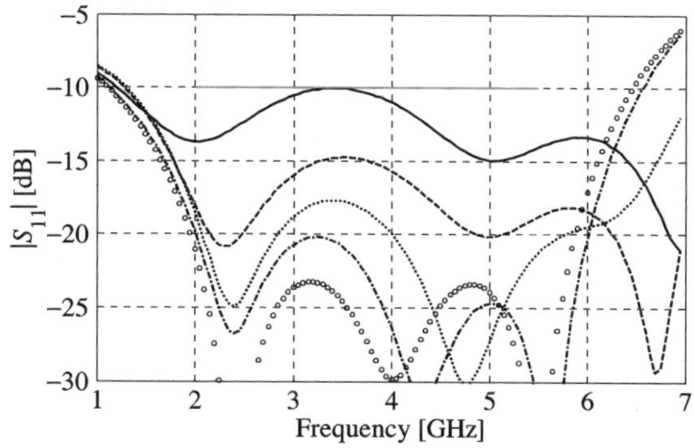

Fig. 11.5. Frequency responses of the high-fidelity model designs listed in Table 11.1: $\boldsymbol{x}_f^{(1)}$ – (—), $\boldsymbol{x}_f^{(3)}$ – (– –), $\boldsymbol{x}_f^{(5)}$ – (···), $\boldsymbol{x}_f^{(7)}$ – (–·), $\boldsymbol{x}_f^{(9)}$ – (∘).

cost of SDP is lower than that of Sec. 7.1, but the quality of the Pareto set is slightly better for the metaheuristic approach. Still, both methods work well, yet SDP is simpler to implement. As mentioned earlier in the book, the assumption of SDP is that the Pareto front is a connected set. While this assumption does not hold in general, it is normally satisfied for majority of microwave structures including the class of impedance transformers considered here.

11.2 Multi-Objective Optimization of Compact Couplers

In this section, we illustrate application of Pareto front exploration method for multi-objective design of compact microstrip couplers. Design of such circuits is challenging for similar reasons as elaborated in Sec. 11.1. An additional difficulty is the necessity of handling quite a few objectives (including the structure size, BW, power split ratio, operating frequency) defined for several responses (matching, transmission, isolation). The specific circuit under consideration is compact RRC designed for the operating frequency of 1 GHz.

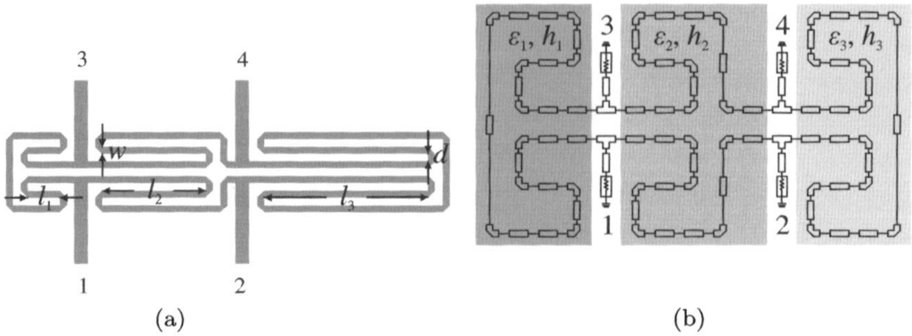

Fig. 11.6. Folded RRC (Koziel *et al.*, 2015b): (a) layout; (b) equivalent circuit model (Agilent ADS). Highlighted regions correspond to different sets of implicit space mapping (ISM) parameters $\boldsymbol{p}$.

11.2.1 *Coupler Structure and Problem Formulation*

One of the most popular microwave circuits designed for space-limited applications is a RRC (Pozar, 2012). Here, a specific example is considered, which is a folded equal-split RRC (Koziel *et al.*, 2015b). By folding each 70.7 Ohm section of a conventional circuit, a significant RRC size reduction can be expected. We choose Taconic RF-35 dielectric substrate ($\varepsilon_r = 3.5$, $h = 0.762$ mm, $\tan\delta = 0.018$), 50 Ohm port impedance and 1 GHz operating frequency for the prototype design. A parameterized layout of the RRC is shown in Fig. 11.6(a). Designable parameters are $\boldsymbol{x} = [l_1\ l_2\ l_3\ d\ w]^T$, with $w_0 = 1.7$, $l_0 = 15$ fixed (all dimensions in mm). The low- and high-fidelity models of the structure are implemented in Agilent ADS (ADS, 2011, Fig. 11.6(b)) and CST Microwave Studio (CST, 2013) (~220,000 mesh cells, ~15 min simulation time per design), respectively. The initial design is $\boldsymbol{x} = [5\ 14\ 21\ 0.7\ 0.9]^T$. The lower/upper bounds $\boldsymbol{l}/\boldsymbol{u}$ of the design space are $\boldsymbol{l} = [2\ 10\ 17\ 0.2\ 0.5]^T$ and $\boldsymbol{u} = [8\ 18\ 25\ 1.2\ 1.3]^T$.

The initial design has been chosen so that the lengths and widths of respective sections of the folded RRC correspond to their counterparts of the conventional coupler. The range of each parameter should be sufficiently large for the Pareto set to be located inside of the search space. The following design objectives are considered: F_1 — BW maximization (defined as intersection of $|S_{11}|$ and $|S_{41}|$, both below –20 dB) and F_2 — minimization of coupler size (layout area) defined as $S(\boldsymbol{x}) = A \times B$ rectangle, where $A = 6w + 5d$ and $B = 8w+7d+l_1+l_2+l_3$. Equal power split between the output ports at the operating frequency is ensured by adding a suitably defined penalty term (Koziel *et al.*, 2016c).

11.2.2 *Low-Fidelity Model Space Mapping Surrogate*

As mentioned before, an equivalent circuit is used here an a low-fidelity model. The low-fidelity model is very fast but it is not an accurate representation of $\boldsymbol{f}$. The optimization process utilizes a space mapping-corrected model $\boldsymbol{c}$, denoted as $\boldsymbol{s}$. Here, we use implicit and frequency space mapping (Koziel *et al.*, 2008a). The surrogate is defined as

$$\boldsymbol{s}(\boldsymbol{x}) = \boldsymbol{c}_F(\boldsymbol{x}; \boldsymbol{F}, \boldsymbol{p}), \tag{11.1}$$

where $\boldsymbol{c}_F$ is a frequency-scaled low-fidelity model, whereas $\boldsymbol{F}$ and $\boldsymbol{p}$ are frequency space mapping and ISM parameters, respectively. Let $\boldsymbol{c}(\boldsymbol{x}) = [c(\boldsymbol{x},\omega_1)\ c(\boldsymbol{x},\omega_2) \ldots c(\boldsymbol{x},\omega_m)]^T$, where $c(\boldsymbol{x},\omega_j)$ is evaluation of the circuit model at a frequency ω_j. Then, $\boldsymbol{c}_F(\boldsymbol{x};\boldsymbol{F},\boldsymbol{p}) = [R_c(\boldsymbol{x}, f_0 + \omega_1 \cdot f_1, \boldsymbol{p}) \ldots R_c(\boldsymbol{x}, f_0 + \omega_m \cdot f_1, \boldsymbol{p})]^T$, with f_0 and f_1 being frequency scaling parameters. Here, ISM parameters $\boldsymbol{p}$ are dielectric permittivity and substrate thickness of the microstrip components of the circuit corresponding to $\boldsymbol{p} = [\varepsilon_1\ \varepsilon_2\ \varepsilon_3\ h_1\ h_2\ h_3]^T$. Space mapping parameters are extracted to minimize misalignment between $\boldsymbol{s}$ and $\boldsymbol{f}$ as follows

$$[\boldsymbol{F}^*, \boldsymbol{p}^*] = \arg\min_{\boldsymbol{F},\boldsymbol{p}} ||\boldsymbol{f}(\boldsymbol{x}) - \boldsymbol{c}_F(\boldsymbol{x}; \boldsymbol{F}, \boldsymbol{p})||. \tag{11.2}$$

The responses of the high- and low-fidelity model at a selected design $\boldsymbol{x}$, as well as the response of the surrogate model $\boldsymbol{s}$ at the same design are shown in Fig. 11.7(a). Although the model alignment is greatly improved, its generalization capability is limited (cf. Fig. 11.7(b)). Consequently, it is not possible to find a single set of space mapping parameters that would ensure surrogate model accuracy across the entire design space. Thus, the surrogate has to be iteratively refined during the optimization process.

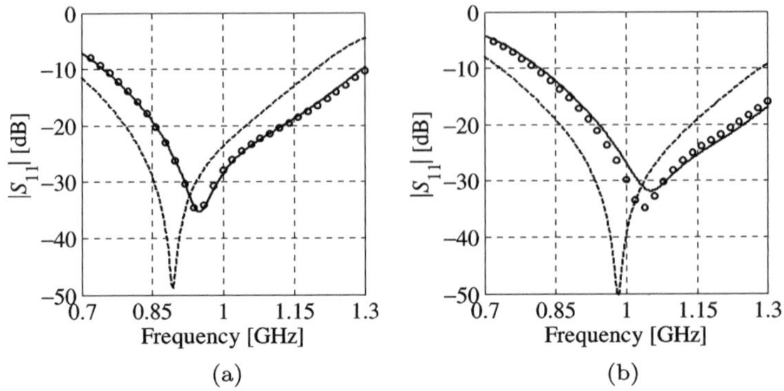

Fig. 11.7. Responses of the high- (—) and low-fidelity (– –) RRC models and the space mapping surrogate (○): (a) at certain design $\boldsymbol{x}$ (at which the surrogate is extracted); (b) at some other design. Plot (b) indicates limited generalization capability of $\boldsymbol{s}$ (Koziel *et al.*, 2015b).

11.2.3 *Optimization Algorithm*

The optimization approach used here is based on Pareto front exploration algorithm of Sec. 7.2 (Koziel *et al.*, 2015b). Starting from the first Pareto optimal design $\boldsymbol{x}^{(1)}$ found by optimizing the high-fidelity model with respect to the first objective (cf. (7.5)) using space mapping, the subsequent designs are found by solving (7.6).

Here, (7.6) is solved using the space mapping surrogate model as described in Sec. 11.2.2 and it is itself realized as an iterative process

$$\boldsymbol{x}^{(j.k)} = \arg \min_{\boldsymbol{x},\ F_2(\boldsymbol{s}^{(j.k)}(\boldsymbol{x})) \le F_2^{(j)}} F_1(\boldsymbol{s}^{(j.k)}(\boldsymbol{x})), \tag{11.3}$$

where

$$\boldsymbol{s}^{(j.k)}(\boldsymbol{x}) = \boldsymbol{c}_F(\boldsymbol{x}; \boldsymbol{F}^{(j.k)}, \boldsymbol{p}^{(j.k)}) \tag{11.4}$$

and

$$[\boldsymbol{F}^{(j.k)}, \boldsymbol{p}^{(j.k)}] = \arg \min_{\boldsymbol{F},\boldsymbol{p}} ||\boldsymbol{f}(\boldsymbol{x}^{(j.k)}) - \boldsymbol{c}_F(\boldsymbol{x}^{(j.k)}; \boldsymbol{F}, \boldsymbol{p})||. \tag{11.5}$$

The starting point for (11.3) is $\boldsymbol{x}^{(j-1)}$ (the previously obtained Pareto set point). Equal power split between the output ports at the operating frequency is secured by adding to F_1 in (11.3) the terms proportional to $(|S_{21}| + 3)^2$ and $(|S_{31}| + 3)^2$ (at 1 GHz) to penalize violations of the aforementioned requirement.

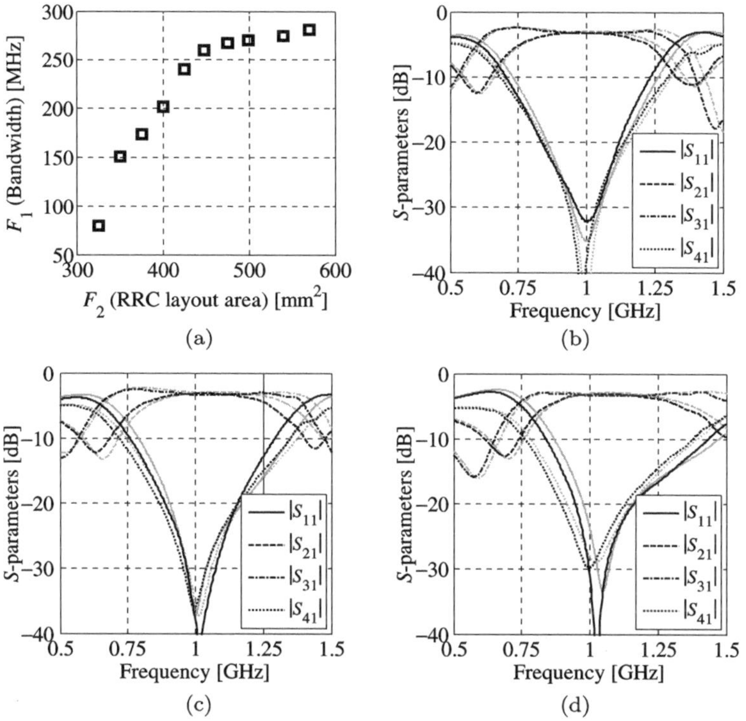

Fig. 11.8. RRC optimization results (Koziel *et al.*, 2015b): (a) a Pareto-optimal set, obtained using Pareto front exploration technique, as well as simulated (gray lines) and measured (black lines) frequency characteristics for selected designs, corresponding to the layout areas of: (b) 570 mm^2; (c) 448 mm^2; (d) 375 mm^2.

11.2.4 *Numerical Results and Experimental Validation*

The first design obtained by optimizing BW without any area constraints resulted in BW of 281 MHz and the layout area of 570 mm^2. Nine other designs have been obtained by setting up $F_2^{(j)}$ to 540, 500, 475, 450, 425, 400, 375, 350 and 325 mm^2, respectively. Figure 11.8(a) shows the obtained high-fidelity representation of the Pareto front. For the layout area of 300 mm^2, it was impossible to obtain a design with positive value of BW. Table 11.2 and Figs. 11.8(b)–11.8(d) show the numerical data as well as simulated and measured frequency characteristics for the selected designs. The photograph of the fabricated

Table 11.2 Multi-objective RRC optimization: selected results

Design variables [mm]					Objectives		
l_1	l_2	l_3	d	W	BW [MHz]	Layout area [mm^2]	Miniaturization* %
4.18	13.20	20.68	0.994	0.865	281	570	90.8
3.83	11.76	20.44	0.825	0.877	270	500	91.9
4.10	13.78	21.14	0.581	0.887	260	450	92.7
4.25	12.17	22.12	0.400	0.923	202	400	93.5
3.95	10.87	21.71	0.350	0.936	174	375	93.9
4.37	12.33	22.52	0.350	0.820	151	350	94.3

*With respect to conventional RRC (radius = 44.39 mm, size = 6,190 mm^2).

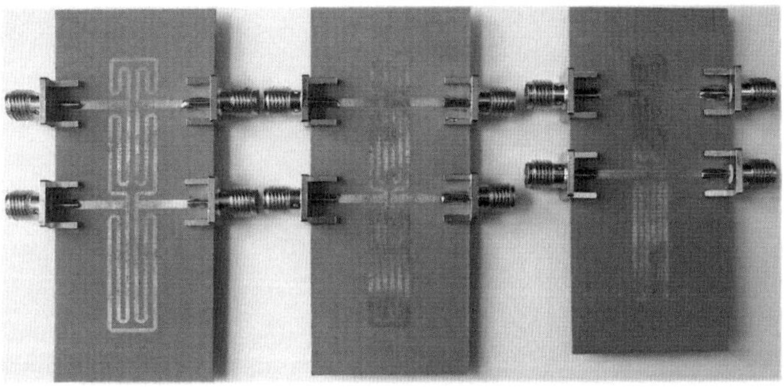

(a) (b) (c)

Fig. 11.9. A photograph of the fabricated RRC prototypes. Designs characterized by footprint of: (a) 570 mm^2; (b) 448 mm^2; and (c) 375 mm^2.

prototypes is shown in Fig. 11.9. The range of sizes of designs in Fig. 11.8(a) is from 570 mm^2 down to 325 mm^2. At the same time, the Pareto-optimal designs offer around 91–94 percent size reduction with respect to the conventional RRC.

The total cost of multi-objective design optimization corresponds to less than 30 high-fidelity model evaluations (~7.5 h of the CPU-time), including the overhead related to multiple evaluations of the circuit model $\boldsymbol{c}$ (the latter does not exceed 20 percent of the overall EM-simulation cost). This cost is orders of magnitude lower than what would be required by any type of metaheuristic algorithm applied for direct optimization of the RRC electromagnetic model (EM) model. For example, assuming 15 min for the RRC fine model evaluation and 100 iterations of the metaheuristic algorithm with population size of 100 individuals, one obtains 10,000 evaluations of $\boldsymbol{f}$ model, which translates into 100 days of the CPU-time.

11.3 Multi-Objective Optimization of Transonic Airfoils

Design of aerodynamic surfaces such as aircraft wings and turbine blades is largely based on computational fluid dynamic (CFD) models (Leoviriyakit *et al.*, 2003; Braembussche, 2008). Similarly as in computational EM, CFD models are expensive to evaluate so that their direct optimization is challenging (Mader and Martins, 2012; Leung and Zingg, 2012). Here, multi-objective design of transonic airfoils using the surrogate-assisted technique of Sec. 7.1 is discussed. In order to make the section self-contained, we briefly outline formulation of the airfoil design problem, airfoil parameterization, CFD models utilized in the optimization process, and, finally, present the results for a specific case study.

11.3.1 *Transonic Airfoil Shape Problem Formulation*

We aim at re-designing the RAE 2822 transonic airfoil shape (Koziel *et al.*, 2016d) having in mind two objectives: maximization of the section lift coefficient C_l and minimization of the section drag coefficient C_d. In other words, we have, $F_1(\boldsymbol{x}) = -C_{l.f}$ and $F_2(\boldsymbol{x}) = C_{d.f}$, where index f indicates that the aerodynamic forces are to be evaluated using a high-fidelity CFD model. More specifically, we are interested in the possible design trade-offs between these two objectives. In our demonstration case, the free-stream Mach number and the angle of attack are fixed at $M_\infty = 0.734$ and to $\alpha = 2.79°$, respectively.

The airfoil shape is parameterized using B-spline curves. The design variable vector is $\boldsymbol{x} = \boldsymbol{p}$, where $\boldsymbol{p}$ is a vector of the size $m \times 1$, with m being the total number of control parameters. The airfoil surfaces can be written in a parametric form as

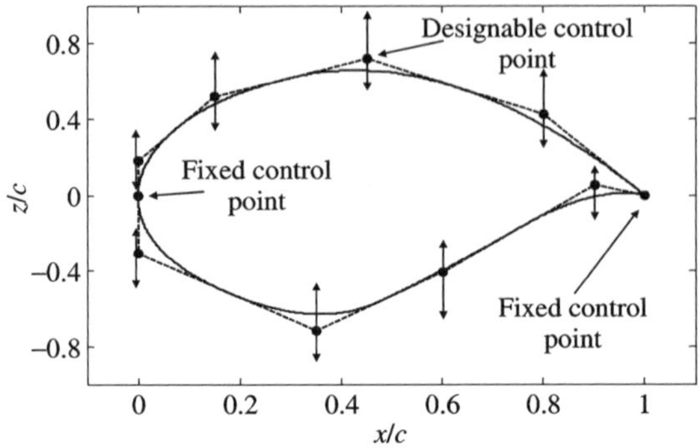

Fig. 11.10. Example B-spline parameterization of an airfoil. The designable control points are restricted to vertical movements only (Koziel *et al.*, 2016d).

(Farin, 1993)

$$x(t) = \sum_{i=1}^{n+1} X_i N_{i,k}(t), \quad z(t) = \sum_{i=1}^{n+1} Z_i N_{i,k}(t), \tag{11.6}$$

where (x,z) are the Cartesian coordinates of the airfoil surface, $N_{i,k}$ is the B-spline basis function of order k, (X_i, Z_i) are the coordinates of the B-spline control polygon; $m = n + 1$ is the total number of control points. These points are our design variables and allowed only to move freely vertically as shown in Fig. 11.10. Thus, we have $\boldsymbol{x} = [Z_1\ Z_2 \ldots Z_{n+1}]^T$ and the corresponding X_i coordinates are fixed during the optimization process.

11.3.2 *Computational Models*

The computational grids used in this study are generated by the hyperbolic C-mesh of Kinsey and Barth (1984). The far field is set 100 chords away from the airfoil surface and the grid points are clustered at the trailing and leading edge of the airfoil to give a minimum stream-wise spacing of $0.001 \times$ chord length. The distance from the airfoil surface to the first node is $5 \cdot 10^{-5} \times$ chord length. An example grid is shown in Fig. 11.11. The free-stream Mach number, static pressure and angle of attack are prescribed at the far field boundary.

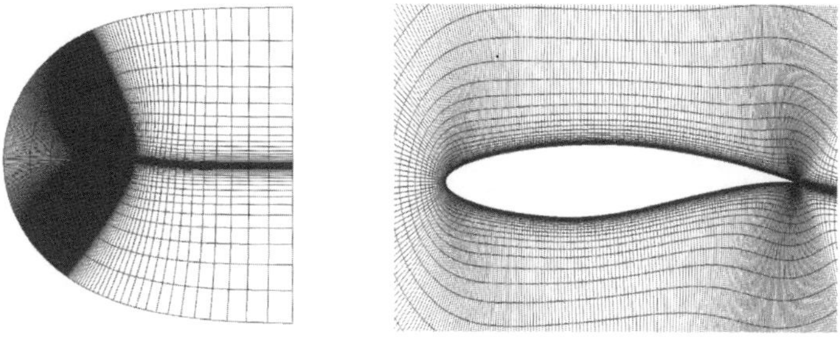

(a) (b)

Fig. 11.11. Example hyperbolic C-grid (Koziel *et al.*, 2016d): a farfield view (a) and a view close to the airfoil surface (b).

The flow is assumed to be steady, viscous and turbulent. The compressible Reynolds-averaged Navier–Stokes (RANS) equations are taken to be the governing fluid flow equations with the Spalart–Allmaras turbulence model (Spalart and Allmaras, 1992). The flow solver is of implicit density-based formulation and the fluxes are calculated by an upwind-biased second-order spatially accurate Roe flux scheme. Asymptotic convergence to a steady-state solution is obtained for each case. The solution convergence criterion for the high-fidelity model is the one that occurs first of the following: a reduction of the residuals by six orders of magnitude, or a maximum number of iterations of 1,800. Numerical fluid flow simulations are performed using FLUENT (FLUENT, 2012).

A grid convergence study was performed using the RAE 2822 airfoil at a Mach number of $M_\infty = 0.734$ and an angle of attack of $\alpha = 2.79°$. The study (cf. Table 11.3) indicates that roughly 520,000 mesh cells are needed for mesh convergence, and that particular mesh was used for the high-fidelity model. The grid has 1,280 points in the streamwise direction and 512 points in the direction normal to the airfoil surface. The region behind the airfoil to the far field contains 256 points. Although the flow equation residuals are not converged after 2,000 iterations, the lift and the drag coefficients are converged within 1,800 iterations. Therefore, the maximum number of iterations is set to 1,800. The overall simulation time of the high-fidelity model is 2 h.

Table 11.3 Grid convergence study of the high-fidelity CFD model for RAE 2822 at $M_\infty = 0.734$ and $\alpha = 2.79°$

Grid size	C_l (l.c)	C_d (d.c)
255 × 32 (∼8,000)	78.64	185.95
511 × 64 (∼33,000)	79.84	187.86
1,023 × 128 (∼132,000)	80.40	190.20
2,047 × 256 (∼526,000)	80.56	190.78
4,095 × 512 (∼2,101,000)	80.53	190.58

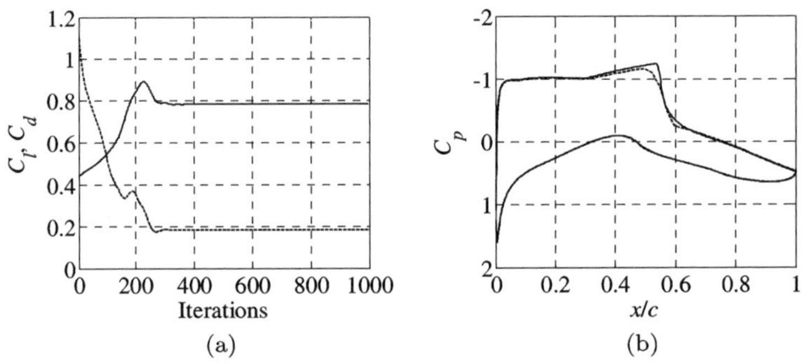

Fig. 11.12. Simulation results for RAE 2822 at a Mach number of $M_\infty = 0.734$ and an angle of attack of $\alpha = 2.79°$: (a) the evolution of the lift C_l (—) and drag coefficients C_d (– –) obtained by the low-fidelity model; (b) a comparison of the pressure distributions obtained by the high- (—) and low-fidelity (– –) models (Koziel *et al.*, 2016d).

The low-fidelity CFD model is constructed in the same way as the high-fidelity model, but with a coarser computational mesh and relaxed convergence criteria. For the low-fidelity model, we use the coarse mesh in the grid study presented in Table 11.3, with ∼8,000 mesh cells. The flow solution history for the low-fidelity model, shown in Fig. 11.12(a), indicates that the lift and drag coefficients are nearly converged after 400–500 iterations. The maximum number of iterations is set to 500 for the low-fidelity model. Hence, the overall simulation time is around 4 min on the same four processors as the high-fidelity model.

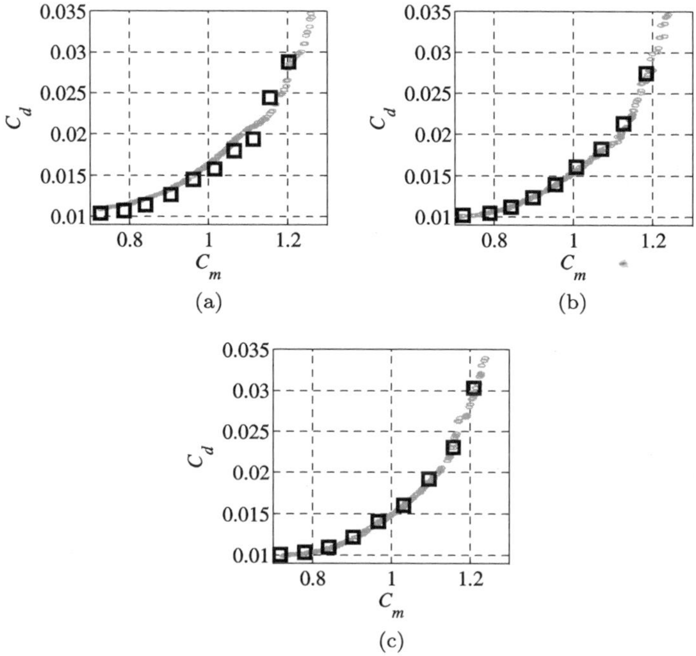

Fig. 11.13. Pareto sets for 8-parameter transonic airfoil example: Pareto front predicted by the optimized co-Kriging surrogate model (○), and high-fidelity verification samples (□): (a) initial iteration, (b) second iteration, (c) fourth and final iteration (Koziel *et al.*, 2016d).

A comparison of the pressure distributions (Fig. 11.13(b)) indicates that the low-fidelity model captures the main features of the high-fidelity model pressure distribution quite well. The biggest discrepancy in the distributions is around the shock on the upper surface, leading to an under estimation of both the lift and drag coefficients.

Note that the lift and drag coefficients are presented in counts, i.e., one lift count (l.c.) is $\Delta C_l = 0.01$ and one drag count (d.c.) is $\Delta C_d = 0.0001$.

11.3.3 *Case Study and Results*

In the design case study presented here, B-spline-parameterized airfoil is considered with eight design variables. The angle of attack is

$\alpha = 2.79°$. The upper and lower bounds for the design variables are set 0.1 and –0.1, respectively. The extreme points of the Pareto front obtained at the level of the low-fidelity CFD model are as follows: drag minimization, $\boldsymbol{x}_c^{*(1)} = [0.0131\ 0.0453\ 0.0621\ 0.0406\ -0.0224\ -0.0487\ -0.0540\ -0.0287]^T$, $C_l = 0.4244$, $C_d = 0.0098$; and lift maximization $\boldsymbol{x}_c^{*(2)} = [0.0597\ 0.0498\ 0.0681\ 0.0814\ -0.0291\ -0.0679\ -0.0310\ 0.0654]^T$, $C_l = 1.0387$, $C_d = 0.0791$. The reduced design space is therefore defined by the following bounds: $\boldsymbol{l}^* = [0.0131\ 0.0453\ 0.0621\ 0.0406\ -0.0291\ -0.0679\ -0.0540\ -0.0287]^T$ and $\boldsymbol{u}^* = [0.0597\ 0.0498\ 0.0681\ 0.0814\ -0.0224\ -0.0487\ -0.0310\ 0.0654]^T$.

It should be noted that the reduced space is over 10^8 times smaller (volume-wise) than the original one. The extreme points were obtained using the pattern search algorithm (Kolda *et al.*, 2003) at the cost of 250 low-fidelity model evaluations per point.

In the next step, a Kriging interpolation model $\boldsymbol{s}_{\text{KR}}$ has been set up using 1,056 training points, including 256 corner points and 800 samples allocated using Latin Hypercube Sampling (Beachkofski and Grandhi, 2002).

Subsequently, the procedure of Sec. 7.1.4 was performed using 10 high-fidelity verification points uniformly sampled along the Pareto front predicted by the surrogate model. The process was completed in four iterations. Figure 11.13 shows the predicted Pareto fronts and high-fidelity verification samples at the four iterations of the algorithm. It can be observed that the agreement between the predicted front and the verification samples is improving between iterations. Also, the front is pushed slightly down (i.e., it improves) towards the end of the process.

The total design optimization cost corresponds to only 92 high-fidelity model evaluations (including all necessary low-fidelity model samples as reported above and 40 high-fidelity verification samples). The cost of multi-objective optimization of the Kriging surrogate can be neglected. The Pareto front, shown in Fig. 11.13, illustrates the best possible trade-offs between the drag and lift coefficients at subsequent design refinements.

For further analysis, we only consider the part of the Pareto front (Fig. 11.13(c)) for which the drag coefficient is lower than 0.02, and

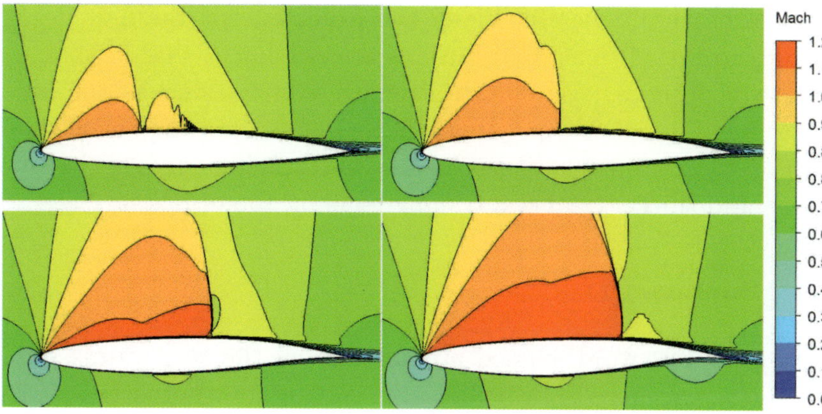

Fig. 11.14. Mach contours for the selected designs from the top left corner and to the bottom right corner (clockwise) — first design, third design, fifth design and seventh design (Koziel *et al.*, 2016d).

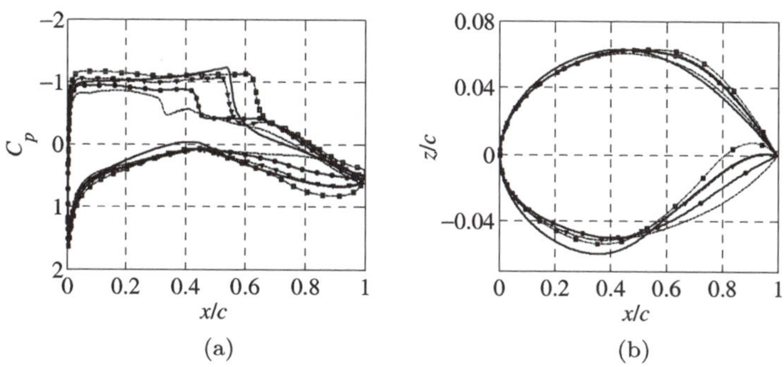

Fig. 11.15. Pressure coefficient distributions (a) and airfoil shapes (b) of the reference (—) and Pareto designs of Fig. 11.13(c): first (– –), third (○), fifth (∇) and seventh (□) (Koziel *et al.*, 2016d).

contains seven designs in that range. Starting from the lower lift as the first design, we consider the first, third, fifth and the seventh design and analyze them in more detail. Their corresponding Mach contours and pressure coefficient distributions and airfoil shapes are shown in Figs. 11.14 and 11.15, respectively. As we move along the front it starts with an almost shock-free design and then in the consecutive designs the shock strength is increasing and the shock location is moving towards the trailing edge. Starting from the lower lift towards the higher lift along the front, except the first design, the pressure coefficient distributions are similar in shape. As it can be clearly observed, the shock strength is increasing as the lift is increasing. The other trend that can be seen is that the airfoil thickness is getting thinner near the leading and trailing edges in the lower surface and thicker near the trailing edge on the upper surface.

11.4 Summary

In this chapter, we discussed applications of the surrogate-assisted techniques considered in this book for solving multi-objective design optimization problems in microwave and aerospace engineering. The results obtained in all cases are consistent with those demonstrated in the case of antenna structures, i.e., the reliable representations of the Pareto front can be obtained at the computational cost corresponding to a few dozen of high-fidelity model evaluations (for lower-dimensional problems) to about 200 of model evaluations (for higher-dimensional problems). If available, other alternative types of low-fidelity model can be used (here, equivalent circuits in case of coupler structures, pre-conditioned using space mapping). In general, it seems that combination of variable-fidelity/variable-resolution simulations, data-driven modeling and global optimization techniques (or local, depending on a particular solution approach) permit robust handling of multi-objective problems across various engineering disciplines.

Chapter 12

Applications of Multi-Objective Optimization

The goal of multi-objective optimization — as understood in this book — is to find comprehensive information about a given structure in the form of a set of alternative designs representing the best possible trade-offs between conflicting objectives. The numerical methods presented in Chapters 7 and 8, and demonstrated in Chapters 9 and 11, allow for finding such Pareto sets in reasonable timeframes, even for relatively expensive high-fidelity electromagnetic (EM)-simulation models. In this chapter, we demonstrate utilization of Pareto sets for comprehensive comparison of various antenna structures and microwave circuits. Three case studies are considered. The first one involves three structures of compact ultra-wideband (UWB) antennas, where information about available size-performance trade-offs allow for conclusive comparison of the competing antenna topologies. The second case is a comparison of two structures of UWB monopole antennas that are topologically similar, yet a small modification leads to considerable performance improvement in terms of attainable miniaturization rate. The last study investigates selection of the best possible architecture for compact impedance matching transformers, specifically, the optimum type and arrangement of the compact cells utilized as the basic building blocks of the transformer structure. In all cases, the knowledge of Pareto fronts is indispensable to make application-driven design decisions. At the same time, it is pointed out that the antenna dimensions reported in the literature are often far from the optimum ones; consequently, various comparisons of antenna structures provided in the published works might be of limited use.

12.1 Performance Comparison of Ultra-Wideband Antennas

In this section, we utilize multi-objective optimization as a means for comparing various UWB antenna structures with respect to their miniaturization capability as well as electrical performance. The knowledge of the Pareto fronts is used here as a tool to conduct such a comparison. Clearly, from a practical point of view, it is important that the design trade-offs are generated at low computational cost. Here, a sequential domain patching (SDP) algorithm is utilized (Sec. 7.3).

12.1.1 *Antenna Comparison Using Pareto Sets*

For compact UWB antennas, the two most important figures of merit that are normally subjected to optimization process are reflection characteristic and the antenna size. In most cases, reflection response is optimized directly so that the acceptable antenna performance can be achieved. At the same time, the small size is normally a result of the introduced geometry modifications (Li *et al.*, 2012; Wu *et al.*, 2014). Methods for explicit minimization of the antenna size have been recently reported (Bekasiewicz and Koziel, 2015a; Koziel and Bekasiewicz, 2016a). Regardless of the design problem formulation, usually a single design is produced as the outcome of the optimization algorithm.

More comprehensive information about capabilities of a given antenna, e.g., in terms of the best possible trade-offs between the structure size and the maximum in-band reflection, can be obtained by means of multi-objective optimization, specifically in the form of the Pareto set as illustrated in Fig. 12.1 for typical objectives

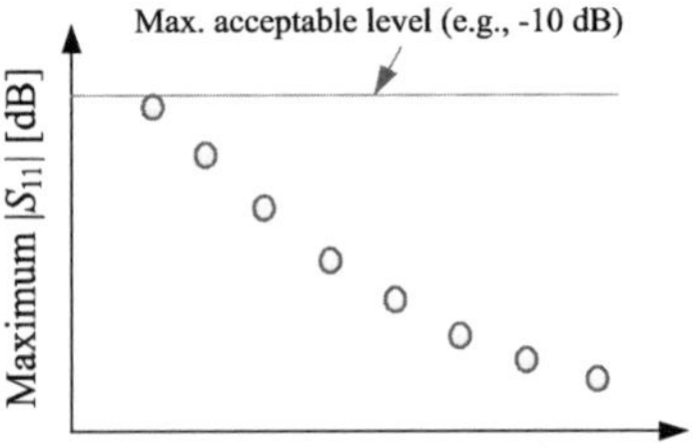

Fig. 12.1. Pareto set illustration for a compact UWB antenna with the two design objectives being minimization of the maximum in-band antenna reflection and minimization of the antenna footprint.

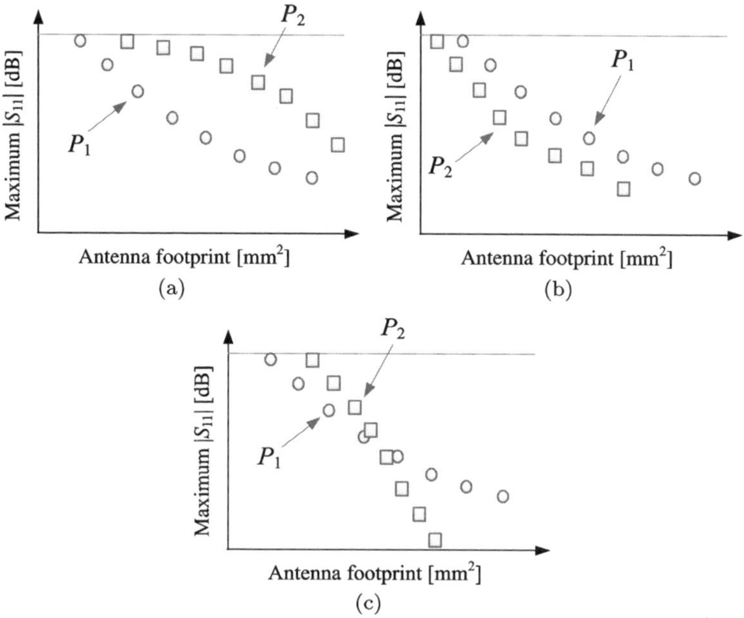

Fig. 12.2. Comparison of antenna structures A_1 and A_2 based on their Pareto sets P_1 and P_2. Some of possible scenarios include: (a) complete domination of P_1 over P_2, (b) complete domination of P_2 over P_1, (c) partial domination of P_1 over P_2.

important for compact antennas. A conclusive assessment of several antenna structures can be realized by comparing their corresponding Pareto sets. This is illustrated in Fig. 12.2 which shows some of possible scenarios concerning the Pareto set allocation. Domination of one of the Pareto sets over the other indicates that the corresponding antenna structure outperforms its competitor in both aspects taken into account (size and reflection response). Partial domination indicates that one of the antenna exhibits better miniaturization potential whereas the other can be designed for superior electrical performance. The knowledge of the Pareto set also allows for selecting a design that ensures a specific maximum in-band reflection level (e.g., –12 dB) that might be important (as a safety margin) to account for possible manufacturing tolerances.

A practical problem is that multi-objective optimization of antennas is a very challenging task when EM analysis is used for structure

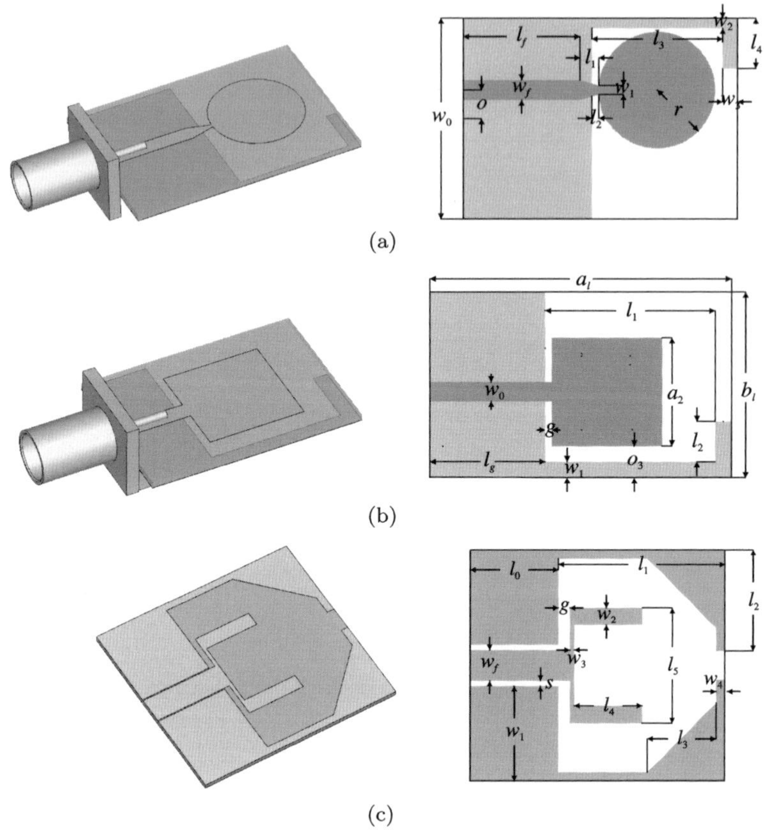

Fig. 12.3. Geometries of UWB antennas considered in our comparative study: (a) Antenna I (Li *et al.*, 2012), (b) Antenna II (Koziel and Bekasiewicz, 2016j) and (c) Antenna III (Qing and Chen, 2009).

evaluation. However, the design process can be considerably sped up by using surrogate-assisted techniques. A specific method exploited here is SDP of Sec. 7.3.

12.1.2 *Antenna Structures*

For the sake of example, we compare three structures of compact UWB antennas recently proposed in the literature (Li *et al.*, 2012; Koziel and Bekasiewicz, 2016j; Qing and Chen, 2009). The antenna geometries are shown in Fig. 12.3. The structures are designed on a

0.762 mm thick Taconic RF-35 dielectric substrate ($\varepsilon_r = 3.5$, $\tan\delta = 0.0018$). Antenna models are implemented in CST Microwave Studio and simulated using its time domain solver (CST, 2013).

The first structure (Antenna I) consists of a circular patch fed through a 50 Ohm microstrip line and a modified ground plane with L-shaped strip aimed at enhancement of the current path and, consequently, miniaturization (Li *et al.*, 2012). The design variables are: $\boldsymbol{x}_I = [w_0\ l_1\ l_2\ l_3\ l_4\ w_1\ w_2\ w_3\ r\ o_r]^T$, where parameters $w_f = 1.7$, $l_f = 10$ and $o = 0.5w_0 \cdot r \cdot o_r$ (all dimensions in mm). The high-fidelity antenna model $\boldsymbol{f}$ consists of ~2,600,000 mesh cells (simulation time 24 min). The low-fidelity model $\boldsymbol{c}$ contains ~370,000 cells (simulation time 150 s). The models include the SMA connector to ensure reliability of antenna evaluation. The ranges of design variables for considered design are: $\boldsymbol{l}_I = [15\ 2\ {-2}\ 10\ 3\ 0.2\ 0.2\ 0.2\ 4.5\ {-1}]^T$ and $\boldsymbol{u}_I = [30\ 8\ 2\ 15\ 8\ 1.7\ 1.2\ 1.2\ 8.5\ 1]^T$.

The second structure (Antenna II) consists of a rectangular radiator and a ground plane with L-shaped strip for current path enhancement. The structure is based on a radiator of the multi-input and multi-output (MIMO) antenna proposed in (Liu *et al.*, 2013) and extracted as a single antenna for the purpose of this study. The vector of design variables is $\boldsymbol{x}_{II} = [l_g\ g\ a_2\ l_1\ l_2\ w_1\ o_3]^T$. Parameter $w_0 = 1.7$ to ensure 50 Ohm input impedance. All dimensions are in mm. The EM antenna models are implemented in CST Microwave Studio ($\boldsymbol{f}$: ~4,600,000 mesh cells, simulation time 40 min, and $\boldsymbol{c}$: ~850,000 cells, 2 min). Similarly as for Antenna I, the models include the SMA connector to ensure reliability of antenna evaluation. The ranges of design variables for considered design are: $\boldsymbol{l}_{\text{II}} = [4\ {-2}\ 4\ 5\ 1\ 0.5\ 0.5]^T$ and $\boldsymbol{u}_{\text{II}} = [15\ 2\ 15\ 20\ 10\ 3.5\ 5.5]^T$.

The third design (Antenna III) is a uniplanar structure composed of a driven element in the form of fork-shaped radiator fed through a coplanar waveguide (CPW) and an open slot (Qing and Chen, 2009). The design variables are: $\boldsymbol{x}_{\text{III}} = [l_0\ l_1\ l_{2r}\ l_{3r}\ l_4\ l_5\ w_1\ w_2\ w_3\ w_4\ g]^T$. Parameters $w_f = 3.5$ and $s = 0.16$ to ensure 50 Ohm input impedance, where variables $l_2 = (0.5w_f + s + w_1) \cdot \max\{l_{2r}, l_{3r}\}$ and $l_3 = (0.5w_f + s + w_1) \cdot l_{3r}$. The low-fidelity antenna model consists of ~215,000 cells (simulation time 67 s), whereas the high-fidelity one is composed of ~1,600,000 cells (simulation time 10 min). The ranges of parameters are: $\boldsymbol{l}_{\mathrm{III}} = [5\ 15\ 0.2\ 0.2\ 4\ 8\ 7\ 0.5\ 0.2\ 0.2\ 0.2]^T$ and $\boldsymbol{u}_{\mathrm{III}} = [15\ 25\ 1\ 0.8\ 11\ 16\ 15\ 3.5\ 2.5\ 2\ 2]^T$.

Based on the structure dimensions from (Li *et al.*, 2012; Liu *et al.*, 2013; Qing and Chen, 2009) one can conclude that Antenna III is noticeably larger (766 mm^2) than Antenna I (673 mm^2). At the same time, the size of Antenna II is only 390 mm^2.

The following two design objectives are considered to compare the structures: F_1 — minimization of reflection in 3.1–10.6 GHz band and F_2 — minimization of footprint. The latter is defined as $A_I(\boldsymbol{x}_I) = w_0 \cdot \max\{l_f + l_1 + 2r - 0.05r,\ l_f + l_1 - 0.05r + l_2 + l_3 + w_3\}$, $A_{\mathrm{II}}(\boldsymbol{x}_{\mathrm{II}}) = (a_2 + 2o_3) \cdot (l_g + l_1 + w_1)$ and $A_{\mathrm{III}}(\boldsymbol{x}_{\mathrm{III}}) = (l_0 + l_1) \cdot (2s + 2w_1 + w_f)$ for Antennas I–III, respectively.

12.1.3 *Pareto Fronts Identification Using Sequential Domain Patching*

Figure 12.4 shows the Pareto fronts obtained using the SDP algorithm for the three antenna structures considered. Figure 12.5 shows reflection characteristics of the selected Pareto-optimal designs. Corresponding dimensions have been gathered in Tables 12.1–12.3 for Antennas I, II and III, respectively.

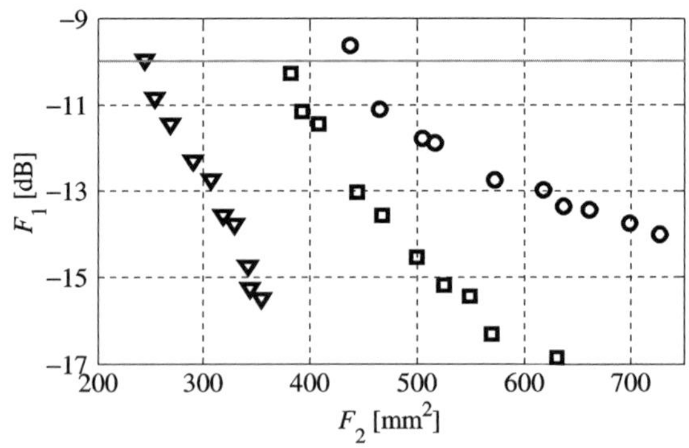

Fig. 12.4. Comparison of the high-fidelity Pareto sets for Antenna I (□), Antenna II (▽) and Antenna III (○).

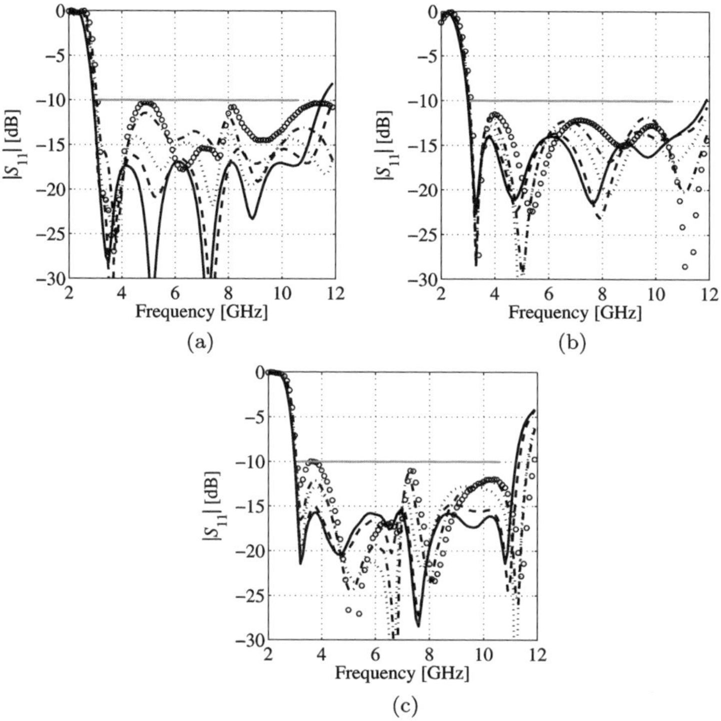

Fig. 12.5. Frequency responses of the high-fidelity model of: (a) Antenna I for the designs listed in Table 12.1; (b) Antenna II for the designs of Table 12.2 and (c) Antenna III for the designs from Table 12.3. For all structures the designs are as follows: $\boldsymbol{x}_f^{(1)}$ — (—), $\boldsymbol{x}_f^{(3)}$ — (- -), $\boldsymbol{x}_f^{(6)}$ — (· · ·), $\boldsymbol{x}_f^{(8)}$ — (- ·), $\boldsymbol{x}_f^{(10)}$ — (○).

Table 12.1 Antenna I: selected Pareto-optimal designs

			Design variables [mm]									
	F_1	F_2	w_0	l_1	l_2	l_3	l_4	w_1	w_2	w_3	r	o_r
$\boldsymbol{x}_f^{(1)}$	–17.0	577	20.33	4.05	–0.09	13.52	5.47	0.23	1.06	1.19	5.86	0.07
$\boldsymbol{x}_f^{(3)}$	–15.2	536	20.12	3.65	–0.36	12.48	5.43	0.20	0.85	1.14	5.52	–0.11
$\boldsymbol{x}_f^{(6)}$	–13.8	475	17.90	3.40	–0.25	12.53	5.40	0.20	0.68	1.13	5.44	–0.05
$\boldsymbol{x}_f^{(8)}$	–12.1	448	17.17	2.91	–0.30	12.61	5.44	0.24	0.76	1.15	5.39	–0.05
$\boldsymbol{x}_f^{(10)}$	–9.9	424	16.95	2.78	–0.84	12.24	5.69	0.20	0.69	1.09	5.02	–0.15

Table 12.2 Antenna II: selected Pareto-optimal designs

			Design variables [mm]						
	F_1	F_2	l_g	g	a_2	l_1	l_2	w_1	o_3
$\boldsymbol{x}_f^{(1)}$	–15.5	355	4.56	1.19	9.80	14.04	4.57	2.96	3.32
$\boldsymbol{x}_f^{(3)}$	–14.7	341	4.54	1.20	9.60	13.62	5.08	2.96	3.28
$\boldsymbol{x}_f^{(6)}$	–12.7	307	4.39	1.20	9.09	12.95	6.10	2.78	3.09
$\boldsymbol{x}_f^{(8)}$	–11.4	269	4.34	1.29	9.04	13.45	6.03	1.86	2.32
$\boldsymbol{x}_f^{(10)}$	–10.0	245	4.39	1.20	8.88	12.68	6.91	1.55	2.13

Table 12.3 Antenna III: selected Pareto-optimal designs

			Design variables [mm]										
	F_1	F_2	l_0	l_1	l_{2r}	l_{3r}	l_4	l_5	w_1	w_2	w_3	w_4	g
$\boldsymbol{x}_f^{(1)}$	–14.0	727	9.04	18.28	0.89	0.61	8.40	14.65	11.40	2.89	0.49	1.46	0.52
$\boldsymbol{x}_f^{(3)}$	–13.4	661	9.04	17.53	0.90	0.69	8.31	14.57	10.54	2.86	0.42	1.04	0.48
$\boldsymbol{x}_f^{(6)}$	–12.7	573	8.76	17.50	0.82	0.76	8.34	14.46	9.00	2.91	0.42	0.61	0.46
$\boldsymbol{x}_f^{(8)}$	–11.8	505	9.02	16.61	0.87	0.77	8.13	14.54	7.95	2.91	0.42	0.35	0.45
$\boldsymbol{x}_f^{(10)}$	–9.6	438	7.80	16.71	0.71	0.83	8.17	14.46	7.02	2.91	0.52	0.22	0.44

The computational cost of the multi-objective optimization process is low and corresponds to only 128 high-fidelity model simulations for Antenna I (including 944 evaluations of $\boldsymbol{c}$ and 30 evaluations of $\boldsymbol{f}$). The cost for Antenna II corresponds to 72 evaluations of the high-fidelity model (831 $\times$ $\boldsymbol{c}$ and 30 $\times$ $\boldsymbol{f}$). Finally, the cost for Antenna III corresponds to 132 evaluations of the high-fidelity model (914 $\times$ $\boldsymbol{c}$ and 30 $\times$ $\boldsymbol{f}$). Note that the costs are similar for all structures.

12.1.4 *Structure Comparison*

The results presented in Figs. 12.4 and 12.5 as well as in Tables 12.1–12.3 indicate that Antenna I dominates over Antenna III. At the same time, Antenna I is dominated by Antenna II (consequently, Antenna II dominates over Antenna III as well) in the multi-objective sense. Antenna I allows for achieving the lowest reflection level (of around –17 dB in the UWB range) with the minimum reflection levels for Antenna II and III being around –15.5 dB and –14 dB. Antenna II can be designed for the smallest size of about 240 mm^2, which is much smaller than for the other structures. Antenna III offers the largest variability of size along the Pareto front (290 mm^2), whereas sizes changes for Antennas I and II are 249 mm^2 and 110 mm^2, respectively.

An important aspect is that in all cases, the initial designs taken from the literature are dominated by the designs corresponding to the Pareto sets found for the considered antennas. In other words, the designs provided in the literature are not optimal and could have been improved by applying appropriate numerical techniques. At the same time, non-optimality of these designs (which is actually the case for majority of the designs presented in the literature) makes various comparisons involving the literature data of limited use.

12.2 Performance Comparison of Rectangular Ultra-Wideband Monopoles

In this section, we consider two similar structures of UWB monopole antennas with the only difference being the slit below the microstrip feeding line. As indicated using Pareto front comparison, this small difference has a significant effect on the attainable miniaturization rate of the antenna.

12.2.1 *Antenna Description*

We consider two realizations of a UWB monopole antenna. The basic realization — denoted as Antenna I — consists of a square-shaped radiator with ground plane modification in the form of L-shaped strip. Design variables are $\boldsymbol{x}_I = [b_l\ g\ a_2\ l_1\ l_2\ w_1\ o_3]^T$, whereas $o_1 = 0.5b_l$, $l_g = 6.79$, $w_0 = 1.7$ is fixed. The modified geometry (Antenna II) comprises ground plane slot below the microstrip feed line and features increased number of degrees of freedom. The parameter vector is $\boldsymbol{x}_{II} = [l_g\ g\ a_1\ a_2\ l_1\ l_2\ w_1\ s_1\ s_2\ o_1\ o_3]^T$ (all dimensions are in mm). Geometries of both structures are shown

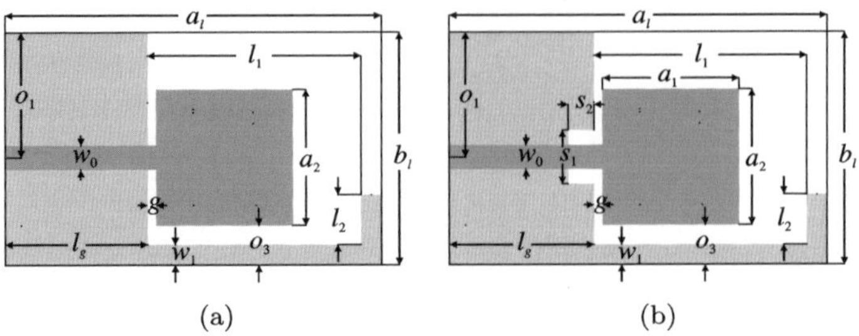

Fig. 12.6. Two versions of the considered monopole antenna with rectangular radiator: (a) seven-parameter structure (Koziel and Bekasiewicz, 2016j); and (b) 11-parameter design with the slit below the microstrip feed line.

in Fig. 12.6. We would like to point out that EM models of the considered antennas are supplemented with identical SMA connectors (see Figs. 9.29(a) and 12.3(b)).

The design requirements are: minimization of antenna reflection within 3.1–10.6 GHz frequency band of interest (objective F_1) and reduction of antenna footprint (objective F_2) defined as $S(\boldsymbol{x}) = a_l \times b_l$, where $a_l = 0.5a_2 + o_1 + o_3$. For Antenna II $b_l = l_g + l_1 + w_1$. More comprehensive description of considered antennas can be found in Secs. 9.6 and 12.1, respectively.

12.2.2 *Multi-Objective-Based Performance Comparison*

A comparison of the high-fidelity Pareto designs obtained for both antennas is shown in Fig. 12.7. Their corresponding dimensions can be found in Tables 9.8 and 12.2, respectively. The obtained results indicate that geometrical modifications introduced to Antenna II have significant impact on its size (upon proper optimization of antenna dimensions). As a consequence, Antenna I is entirely dominated by Antenna II.

We would like to emphasize that, despite topological changes, considered structures feature comparable performance along objective F_1 with minimum reflection responses being around –16 dB. At the same time, the distance between the obtained Pareto fronts along F_2 is more or less stable along F_1. On average, all the Pareto

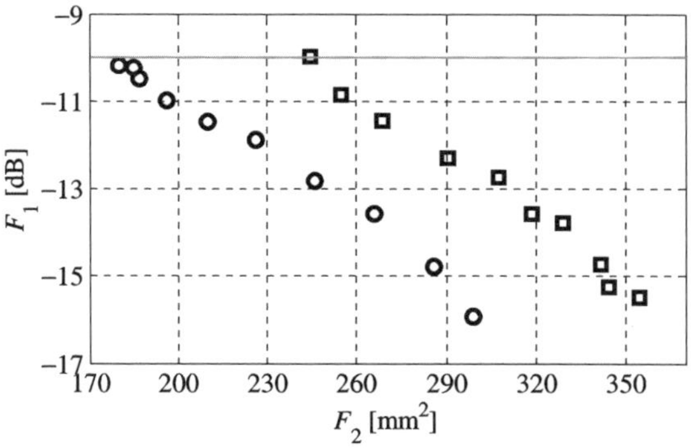

Fig. 12.7. Comparison of the high-fidelity Pareto sets for Antenna I (□) and Antenna II (○).

designs of the structure with modified geometry are 59 mm^2 smaller in comparison to the reference antenna.

12.3 Optimum Architecture Selection of Compact Impedance Matching Transformers

In this section, we demonstrate utilization of multi-objective optimization for optimum selection of the architecture for decomposable compact microwave structures. A specific class of circuits considered here are impedance matching transformers. Their miniaturized versions are composed of so-called compact microstrip resonant cells (CMRCs) (Bekasiewicz *et al.*, 2012; Nie *et al.*, 2014). The knowledge of Pareto sets obtained for various configurations of CMRCs permits selection of the best arrangement of the cells that permits optimum electrical performance or minimum structure size.

12.3.1 *CMRC-Based Miniaturization: Architecture Selection Problem*

Multi-objective design of impedance matching transformers has been considered — as one of the demonstration examples — in Chapter 11. Section 11.1.1 contains information about the miniaturization process based on replacing conventional transmission lines (TLs) by

Fig. 12.8. Conceptual illustration of impedance transformer miniaturization using CMRC.

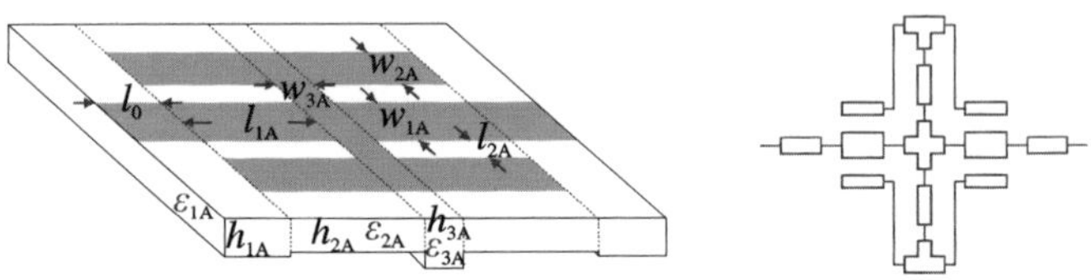

(a) (b)

Fig. 12.9. CMRC component (Cell A): (a) parameterized geometry and the pre-assigned parameters used for implicit space mapping (ISM) correction of the equivalent model; (b) equivalent circuit model.

CMRCs, as well as related design challenges. Here, we only recall, that such a replacement may not only lead to considerable reduction of the circuit size (cf. Fig. 12.8) but also increase the number of geometry parameters to be adjusted and considerably increase of the computational cost of the EM-simulation model.

In practice, a designer develops or selects CMRCs from available topologies (Kurgan *et al.*, 2012; Bekasiewicz and Kurgan, 2014) so that the subject of subsequent simulation-driven optimization (or tuning) procedure are only dimensions of the circuit at hand. Given a large variety of cells, the problem of selecting CMRCs that would ensure higher miniaturization rate or the best electrical performance of the circuit is a non-trivial one. Moreover, the optimum circuit architecture may consist of cells of several types in a specific arrangement. The only way of finding out which configuration is the most advantageous is finding optimum circuit dimensions (for all alternative architectures) and making a comparison. Obviously, such an approach is infeasible when using conventional numerical techniques.

In this section, we demonstrate comparison of various circuit architectures based on their Pareto sets generated for a specific set of cell arrangements. The considered CMRCs are shown in Figs. 12.9

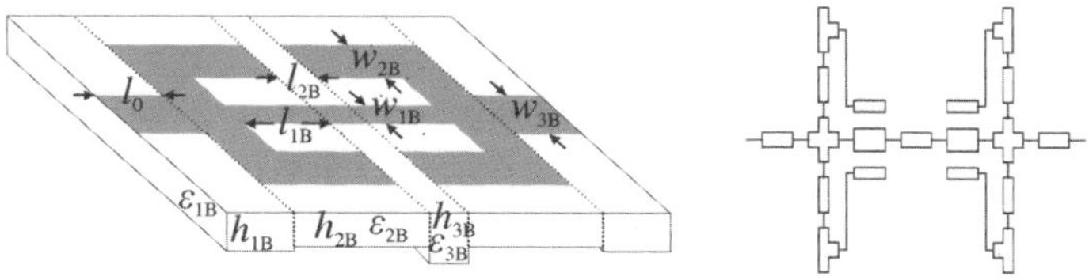

(a) (b)

Fig. 12.10. CMRC component (Cell B): (a) parameterized geometry and the preassigned parameters used for ISM correction of the equivalent model; (b) equivalent circuit model.

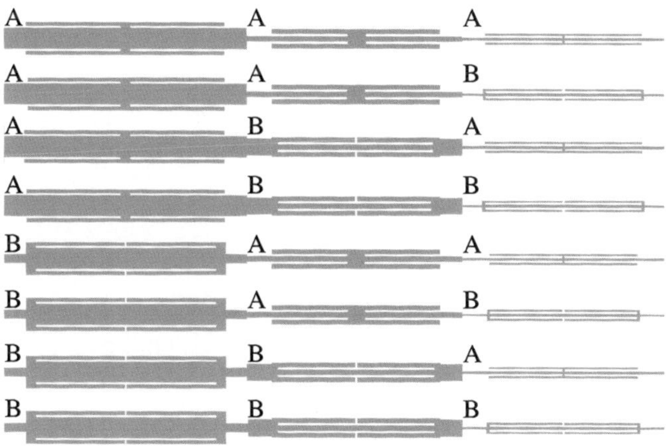

Fig. 12.11. Eight realizations of a three-section wideband impedance transformer using CMRC components of Figs. 12.9 (cell A) and 12.10 (cell B). These realizations are considered in the comparative study aimed at selecting the best transformer architecture for a given application (here, 50–130 Ohm transformer working in 2–5.5 GHz frequency range).

and 12.10, respectively. Figure 12.11 shows eight impedance transformer architectures that will be compared so that the most suitable one can be selected to realize a 50–130 Ohm transformer for 2–5.5 GHz frequency range. In the remaining part of this section, we explain the process of generating Pareto sets as well as present numerical results and their discussion.

12.3.2 *Generation of Pareto Fronts*

The Pareto fronts for the considered realizations of the transformer (cf. Fig. 12.11) are obtained using the multi-objective evolutionary algorithm (MOEA)-based optimization algorithm of Sec. 7.1. However, the no data-driven surrogate model is utilized. Instead, the surrogate model of the transformer is obtained from equivalent circuit models of the respective CMRCs (see Figs. 12.9(b) and 12.10(b)). In the following paragraphs, we describe the process of surrogate model construction in detail.

The first stage of the surrogate modeling process is to obtain alignment between the equivalent circuit and the EM-simulation models for the CMRC structure. For this purpose, we utilize implicit space mapping (ISM) (Bandler *et al.*, 2004b) with the substrate parameters playing the role of the preassigned parameters.

The ISM is chosen because of its flexibility (any number of variables, even for individual microstrip components, can be used if necessary) and the fact that it does not alter (as opposed to the input space mapping, Koziel *et al.*, 2008a) the model variable space. Here, we use six preassigned parameters (three substrate heights and three permittivities) $\boldsymbol{p} = [h_1\ h_2\ h_3\ \varepsilon_1\ \varepsilon_2\ \varepsilon_3]^T$ which have been selected based on engineering experience (cf. Figs. 12.9 and 12.10). The space mapping surrogate $\boldsymbol{s}_{\text{cell}}$ of a CMRC component is obtained as ($\boldsymbol{y}$ represents geometry parameters of the cell)

$$\boldsymbol{s}_{\text{cell}}(\boldsymbol{y}) = \boldsymbol{s}_{I.\text{cell}}(\boldsymbol{y}, \boldsymbol{p}^*). \tag{12.1}$$

Here, $\boldsymbol{s}_{I.\text{cell}}(\boldsymbol{y},\ \boldsymbol{p})$ denotes an equivalent circuit cell model with explicit dependence on its preassigned parameters $\boldsymbol{p}$. The optimum variable vector $\boldsymbol{p}^*$ is obtained by solving the nonlinear parameter extraction process,

$$\boldsymbol{p}^* = \arg\min_{\boldsymbol{p}} \sum_{k=1}^{N_{\text{cell}}} ||\boldsymbol{s}_{I.\text{cell}}(\boldsymbol{y}^{(k)}, \boldsymbol{p}) - \boldsymbol{f}_{\text{cell}}(\boldsymbol{y}^{(k)})||. \tag{12.2}$$

The vectors $\boldsymbol{y}^{(k)}$, $k = 1, \ldots, N_{\text{cell}}$, denote the training (or base) designs. Here, we use the star-distributed (Koziel *et al.*, 2011a) base with $N_{\text{cell}} = 2n + 1$, where n is the number of components in $\boldsymbol{y}$. The

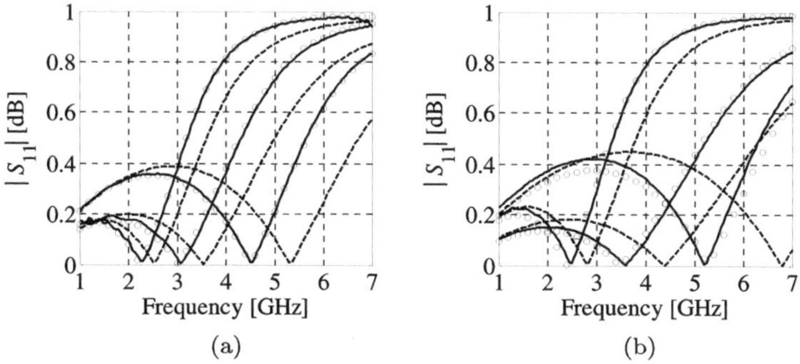

Fig. 12.12. ISM correction of the type A CMRC structure of Fig. 12.9. Equivalent circuit model (– –), EM-simulation model (—) and ISM-corrected circuit model (◯) at: (a) selected training points, (b) random test points.

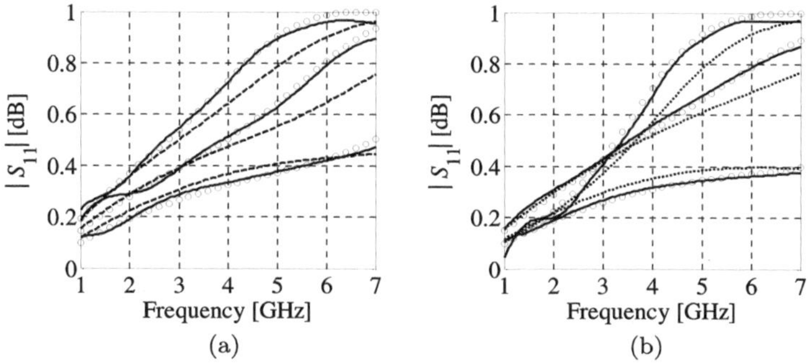

Fig. 12.13. ISM correction of type B CMRC structure (Fig. 12.10). Equivalent circuit model (– –), EM-simulation model (—) and ISM-corrected circuit model (◯) at: (a) selected training points, (b) random test points.

training points are allocated at the center of the parameter space and the centers of its faces.

Figures 12.12 and 12.13 show the responses of the CMRC components (Cells A and B): their equivalent circuit models before and after ISM correction, as well as the corresponding responses of the EM-simulation model of the cell. Verification performed using the random testing points indicates good generalization capability of $\boldsymbol{s}_{\text{cell}}$.

A surrogate model $\boldsymbol{s}(\boldsymbol{x})$ of the entire compact structure is assembled from individual CMRC components corrected using ISM as described above. The geometry parameter vector is $\boldsymbol{x} = [\boldsymbol{y}_1^T \ldots \boldsymbol{y}_p^T]^T$,

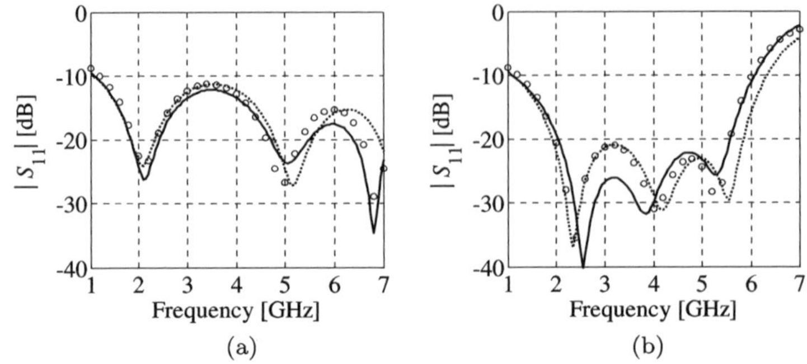

Fig. 12.14. Wideband impedance transformer in AAB configuration (cf. Fig. 12.10). The plots show the responses of the equivalent circuit surrogate model assembled using ISM-corrected CMRC components (– –), the high-fidelity EM transformer model (—), and the frequency scaled surrogate (○) at two different designs denoted as (a) and (b).

where p is the number of CMRC structures and $\boldsymbol{y}_k$ is the parameter vector of the kth cell. Each cell has its own parameter vector $\boldsymbol{p}_k^*$ determined using (12.2).

In order to improve the alignment between $\boldsymbol{s}(\boldsymbol{x})$ and the high-fidelity model of the entire structure, the surrogate can be additionally corrected using frequency space mapping (Koziel and Ogurtsov, 2014a), which is a simple operation and requires only two parameters (here, we use affine frequency space mapping being a composition of a shift and frequency scaling, Bandler *et al.*, 2004a). Moreover, the cross-couplings between CMRC components not accounted for at the level of individual cells usually lead to frequency shifts between $\boldsymbol{s}$ and the high-fidelity model $\boldsymbol{f}$.

The advantages of this approach can be observed in Fig. 12.14 showing the response of a three-section impedance matching transformer consisting of two type A cells and one type B cell (AAB configuration; cf. Fig. 12.11) at two different designs before and after frequency space mapping is applied. Note that frequency scaling improves model alignment in terms of the frequency allocation of the steep parts of the response (obviously, it does not affect the response levels). This is helpful for design refinement.

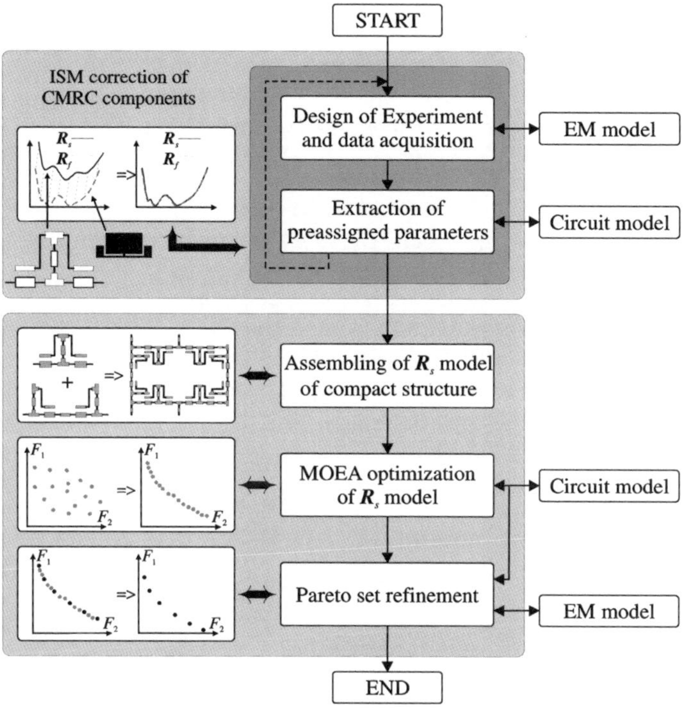

Fig. 12.15. Flow diagram of multi-objective design optimization of decomposable compact structures. Note that ISM correction of the cell is performed only once and its results can be reused for other projects. Dashed line indicates that the ISM step is repeated for all CMRCs considered for construction of a given compact structure.

Generation of the Pareto front is realized, as mentioned before, by optimizing the surrogate model using MOEA and refining selected designs as described in Sec. 7.1.3. Figure 12.15 shows the flow diagram of the optimization process.

12.3.3 *Numerical Results and Comparisons*

Our numerical study concerns comparison of the eight realizations of a three-section wideband 50–130 Ohm impedance matching transformer, implemented on a 0.762 mm thick Taconic RF-35 substrate. The realizations will be referred to as AAA, AAB, ABA, ABB, BAA, BAB, BBA and BBB as shown in Fig. 12.11. Each structure has been designed in a multi-objective sense as described in Sec. 12.3.2. Their respective Pareto fronts are subsequently compared with respect to the structure size and its electrical performance.

The first stage of the process is to carry out ISM correction of the CMRC components. The type A and B cells have five geometry parameters $\boldsymbol{y} = [w_{1A}\ w_{2A}\ w_{3A}\ l_{1A}\ l_{2A}]^T$ (cell A) and $\boldsymbol{y} = [w_{1B}\ w_{2B}\ w_{3B}\ l_{1B}\ l_{2B}]^T$ (cell B). The lower and upper bounds for these parameters are identical for both cell types: $\boldsymbol{l} = [0.1\ 0.1\ 0.1\ 2.0\ 0.1]^T$ mm, and $\boldsymbol{u} = [1.0\ 1.2\ 1.0\ 0.5\ 1.0]^T$ mm. Detailed dimensions of considered CMRCs are marked in Figs. 12.9 and 12.10. The ISM correction is realized using 11 star-distributed high-fidelity training points and results in the following parameter values $\boldsymbol{p}_A^* = [h_1\ h_2\ h_3\ \varepsilon_1\ \varepsilon_2\ \varepsilon_3]^T = [0.979\ 0.158\ 0.219\ 4.28\ 2.00\ 2.00]^T$ and $\boldsymbol{p}_B^* = [1.023\ 0.172\ 0.219\ 4.48\ 2.32\ 2.61]^T$ (substrate height in mm).

The cell responses before and after ISM corrections are shown in Fig. 12.12 (for the A cell) and in Fig. 12.13 (B cell). It can be observed that ISM performs well for both cells ensuring good approximation and generalization of the surrogate.

The initial Pareto front approximations have been obtained by optimizing the transformer surrogates using MOEA. The design objectives are:

- F_1 — circuit size defined as the area of the smallest rectangle encompassing the structure;
- F_2 — maximum in-band reflection, $\max(|S_{11}|)$, for 2–5.5 GHz frequency range.

Figure 12.16 shows the comparison of the Pareto fronts obtained for all eight transformer structures. The cost of evolutionary optimization was up to 20,000 surrogate model evaluations (population size 200, maximum number of iterations 100), which takes about 20 min. In all cases, the evolutionary algorithm was initialized in part randomly and in part using the optimum design of the surrogate model with respect to its electrical performance (design objective F_2). This was done in order to ensure that the Pareto front will contain the F_2-optimized design (which is not guaranteed without appropriate initialization because of complexity of the optimization problem at hand).

The plots in Fig. 12.16 indicate some fundamental differences between various configurations of the impedance transformer of Fig. 12.11. While the range of maximum in-band reflection along the Pareto fronts is similar for all cases (from about –8 dB to about –23 dB), the structures differ considerably with respect to the size. It can be observed that the three realizations, specifically, AAA, AAB and ABB exhibit similar performance. On the other hand, all configurations with the type B cell at the beginning of the structure are characterized by significantly larger size. Clearly, it is expected that the situation may be slightly different after design refinement, however, the general conjecture about advantages of the Axx configurations versus Bxx ones seems to be reasonable.

In the last stage of the design process, 10 designs selected from the initial Pareto front representation have been refined using the algorithm of Sec. 7.1.3. The final Pareto sets obtained this way for all eight realizations of the impedance transformer are shown in Fig. 12.17. The cost of the refinement stage is 50 high-fidelity transformer model evaluations (five per design) for each configuration.

The Pareto sets presented in Fig. 12.17 indicate a general agreement in terms of the front arrangements compared to what was shown in Fig. 12.16 (except the configurations BAB and BBA that perform better than at the level of the surrogate model). The maximum in-band reflection is lower than for the surrogate models (from around –15 dB to –23 dB), which is partially due to the fact that the refinement procedure is focused on more precise local optimization which allows to improve the electrical performance of the structure more than in case of evolutionary optimization.

It can be observed that the two configurations, specifically, AAB and ABB are superior compared to others with ABB being an overall winner (its corresponding Pareto front dominates over all other configurations). This means that ABB configuration is the most advantageous for realizing the considered structure of the three-section impedance matching transformer (for the selected frequency range).

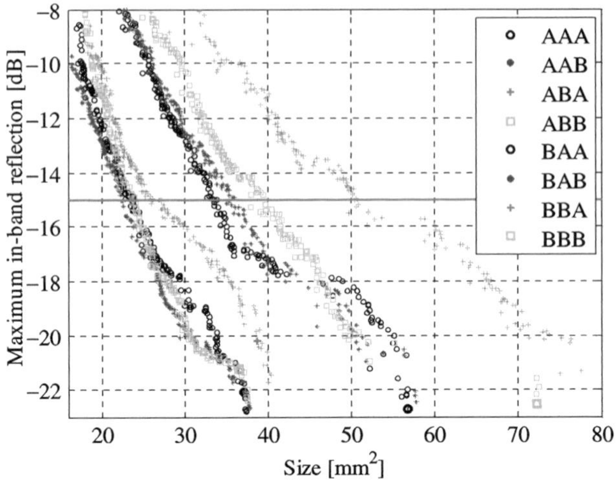

Fig. 12.16. Initial approximations of the Pareto fronts for eight realizations of the three-section wideband impedance matching transformer shown in Fig. 12.11. Acceptable level of the maximum in-band reflection marked using horizontal line.

Table 12.4 Performance comparison of selected Pareto-optimal transformer designs (ABB configuration)

	Design		
Performance parameter	$\boldsymbol{x}^{(1)}$	$\boldsymbol{x}^{(2)}$	$\boldsymbol{x}^{(3)}$
Circuit area [mm^2]	18.2	22.4	34.1
Maximum in-band reflection [dB]	–16.2	–19.3	–22.9

In particular, it offers the best trade-offs between the structure size and its electrical performance.

Figure 12.18 shows frequency responses at the three designs selected from the Pareto front of the ABB transformer configuration. The designs correspond to the smaller transformer size (design $\boldsymbol{x}^{(1)}$), the best electrical performance (design $\boldsymbol{x}^{(3)}$), and the intermediate solution (design $\boldsymbol{x}^{(2)}$). The sizes and maximum in-band reflection levels are given in Table 12.4. The detailed transformer dimensions are gathered in Table 12.5.

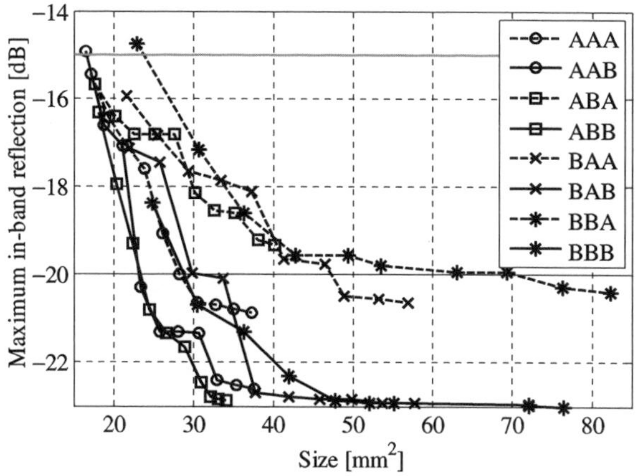

Fig. 12.17. Final Pareto set representations (at the high-fidelity EM-simulation model level) for eight realizations of the three-section wideband impedance matching transformer shown in Fig. 12.11. Acceptable level of the maximum in-band reflection marked using horizontal line.

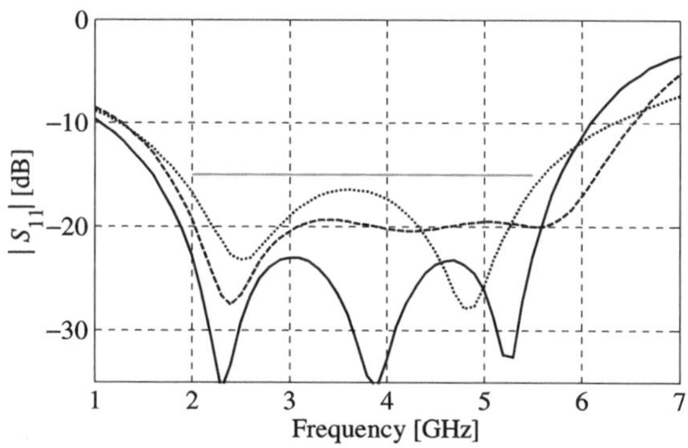

Fig. 12.18. Frequency responses of the selected designs of the optimized transformer (ABB configuration): design $\boldsymbol{x}^{(1)}$ $(\cdots\cdots)$, design $\boldsymbol{x}^{(2)}$ (- - -) and design $\boldsymbol{x}^{(3)}$ (—).

It can be noted that the range of the transformer sizes is from about 18 mm^2 to over 34 mm^2 (almost 50 percent of size reduction from the largest to the smallest design), whereas the corresponding variation of maximum in-band reflection levels is from around -16 dB to around -23 dB.

Table 12.5 Parameters of selected Pareto-optimal designs of the transformer (ABB configuration)

		Design		
Cell	Parameter	$\boldsymbol{x}^{(1)}$	$\boldsymbol{x}^{(2)}$	$\boldsymbol{x}^{(3)}$
A	$w_{1\text{A}}$	0.417	0.489	0.841
	$w_{2\text{A}}$	0.243	0.251	0.198
	$w_{3\text{A}}$	1.000	1.000	0.978
	$l_{1\text{A}}$	2.887	3.107	3.182
	$l_{2\text{A}}$	0.100	0.100	0.119
B	$w_{1\text{B}}$	0.134	0.232	0.422
	$w_{2\text{B}}$	0.204	0.179	0.138
	$w_{3\text{B}}$	1.000	0.485	0.313
	$l_{1\text{B}}$	2.000	2.000	2.461
	$l_{2\text{B}}$	0.100	0.254	0.343
B	$w_{1\text{B}}$	0.100	0.107	0.196
	$w_{2\text{B}}$	0.100	0.103	0.104
	$w_{3\text{B}}$	0.312	0.240	0.223
	$l_{1\text{B}}$	2.000	2.776	3.477
	$l_{2\text{B}}$	0.152	0.183	0.172

As mentioned before, configuration AAB exhibits similar (although not as good) performance. Configurations relying on identical cells (AAA and BBB) are comparable to each other in terms of their size-performance trade-offs although the BBB configuration offers lower reflection. Also, these configurations are considerably larger (29 mm^2 versus 23.5 mm^2, i.e., about 25 percent difference) than the ABB configuration for maximum in-band reflection of -20 dB.

The examples considered here demonstrate that fast design optimization methods allow systematic and comprehensive comparison of various CMRC components in terms of their suitability to implement compact microwave structures (here, an impedance matching transformer).

Chapter 13

Discussion and Recommendations

Simulation-driven design of antenna structures is an exciting area of research yet of paramount importance for engineering practice. One of its fundamental challenges is high computational cost of full-wave electromagnetic (EM) antenna analysis. High-fidelity EM simulation models are necessary to ensure accurate evaluation of the antenna performance but their optimization using conventional numerical techniques is impractical or — in extreme cases — even prohibitive. Multi-objective optimization, the main focus of this book, is considerably more challenging because the aim is to find the entire set of designs that represent the best possible trade-offs between several (and often conflicting) objectives. In multi-objective design, the high cost of evaluating an antenna structure becomes a more serious bottleneck than for single-objective optimization. Another important issue is multi-dimensionality of the design space and potentially wide ranges of geometry parameters of the antenna. In particular, global search needs to be involved, first to identify a relevant portion of the design space (the one that contains Pareto-optimal designs), then to find a representation of the Pareto front. None of these can be executed directly at the level of EM-simulation models, especially if optimization is conducted using population-based metaheuristics.

The purpose of this book was twofold. In the broad sense, it was supposed to provide introduction to multi-objective optimization as well as an outline of the fundamental solution techniques. For the sake of being self-contained, the book also contains some background material concerning single-objective optimization, both conventional (gradient-based as well as derivative-free) and surrogate-assisted. Consequently, the presented material can be considered as a more or less complete — yet basic — introduction to numerical optimization including multi-objective design (with the emphasis on metaheuristics)

The second purpose of the book was to introduce the subject of multi-objective design optimization of antenna structures and demonstrate techniques that allow for conducting such optimization in a computationally efficient manner. We focused on simulation-driven optimization and formulated methods for generating Pareto-optimal designs at a practically acceptable cost. The approach considered here combine design space reduction, variable-fidelity EM-simulations, physics-based and data-driven surrogate modeling, as well as response correction schemes. The three basic algorithms were presented. The first technique utilizes multi-objective evolutionary algorithms (MOEAs) to find an initial approximation of the Pareto front using an auxiliary data-driven surrogate, constructed from coarse-discretization EM-simulations of the antenna at hand. The final Pareto set (at the level of the high-fidelity EM model) was obtained using response correction methods. In order to permit computationally feasible construction of the surrogate, the ranges of antenna geometry parameters have to be restricted by means of design space reduction methods. This MOEA-based technique is quite generic because no specific assumptions are made about the Pareto front geometry. On the other hand, it still relies on population-based metaheuristics.

The second sequential domain patching (SDP) and the third algorithms (point-by-point Pareto front exploration) are fully deterministic. SDP creates a path connecting the extreme Pareto-optimal solutions (found by independent single-objective optimization runs), and the initial Pareto set is further refined using response correction methods. On the other hand, Pareto front exploration approach finds subsequent trade-off solutions by executing a series of constrained local optimization runs aided by local response surface approximation models. The computational cost of these methods is typically lower than for the MOEA-based technique but an implicit assumption is that the Pareto front is a connected set. Fortunately, this assumption normally holds for majority of antenna-related design problems. Also, both deterministic methods presented here are formulated for two-objective problems only. Reformulation for higher-dimensional problem is possible and rather straightforward but would increase computational complexity of the respective algorithms.

The optimization methods presented in the book have been comprehensively demonstrated using several antenna structures of various complexities from a few to over 20 adjustable parameters. Furthermore, applications of the technique for solving multi-objective design problems in other engineering areas including microwave and aerospace have been discussed. Some of the test cases have been solved using several methods which permit direct comparison of their advantages and disadvantages. A separate chapter (Chapter 10) was focused on investigating practically relevant properties of the optimization algorithms such as their scalability (with respect to dimensionality of the design space), the influence of MOEA performance on the quality of the Pareto front produced by the algorithm, as well as performance/cost trade-offs of the SDP algorithm with respect to the patch size setup.

The final outcome of multi-objective optimization (as understood in this book, i.e., from the point of view of posteriori preference articulation, Coello Coello *et al.*, 2007) is to find a set of trade-off designs. In Chapter 12, we demonstrated how such information can be utilized for performance comparison of antenna and microwave devices, either to find out the limits of a given structure in terms of its electrical performance or minimum attainable size (while maintaining acceptable characteristics), or to find the optimum circuit architecture. Clearly, a prerequisite for this kind of comparisons is availability of methods for fast multi-objective optimization.

There are several conclusions that can be drawn from the numerical (in some cases also experimental) results presented in the book:

- Surrogate-assisted methods allow for reliable and low-cost multi-objective optimization of antenna structures. Depending on a problem complexity (within the range considered in the book), the cost of generating a 10-element Pareto set corresponds to anything between a few dozen to a few (two–three) hundreds of high-fidelity EM-simulations of the antenna structure under design.
- The three basic methods considered (MOEA-based optimization with data-driven surrogate, SDP, and point-by-point Pareto front exploration) are comparable in terms of performance. The first method is more expensive yet more generic.
- The scaling properties of the methods, specifically, computational cost versus problem dimensionality, are practically attractive: at most quadratic for MOEA-based algorithm as found based on numerical experiments; and similar or better for deterministic methods, as it follows from the very formulations of the respective techniques. This means that all of these methods can potentially be applied for even more complex antenna design problems.
- The methods are robust; in particular, the Pareto sets produced are rather insensitive to control parameters setups SDP or performance of some key components (MOEA).

Computational cost of the surrogate-assisted optimization algorithms can be further reduced by utilizing more efficient local optimization methods (for finding the extreme points of the Pareto front in case of all considered techniques, or for moving along the Pareto front for the point-by-point technique). One of the possibilities is utilization of adjoint sensitivity techniques whenever available (Ghassemi *et al.*, 2013; Koziel and Bekasiewicz, 2015a).

Majority of the test cases presented in the book involve two-objectives, which is the most typical case in practical applications (e.g., electrical performance versus size for compact antennas). The complexity of multi-objective optimization problem increases with the increase of the number of objectives (in particular, for more than three- or four-objectives, the very concept of Pareto domination becomes problematic as a discriminative factor driving the solutions towards the Pareto front, Yang *et al.*, 2013). As mentioned before, the deterministic techniques presented here can be reformulated to handle three or more objectives but their computational complexity would increase considerably.

From the point of view of practical applications of the surrogate-assisted multi-objective algorithms presented in this book, one needs to consider several factors such as implementation details, generality, computational complexity, potential for further developments, and, of course, suitability for specific design problems. Deterministic methods (Secs. 7.2 and 7.3) are simpler to implement and their computational complexity is slightly lower than for MOEA-based approaches. Yet, as formulated in the book, they only allow handling of two-objectives. Although this is the most typical situation, three- or more-objective designs might be necessary in some cases. On the other hand, it is straightforward to improve computational efficiency of these techniques by means of, e.g., adjoint sensitivities. MOEA-based methods (Sec. 7.1), both with response-correction-based or co-Kriging-based Pareto front refinement, are more general and can handle several objectives using the same formulation. However, they are more difficult to implement because several components are required there such as local optimization algorithms, data-driven surrogate modeling procedures, response correction algorithms, etc. It should be noted that some of these building blocks are freely available over the internet (e.g., surrogate-modeling toolboxes, Lophaven *et al.*, 2002; Gorissen *et al.*, 2010) or built in popular programming environments (e.g., MATLAB, 2012). In practice, all presented approaches need to be supplemented by design space reduction techniques (Chapter 8), which — in some cases — may require globalized search.

Again, one needs to reiterate that multi-objective antenna design is a very challenging task. In particular, conventional multi-objective design by means of straightforward handling of high-fidelity EM-simulation models using population-based metaheuristics is prohibitively expensive in vast majority of situations. As demonstrated, design aided by surrogate modeling techniques and variable-fidelity simulations is a way to conduct such design in a computationally tractable manner. Yet, for some problems, individualized approach is necessary (depending on Pareto front geometry, see, e.g., Secs. 9.1, 9.3 or 9.9). The authors of this book do hope that the presented material will be helpful for the readers interested in multi-objective simulation-driven design optimization of antenna structures (but also computationally expensive models in other engineering disciplines), not only in the sense of providing ready-to use algorithms and design procedures but also to serve as an inspiration to develop their own methods.

Finally, a comment is necessary concerning the perspectives and future developments of surrogate-assisted techniques for antenna design, both in single- and multi-objective contexts. Surrogate-based algorithms, particularly those exploiting variable-fidelity EM-simulations, can be extremely expensive with satisfactory designs (whether it is a single solution or the entire Pareto front representation) found at a low computational cost corresponding to a few evaluations of the high-fidelity EM model of the antenna of interest (in case of single-objective design) or a few dozen to a few 100 of evaluations in case of multi-objective optimization. The presented multi-objective design procedures have been demonstrated for antenna structures with over 20 parameters. Still, more work needs to be done in order to handle even more complex cases. Some open problems remain, such as a proper (an automated) choice of the low-fidelity EM model, which may affect both the cost and performance of the optimization algorithm. Successful application of surrogate-assisted methods still partially depends on user experience. The research attempting to fully automate the algorithms is ongoing. Automation, including implementation of appropriate software tools where most of the decisions regarding model and algorithm selection as well as the setup of relevant control parameters could be done without user interaction, is a necessary condition for wider acceptance of surrogate-based methods in antenna and other engineering communities.

References

Abbass, H.A., Sarker, R., Newton, C. (2001). PDE: A Pareto-frontier differential evolution approach for multi-objective optimization problems. *Proc. Congress Evol. Comp.*, **2**, 971–978, New Jersey: IEEE Service Center.

ADS (Advanced Design System) (2011). Agilent (Keysight) Technologies, Fountaingrove Parkway 1400, Santa Rosa, CA, US.

Afshinmanesh, F., Marandi, A., Shahabadi, M. (2008). Design of a single-feed dual-band dual-polarized printed microstrip antenna using a Boolean particle swarm optimization, *IEEE Trans. Ant. Prop.*, **56**, 1845–1852.

Alander, J.T., Zinchenko, L.A., Sorokin, S.N. (2004). Comparison of fitness landscapes for evolutionary design of dipole antennas, *IEEE Trans. Ant. Prop.*, **52**, 2932–2940.

Alba, E., Marti, R. (eds.) (2006). *Metaheuristic procedures for training neural networks*, Springer, New York, US.

Alexandrov, N.M., Lewis, R.M. (2001). An overview of first-order model management for engineering optimization, *Opt. Eng.*, **2**, 413–430.

Alexandrov, N.M., Dennis, J.E., Lewis, R.M., Torczon, V. (1998). A trust region framework for managing use of approximation models in optimization, *Struct. Multidisciplinary Optim.*, **15**, 16–23.

Andrés, E., Salcedo-Sanz, S., Monge, F., Pérez-Bellido, A.M., (2012). Efficient aerodynamic design through evolutionary programming and support vector regression algorithms, *Int. J. Expert Systems with Applications.* **39**, 10700–10708.

Angiulli, G., Cacciola, M., Versaci, M. (2007). Microwave devices and antennas modelling by support vector regression machines, *IEEE Trans. Magnetics*, **43**, 1589–1592.

Antoniou, A., Lu, W.-S. (2007). *Practical Optimization: Algorithms and Engineering Applications.* Springer, US.

Aljibouri, B., Lim, E., Evans, H., Sambell, A. (2000). Multiobjective genetic algorithm approach for a dual-feed circular polarised patch antenna design, *Electron. Lett.*, **36**, 1005–1006.

Aljibouri, B., Sambell, A., Sharif, B. (2008). Application of genetic algorithm to design of sequentially rotated circularly polarised dual-feed microstrip patch antenna array, *Electron. Lett.*, **44**, 708–709.

Almpanis, G., Fumeaux, C., Vahldieck, R. (2008). The trapezoidal dielectric resonator antenna, *IEEE Trans. Ant. Prop.*, **56**, 2810–2816.

Antenna Magus. (2015). Magus (Pty) Ltd, 3 Meson Street, Technopark Stellenbosch 7600, South Africa.

Ares-Pena, F.J., Rodriguez-Gonzalez, J.A., Villanueva-Lopez, E., Rengarajan, S.R. (1999). Genetic algorithms in the design and optimization of antenna array patterns, *IEEE Trans. Ant. Prop.*, **47**, 506–510.

Azaro, R., Zeni, E., Rocca, P., Massa, A. (2008). Innovative design of a planar fractal-shaped GPS/GSM/Wi-Fi antenna, *Microwave Opt. Technol. Lett.*, **50**, 825–829.

Babu, B.V., Jehan, M.M.L. (2003). Di?erential Evolution for Multi-Objective Optimization. *Proc. Congress Evol. Comp.*, **4**, 2696–2703, IEEE Press.

Bäck, T. (1996). *Evolutionary Algorithms in Theory and Practice*, Oxford University Press, New York, US.

Back, T., Fogel, D.B., Michalewicz, Z. (eds.). (2000). *Evolutionary computation 1: basic algorithms and operators*, Institute of Physics Publishing, Bristol, UK.

Bai, X., Qu, S.W., Yang, S., Hu, J., Nie, Z.P. (2016). Millimeter-wave circularly polarized tapered-elliptical cavity antenna with wide axial-ratio beamwidth, *IEEE Trans. Ant. Prop.*, **64**, 811–814.

Bakr, M.H., Bandler, J.W., Biernacki, R.M., Chen, S.H., Madsen, K. (1998). A trust region aggressive space mapping algorithm for EM optimization, *IEEE Trans. Microwave Theory Tech.*, **46**, 2412–2425.

Bakr, M.H., Bandler, J.W., Georgieva, N.K., Madsen, K. (1999). A hybrid aggressive space-mapping algorithm for EM optimization, *IEEE Trans. Microwave Theory Tech.*, **47**, 2440–2449.

Balanis, C.A. (2005). *Antenna theory analysis and design*, 3rd ed., John Wiley & Sons, Hoboken, New York, US.

Bandler, J.W., Seviora, R.E. (1972). Wave sensitivities of networks, *IEEE Trans. Microwave Theory Tech.*, **20**, 138–147.

Bandler, J.W., Cheng, Q.S., Gebre-Mariam, D.H., Madsen, K., Pedersen, F., Søndergaard, J. (2003). EM-based surrogate modeling and design exploiting implicit, frequency and output space mappings. *IEEE Int. Microwave Symp. Digest*, Philadelphia, PA, US.

Bandler, J.W., Biernacki, R.M., Chen, S.H., Grobelny, P.A., Hemmers, R.H. (1994). Space mapping technique for electromagnetic optimization, *IEEE Trans. Microwave Theory Tech.*, **42**, 2536–2544.

Bandler, J.W., Biernacki, R.M., Chen, S.H., Hemmers, R.H., Madsen, K. (1995). Electromagnetic optimization exploiting aggressive space mapping, *IEEE Trans. Microwave Theory Tech.*, **43**, 2874–2882.

Bandler, J.W., Cheng, Q.S., Dakroury, S.A., Mohamed, A.S., Bakr, M.H., Madsen, K., Søndergaard, J. (2004a). Space mapping: the state of the art, *IEEE Trans. Microwave Theory Tech.*, **52**, 337–361.

Bandler, J.W., Cheng, Q.S., Nikolova, N.K., Ismail, M.A. (2004b). Implicit space mapping optimization exploiting preassigned parameters, *IEEE Trans. Microwave Theory Tech.*, **52**, 378–385.

Bansal, J.C., Singh, P.K., Saraswat, M., Verma, A., Jadon, S.S., Abraham, A. (2011). Inertia weight strategies in particle swarm optimization, *World Congress Nature Biologically Inspired Comp.*, 633–640.

Bates, D.M., Watts, D.G. (1988). *Nonlinear Regression and Its Applications*, John Wiley & Sons, New York, US.

Bayderkhani, R., Forooraghi, K., Abbasi-Arand, B. (2015). Gain-enhanced SIW cavity-backed slot antenna with arbitrary levels of inclined polarization, *IEEE Ant. Wireless Prop. Lett.*, **14**, 931–934.

Bayraktar, Z., Komurcu, M., Bossard, J.A., Werner, D.H. (2013). The wind driven optimization technique and its application in electromagnetics, *IEEE Trans. Ant. Prop.*, **61**, 2745–2757.

Beachkofski, B., Grandhi, R. (2002). Improved distributed hypercube sampling. *Structures, Structural Dynamics, Materials Conf.*, Paper AIAA 2002-1274.

Bekasiewicz, A., Koziel, S. (2015a). Structure and computationally-efficient simulation-driven design of compact UWB monopole antenna, *IEEE Ant. Wireless Prop. Lett.*, **14**, 1282–1285.

Bekasiewicz, A., Koziel, S. (2015b). Efficient multi-fidelity design optimization of microwave filters using adjoint sensitivity, *Int. J. RF Microwave Comp. Aid. Eng.*, **25**, 178–183.

Bekasiewicz, A., Koziel, S. (2016a). Cost-efficient design optimization of compact patch antennas with improved bandwidth, *IEEE Ant. Wireless Prop. Lett.*, 270–273.

Bekasiewicz, A., Koziel, S. (2016b). A novel structure and design optimization of compact UWB slot antenna, *Electron. Lett.*, **52**, 681–682.

Bekasiewicz, A., Kurgan, P. (2014). A compact microstrip rat-race coupler constituted by nonuniform transmission lines, *Microwave Opt. Technol. Lett.*, **56**, 970–974.

Bekasiewicz, A., Kurgan, P., Kitlinski, M. (2012). A new approach to a fast and accurate design of microwave circuits with complex topologies, *IET Microwaves, Ant. Prop.*, **6**, 1616–1622.

Bekasiewicz, A., Koziel, S., Zieniutycz, W. (2014a). Design space reduction for expedited multi-objective design optimization of antennas in highly-dimensional spaces, in (S. Koziel, L. Leifsson, X.-S. Yang, eds.) *Solving Computationally Expensive Engineering Problems: Methods and Applications*, 113–147, Springer, New York, US.

Bekasiewicz, A., Koziel, S., Leifsson, L. (2014b). Computationally efficient multi-objective optimization of and experimental validation of Yagi-Uda antenna, *Int. Conf. Simulation Modeling Methodologies, Technologies App.*, 798–805.

Bekasiewicz, A., Koziel, S., Bandler, J. W. (2016a). Low-cost multi-objective design of compact microwave structures using domain patching, *IEEE Int. Microwave Symp.* 1–3.

Bekasiewicz, A., Koziel, S., Zieniutycz, W., Leifsson, L. (2016b). Expedited simulation-driven multi-objective design optimization of quasi-isotropic dielectric resonator antenna, in (Koziel S., Leifsson. L., Yang., X.-S.

eds.) *Simulation-driven modeling and optimization*, 207–231, Springer, New York, US.

Bertsekas, D.P. (1982). *Constrained Optimization and Langrange Multiplier Methods*, Academic Press, New York, US.

Bevelacqua, P.J., Balanis, C.A. (2007). Optimizing Antenna Array Geometry for Interference Suppression, *IEEE Trans. Ant. Prop.*, **55**, 637–641.

Beyer, H.-G. (2001). *The Theory of Evolution Strategies*, Springer-Verlag, Berlin, Germany.

Beyer, H.-G., Schwefel, H.-P. (2002). Evolution strategies: a comprehensive introduction, *Natural Computing*, **1**, 3–52.

Binh, T.T. Korn, U. (1997). Multiobjective evolution strategy for constrained optimization problems, *Proc. IMACS World Congress Scientific Comp., Modelling Applied Math.*, 357–362.

Bischof, C., Corliss, G., Griewank, A. (1993). Structured second- and higher-order derivatives through univariate Taylor series, *Opt. Methods Soft.*, **2**, 211–232.

Booker, A.J., Dennis, J.E., Frank, P.D., Serafini, D.B., Torczon, V., Trosset, M.W. (1999). A rigorous framework for optimization of expensive functions by surrogates, *Structural Opt.*, **17**, 1–13.

Braembussche, R.A. (2008). Numerical optimization for advanced turbomachinery design, in (Thevenin, D., Janiga, G. eds.) *Optimization and Computational Fluid Dynamics*, 147–189, Springer, Berlin, Germany.

Broyden, C.G. (1965). A class of methods for solving nonlinear simultaneous equations, *Math. Comp.*, **19**, 577–593.

Buell, K., Mosallaei, H., Sarabandi, K. (2006). A substrate for small patch antennas providing tunable miniaturization factors, *IEEE Trans. Microwave Theory Tech.*, **54**, 135–146.

Caorsi, S., Massa, A., Pastorino, M. (2000). A computational technique based on a real-coded genetic algorithm for microwave imaging purposes, *IEEE Trans. in Geoscience Remote Sens.*, **38**, 1697–1708.

Chahat, N., Zhadobov, M., Sauleau, R., Ito, K. (2011). A compact UWB antenna for on-body applications, *IEEE Trans. Ant. Prop.*, **59**, 1123–1131.

Chakraborty, U. (2008). *Advances in Differential Evolution.* Studies in Computational Intelligence, Springer, Berlin, Germany.

Chamaani, S., Abrishamian, M.S., Mirtaheri, S.A. (2010) Time-domain design of UWB Vivaldi antenna array using multiobjective particle swarm optimization, *IEEE Ant. Wireless Prop. Lett.*, **9**, 666–669.

Chang, C.S., Xu, D.Y., Quek, H.B. (1999). Pareto-optimal set based multiobjective tuning of fuzzy automatic train operation for mass transit system, *IEE Proc. Electric Power App.*, **146**, 577–583.

Changkong, V., Haimes, Y.Y. (1983). *Multiobjective Decision Making Theory and Methodology.* North-Holland, New York, US.

Cheng, Q.S., Koziel, S., Bandler, J.W. (2006). Simplified space mapping approach to enhancement of microwave device models, *Int. J. RF and Microwave Computer-Aided Eng.*, **16**, 518–535.

Cheng, Q.S., Bandler, J.W., Koziel, S. (2010). Space mapping design framework exploiting tuning elements, *IEEE Trans. Microwave Theory Tech.*, **58**, 136–144.

Chu, Q.-X., Mao, C.-X., Zhu, H. (2013). A compact notched band UWB slot antenna with sharp selectivity and controllable bandwidth, *IEEE Trans Ant. Prop.*, **61**, 3961–3966.

Chung, Y.S., Cheon, C., Park, I.-H., Hahn, S.Y. (2001). Optimal design method for microwave device using time domain method and design sensitivity analysis. II. FDTD case, *IEEE Trans. Magnetics*, **37**, 3255–3259.

Clemens, M., Weiland, T. (2001). Discrete electromagnetism with the finite integration technique, *Prog. EM Research*, **32**, 65–87.

Clerc, M. (2006). *Particle Swarm Optimization.* ISTE, London, UK.

Clerc, M., Kennedy, J. (2002). The particle swarm — explosion, stability, and convergence in a multidimensional complex space, *IEEE Trans. Evol. Comp.*, **6**, 58–73.

Coello Coello, C.A., Lamont, G.B., Van Veldhuizen, D. (2007). *Evolutionary algorithms for solving multi-objective problems.* 2nd ed., Springer, New York, US.

Conn, A.R., Gould, N.I.M., Toint, P.L. (2000). *Trust Region Methods.* MPS-SIAM Series on Optimization, MPS-SIAM, Philadelphia, US.

Conn, A.R., Scheinberg, K., Vincente, L.N. (2009). *Introduction to Derivative-Free Optimization*, MPS-SIAM Series on Optimization, MPS-SIAM, Philadelphia, US.

Couckuyt, I., (2013). *Forward and inverse surrogate modeling of computationally expensive problems.* PhD Thesis, Ghent University.

Crowder, H.P., Wolfe, P. (1972). Linear convergence of the conjugate gradient method, *IBM J. Res. Development*, **16**, 431–433.

CST. (2013). CST Microwave Studio, CST AG, Bad Nauheimer Str. 19, D-64289 Darmstadt, Germany.

Da, Y., Xiurun, G. (2005). An improved PSO-based ANN with simulated annealing technique, *Neurocomputing*, **63**, 527–533.

David, H., (1994). A Markov Chain analysis of genetic algorithms with a state dependent fitness function, *Complex Systems*, **8**, 407–417.

Davidson, D.B. (2010). *Computational Electromagnetics for RF and Microwave Engineering*, 2nd ed., Cambridge University Press, Cambridge, UK.

Davis, C. (1954). Theory of positive linear dependence, *Amer. J. Math.*, **76**, 733–746.

De Jong, A.K. (1975). *An Analysis of the Behavior of a Class of Genetic Adaptive Systems.* PhD thesis, University of Michigan.

Dean, A., Lewis, S. (2006). *Screening: Methods for Experimentation in Industry, Drug Discovery, and Genetics*, Springer-Verlag, New York, US.

Deb, K. (2001). *Multi-Objective Optimization Using Evolutionary Algorithms*, John Wiley & Sons, Chichester, UK.

Deb, K., Pratap, A., Agarwal, S., Meyarivan, T. (2002). A fast and elitist multiobjective genetic algorithm: NSGA-II, *IEEE Trans. Evol. Comp.*, **6**, 182–197.

Deb, K., Goldberg, D.E. (1989). An investigation of niche and species formation in genetic function optimization, *Proc. Int. Conf. Genetic Algorithms*, 42–50.

Ding, D., Wang, G. (2013). Modified multiobjective evolutionary algorithm based on decomposition for antenna design, *IEEE Trans. Ant. Prop.*, **61**, 5301–5307.

Ding, M., Jin, R., Geng, J., Wu, Q., Yang, G. (2008). Auto-design of band-notched UWB antennas using mixed model of 2D GA and FDTD, *Electron. Lett.*, **44**, 257–258.

Director, S.W., Rohrer, R.A. (1969). The Generalized Adjoint Network and Network Sensitivities, *IEEE Trans. Circuit Theory*, **16**, 318–323.

Duffy, A., Zhang, G., Koziel, S., Wang, L. (2015). Objective selection of minimum acceptable mesh refinement for EMC simulations, *IEEE Trans. EM Compatibility*, **57**, 1266–1269.

Echeverría, D., Hemker, P.W. (2005). Space mapping and defect correction, *Comput. Methods Appl. Math.*, **5**, 107–136.

Echeverría, D., Hemker, P.W. (2008). Manifold mapping: a two-level optimization technique, *Computing Vis. Sci.*. **11**, 193–206.

Eiben, A.E., Smith, J.E. (2003). *Introduction to Evolutionary Computing.* Springer-Verlag, Berlin Heidelberg, Germany.

El-Khouly, E., Ghali, H., Khamis, S.A. (2007). High directivity antenna using a modified Peano space-filling curve, *IEEE Ant. Wireless Prop. Lett.*, **6**, 405–407.

Farin G. (1993). *Curves and Surfaces for Computer Aided Geometric Design*, Academic Press, Boston, US.

FEKO. (2012). FEldberechnung für Körper mit beliebiger Oberfläche, Altair Development S.A. (Pty) Ltd, 32 Techno Avenue, Technopark Stellenbosch 7600, South Africa.

Fletcher, R. (1987). *Practical Methods of Optimization*, 2nd ed., John Wiley & Sons, New York, US.

Fletcher R., Reeves, C.M. (1964). Function minimization by conjugate gradients, *Computer J.*, **7**, 149–154.

FLUENT, (2012). ANSYS Inc., Southpointe, 275 Technology Drive, Canonsburg, PA 15317, US.

Fong, S., Wong, R., Vasilakos, A.V. (2016). Accelerated PSO swarm search feature selection for data stream mining big data, *IEEE Trans. Services Computing,* **9**, 33–45.

Fonseca, C.M. (1995). *Multiobjective genetic algorithms with application to control engineering problems.* PhD thesis, Department of Automatic Control and Systems Engineering, University of Sheffield, UK.

Forrester, A.I.J., Keane, A.J. (2009). Recent advances in surrogate-based optimization, *Prog. Aerospace Sci.*, **45**, 50–79.

Forrester, A.I.J., Sobester, A., Keane, A.J. (2007). Multi-fidelity optimization via surrogate modeling, *Proc. Royal Society A*, **463**, 3251–3269.

Forrester, A.I.J., Sobester, A., Keane, A.J. (2008). *Engineering Design via Surrogate Modelling: A Practical Guide*, John Wiley & Sons, Hoboken, New York, US.

Fourman, M.P. (1985). Compaction of symbolic layout using genetic algorithms, in (Grefenstette, J.J. Ed.) *Genetic Algorithms and their Applications: Proc. Int. Conf. Genetic Algorithms*, 141–153. Lawrence Erlbaum, New Jersey, US.

Gautam, A.K., Yadav, S., Kanaujia, B.K. (2013). A CPW-fed compact UWB microstrip antenna, *IEEE Ant. Wireless Prop. Lett.*, **12**, 151–154.

Gazda, C., Couckuyt, I., Rogier, H., Vande Ginste, D., Dhaene, T. (2012). Constrained multiobjective optimization of a common-mode suppression filter, *IEEE Trans. EM Compatibility*, **54**, 704–707.

Gazi, V., Passino, K.M. (2004). Stability analysis of social foraging swarms, *IEEE Trans. Systems, Man, Cybernetics, Part B: Cybernetics*, **34**, 539–557.

Geem, Z.W., Kim, J.H., Loganathan, G.V. (2001). A new heuristic optimization algorithm: harmony search, *Simulation*, **76**, 60–68.

Geisser, S. (1993). *Predictive Inference.* Chapman and Hall.

Gembicki, F. (1974). *Vector Optimization for Control with Performance and Parameter Sensitivity Indices.* PhD thesis, Case Western Reserve University.

Georgieva, N.K., Glavic, S., Bakr, M.H., Bandler, J.W. (2002). Feasible adjoint sensitivity technique for EM design optimization, *IEEE Trans. Microwave Theory Tech.*, **50**, 2751–2758.

Ghassemi, M., Bakr, M., Sangary, N. (2013). Antenna design exploiting adjoint sensitivity-based geometry evolution, *IET Microwaves Ant. Prop.*, **7**, 268–276.

Giannakis, I., Giannopoulos, A., Warren, C. (2016). A realistic FDTD numerical modeling framework of ground penetrating radar for landmine detection, *IEEE J. Selected Topics Applied Earth Observations Remote Sens.*, **9**, 37–51.

Giunta, A.A., (1997). *Aircraft multidisciplinary design optimization using design of experiments theory and response surface modeling methods.* PhD Thesis, Virginia Polytechnic Institute and State University.

Giunta, A.A., Eldred, M.S. (2000). Implementation of a trust region model management strategy in the DAKOTA optimization toolkit. *Proc. Symp. Multidisciplinary Analysis Opt.*, Long Beach, CA, AIAA-2000-4935.

Giunta, A.A., Wojtkiewicz, S.F., Eldred, M.S. (2003). Overview of modern design of experiments methods for computational simulations. *Aerospace Sci. Meeting Exhibit*, paper AIAA 2003–0649.

Goldberg, D.E. (1989). *Genetic Algorithms in Search, Optimization and Machine Learning.* Addison-Wesley Longman Publishing, Boston, MA, US.

Goldberg, D.E., Richardson, J. (1987). Genetic algorithms with sharing for multimodal function optimization, *Proc. Int. Conf. Genetic Algorithms App.*, 41–49.

Golub, G.H., Van Loan, Ch.F. (1996). *Matrix Computations.* 3rd ed., Johns Hopkins University Press, Baltimore, MD, US.

Gorissen, D., Crombecq, K., Couckuyt, I., Dhaene, T., Demeester, P. (2010a). A surrogate modeling and adaptive sampling toolbox for computer based design, *J. Machine Learning Res.*, **11**, 2051–2055.

Gorissen, D., Couckuyt, I., Laermans, E., Dhaene, T. (2010b). Multiobjective global surrogate modeling, dealing with the 5-percent problem, *Eng. with Computers*, **26**, 81–98.

Goudos, S., Zaharis, Z., Kampitaki, D., Rekanos, I., Hilas, C. (2009). Pareto optimal design of dual-band base station antenna arrays using multi-objective particle swarm optimization with fitness sharing, *IEEE Trans. Magnetics*, **45**, 1522–1525.

Gunn, S.R. (1998). Support vector machines for classification and regression. Technical Report. School of Electronics and Computer Science, University of Southampton.

Guney, K., Basbug, S. (2013). Null synthesis of time-modulated circular antenna arrays using an improved differential evolution algorithm, *IEEE Ant. Wireless Prop. Lett.*, **12**, 817–820.

Guo, L., Chen, X., Parini, C. (2012). Miniature ultra-wideband antenna for wireless universal serial bus dongle applications, *IET Microwaves, Ant. Prop.*, **6**, 113–119.

Gutmann, H.-M. (2001). A radial basis function method for global optimization, *J. Global Opt.*, **19**, 201–227.

Hammersley, J.M. (1960). Monte-Carlo methods for solving multivariable problems, *Annals New York Academy Sci.*, **86**, 844–874.

Han, S.P., (1977). A globally convergent method for nonlinear programming, *J. Optim. Theory Appl.*, **22**, 297–309.

Hanninen, I. (2012). Optimization of a reflector antenna system, Whitepaper, Computer Simulation Technology AG.

Hansen, E., Walster, G.W. (2003). *Global Optimization Using Interval Analysis.* 2nd ed., Marcel Dekker, New York, US.

Hansen, N., Kern, S. (2004). Evaluating the CMA evolution strategy on multimodal test functions, in (X. Yao *et al.* eds.) *Parallel Problem Solving from Nature*, 282–291, Springer-Verlag, Berlin Heidelberg, Germany.

Hansen, N., Arnold, D.V., Auger, A. (2015). Evolution Strategies, in (Kacprzyk, J., Pedrycz, W., eds.) *Springer Handbook of Computational Intelligence*, 871–898, Springer-Verlag, Berlin Heidelberg, Germany.

Harrington, R.F. (1993). *Field Computation by Moment Methods.* Hoboken, Wiley-IEEE Press, New Jersey, US.

Haykin, S. (1998) Neural Networks: A Comprehensive Foundation. 2nd ed. Upper Saddle River: Prentice Hall.

Hendrix, E.M.T., Gazdag-Tóth, B. (2010). *Introduction to Nonlinear and Global Optimization*, Springer-Verlag, New York, US.

Hesser, J., Männer, R. (1991). Towards an optimal mutation probability for genetic algorithms, in (Schwefel, H.-P., Männer, R. eds.) *Parallel Problem Solving from Nature*, 23–32, Springer-Verlag, Berlin Heidelberg, Germany.

HFSS (High Frequency Structural Simulator) (2012). ANSYS, Inc., Southpointe 275 Technology Drive, Canonsburg, PA 15317.

Holland, J.H. (1975). *Adaptation in Natural and Artificial Systems: An Introductory Analysis with Applications to Biology, Control, and Artificial Intelligence*, The MIT Press, Cambridge, MA, US.

Hong, J.-S., Lancaster, M. (2001). *Microstrip Filters for RF/Microwave Applications.* John Wiley & Sons, Hoboken, New Jersey, US.

Horn, J. (1993). Finite Markov Chain analysis of genetic algorithms with niching, *Proc. Int. Conf. Gentetic Alg.*, 110–117.

Horn, J., Nafpliotis, N., Goldberg, D. (1994). A niched Pareto genetic algorithm for multiobjective optimization, *Proc. IEEE Conf. Evolutionary Comp.*, **1**, 82–87.

Hornby, G., Globus, A., Linden, D., Lohn, J. (2006). Automated Antenna Design with Evolutionary Algorithms, *AIAA Space Conf. Exposition*, 1–8.

Horst, R., Tuy, H. (1996). *Global Optimization: Deterministic Approaches*, 3rd ed., Springer-Verlag, Berlin Heidelberg, Germany.

Horst, R., Pardalos, P.M., Thoai, N.V. (2000). *Introduction to Global Optimization*, 2nd ed., Kluwer Academic Publishers, Dordrecht, Holland.

Hosder, S., Watson, L.T., Grossman, B., Mason, W.H., Kim, H. (2001). Polynomial response surface approximations for the multidisciplinary design optimization of a high speed civil transport, *Opt. Eng.*, **2**, 431–452.

Huang, L., Gao, Z. (2012). Wing-body optimization based on multi-fidelity surrogate model. *Int. Congress Aeronautical Sci.*, Brisbane, Australia.

Huynh, T., Lee, K.F. (1995). Single-layer single-patch wideband microstrip antenna, *Electron. Lett.*, **31**, 1310–1312.

HyperLynx (2015). Mentor Graphics Inc., 8005 SW Boeckman Road, Wilsonville, OR 97070.

Jacobs, J.H., Etman, L.F.P., van Keulen, F., Rooda, J.E., (2004). Framework for sequential approximate optimization, *Struct. Multidisc. Opt.*, **27**, 384–400.

Jacobs, J.P. (2012). Bayesian support vector regression with automatic relevance determination kernel for modeling of antenna input characteristics, *IEEE Trans. Ant. Prop.*, **60**, 2114–2118.

Jacobs, J.P., Koziel, S. (2014). Cost-effective global surrogate modeling of planar microwave filters using multi-fidelity Bayesian support vector regression, *Int. J. RF Microwave CAE*, **24**, 11–17.

Jacobsson, P., Rylander, T. (2010). Gradient-based shape optimisation of conformal array antennas, *IET Microwaves, Ant. Prop.*, **4**, 200–209.

Jaulin, L., Kieffer, M., Didrit, O., Walter, E. (2001). *Applied Interval Analysis.* Springer-Verlag, London, UK.

Jeong, S., Murayama, M., Yamamoto, K. (2005). Efficient optimization design method using kriging model, *J. Aircraft*, **42**, 413–420.

Jin, J.-M. (2002). *The Finite Element Method in Electromagnetics*, 2nd ed., Wiley, New York, US.

Jin, N., Rahmat-Samii, Y. (2005). Parallel particle swarm optimization and finite-difference time-domain (PSO/FDTD) algorithm for multiband and wide-band patch antenna designs, *IEEE Trans. Ant. Prop.*, **53**, 3459–3468.

Jin, N., Rahmat-Samii, Y. (2007). Advances in particle swarm optimization for antenna designs: real-number, binary, single-objective and multiobjective implementations, *IEEE Trans. Ant. Prop.*, **55**, 556–567.

John, M., Ammann, M. (2009). Antenna optimization with a computationally efficient multiobjective evolutionary algorithm, *IEEE Trans. Ant. Prop.*, **57**, 260–263.

Jolliffe, I. (2002). *Principal Component Analysis*, Springer-Verlag, New York, US.

Jones, D., Schonlau, M., Welch, W. (1998). Efficient global optimization of expensive black-box functions, *J. Global Opt.*, **13**, 455–492.

Journel, A.G., Huijbregts, C.J. (1981). *Mining Geostatistics.* Academic Press, London, UK.

Kabir, H., Wang, Y., Yu, M., Zhang, Q.J. (2008). Neural network inverse modeling and applications to microwave filter design, *IEEE Trans. Microwave Theory Tech.*, **56**, 867–879.

Kao, Y.-T., Zahara, E. (2008). A hybrid genetic algorithm and particle swarm optimization for multimodal functions, *Applied Soft Computing*, **8**, 849–857

Karr, B.A., Durham, T.E., Chryssomallis, M.T., Kralovec, J.A., Gothard, G.K., Christodoulou, C.G. (2004). Investigation of a broad-band cavity-backed array antenna, *IEEE Trans. Antennas Prop.*, **52**, 1913–1916.

Kempel, L.C. (2007). Computational electromagnetics for antennas, in (Volakis, J.L. Ed.) *Antenna Engineering handbook*, 4th ed. McGraw-Hill, New York, US.

Kennedy, J. (1997). The particle swarm: social adaptation of knowledge. *Proc. Int. Conf. Evolutionary Comp.*, 303–308.

Kennedy, J., Eberhart, R.C., Shi, Y. (2001). *Swarm intelligence*, Academic Press, London, UK.

Kennedy, M.C., O'Hagan, A. (2000). Predicting the output from complex computer code when fast approximations are available, *Biometrika*, **87**, 1–13.

Kerkhoff, A.J., Ling, H. (2007). Design of a band-notched planar monopole antenna using genetic algorithm optimization, *IEEE Trans. Ant. Prop.*, **55**, 604–610.

Khamas, S.K. (2008). Circularly polarized dielectric resonator antenna excited by a conformal wire, *IEEE Ant. Wireless Prop. Lett.*, **7**, 240–242.

Kim, Y., Keely, S., Ghosh, J., Hao, L. (2007). Application of artificial neural networks to broadband antenna design based on a parametric frequency model, *IEEE Trans. Ant. Prop.*, **55**, 669–674.

Kinsey, D.W., Barth, T.J. (1984). Description of a hyperbolic grid generation procedure for arbitrary two-dimensional bodies, AFWAL TM 84-191-FIMM.

Kirkpatrick, S., Gelatt, C.D., Vecchi, M.P. (1983). Optimization by simulated annealing, *Science, New Series*, **220**, 671–680.

Kleijnen, J. (2008). *Design and Analysis of Simulation Experiments*, Springer, Berlin, Germany.

Kleijnen, J.P.C. (2009). Kriging metamodeling in simulation: a review, *Eur. J. Oper. Res.*, **192**, 707–716.

Kolda, T.G., Lewis, R.M., Torczon, V. (2003). Optimization by direct search: new perspectives on some classical and modern methods, *SIAM Rev.*, **45**, 385–482.

Koulouridis, S., Psychoudakis, D., Volakis, J. (2007). Multiobjective optimal antenna design based on volumetric material optimization, *IEEE Trans. Ant. Prop.*, **55**, 594–603.

Koziel, S. (2009). Multi-fidelity optimization of microwave structures using response surface approximation and space mapping, *Applied Comp. EM Society J.*, **24**, 601–608.

Koziel, S. (2010a). Shape-preserving response prediction for microwave design optimization, *IEEE Trans. Microwave Theory and Tech.*, **58**, 2829–2837.

Koziel, S. (2010b). Adaptively adjusted design specifications for efficient optimization of microwave structures, *Prog. EM Res. B*, **21**, 219–234.

Koziel, S. (2010c). Shape-preserving response prediction for microwave circuit modeling, *IEEE MTT-S Int. Microwave Symp. Dig.*, Anaheim, CA, 1660–1663.

Koziel, S. (2010d). Efficient optimization of microwave structures through design specifications adaptation, *IEEE Int. Symp. Ant. Prop.*, Toronto, Canada.

Koziel, S. (2010e). Multi-fidelity multi-grid design optimization of planar microwave structures with Sonnet, *Int. Rev. Prog. Applied Comp. EM*, 719-724.

Koziel, S. (2012). Accurate low-cost microwave component models using shape-preserving response prediction, *Int. J. Numerical Modelling*, **25**, 152–162.

Koziel, S., Bandler, J.W. (2007a). Coarse and surrogate model assessment for engineering design optimization with space mapping, *IEEE MTT-S Int. Microwave Symp. Dig.*, Honolulu, HI, 107–110.

Koziel, S., Bandler, J.W. (2007b). Space-mapping optimization with adaptive surrogate model, *IEEE Trans. Microwave Theory Tech.*, **55**, 541–547.

Koziel, S., Bandler, J.W. (2012). Accurate modeling of microwave devices using kriging-corrected space mapping surrogates, *Int. J. Numer. Model.*, **25**, 1–14.

Koziel, S., Bandler, J.W. (2015). Reliable microwave modeling by means of variable-fidelity response features, *IEEE Trans. Microwave Theory Tech.*, **63**, 4247–4254.

Koziel, S., Bekasiewicz, A. (2015a). Fast EM-driven size reduction of antenna structures by means of adjoint sensitivities and trust regions, *IEEE Ant. Wireless Prop. Lett.*, **14**, 1681–1684.

Koziel, S., Bekasiewicz, A. (2015b). Fast multiobjective optimization of narrow-band antennas using RSA models and design space reduction, *IEEE Ant. Wireless Prop. Lett.*, **14**, 450–453.

Koziel, S., Bekasiewicz, A. (2015c). Recent developments in simulation-driven multi-objective design of antennas, *Bulletin Polish Academy of Sci. Tech. Sci.*, **63**, 781–789.

Koziel, S., Bekasiewicz, A. (2015d). Fast simulation-driven feature-based design optimization of compact dual-band microstrip branch-line coupler, *Int. J. RF and Microwave Comp. Aid. Eng.*, **26**, 13–20.

Koziel, S., Bekasiewicz, A. (2015e). Expedited geometry scaling of compact microwave passives by means of inverse surrogate modeling, *IEEE Trans. Microwave Theory Tech*, **63**, 4019–4026.

Koziel, S., Bekasiewicz, A. (2015f). Computationally-efficient multi-objective optimization of antenna structures using point-by-point Pareto set identification and local approximation surrogates, *IEEE MTT-S Int. Conf. Numerical EM Multiphysics Modeling Opt. RF, Microwave, THz App.*

Koziel, S., Bekasiewicz, A. (2016a). A structure and simulation-driven design of compact CPW-fed UWB antenna, *IEEE Ant. Wireless Prop. Lett.*, **15**, 750–753.

Koziel, S., Bekasiewicz, A. (2016b). Fast EM-driven optimization using variable-fidelity EM models and adjoint sensitivities, *IEEE Microwave Wireless Comp. Lett.*, **26**, 80–82.

Koziel, S., Bekasiewicz, A. (2016c). Multi-objective antenna design by means of sequential domain patching, *IEEE Ant. Wireless Prop. Lett.*, **15**, 1089–1092.

Koziel, S., Bekasiewicz, A. (2016d). Fast multi-objective surrogate-assisted design of multi-parameter antenna structures through rotational design space reduction, *IET Microwaves, Ant. Prop.*, **10**, 624–630.

Koziel, S., Bekasiewicz, A. (2016e). Variable-fidelity design optimization of antennas with automated model selection, *European Conf. Ant. Prop.*, Davos, pp. 1–5.

Koziel, S., Bekasiewicz, A. (2016f). Rapid design optimization of antennas using variable-fidelity EM models and adjoint sensitivities, *Eng. Computation.*

Koziel, S., Bekasiewicz, A. (2016g). Low-cost multi-objective optimization and experimental validation of UWB MIMO antenna, *Eng. Computation.*

Koziel, S., Bekasiewicz, A. (2016h). Multi-objective antenna design using sequential domain patching with automated determination of patch size, *European Conf. Ant. Prop.*, Davos, pp. 1–5.

Koziel, S., Bekasiewicz, A. (2016i). Strategies for computationally feasible multi-objective simulation-driven design of compact RF/microwave components, *Eng. Computation*, **33**, 184–201.

Koziel, S., Bekasiewicz, A. (2016j). Low-cost multi-objective optimization of antennas using Pareto front exploration and response features, *IEEE Int. Symp. Ant. Prop.*, Puerto Rico, 2016.

Koziel, S., Bekasiewicz, A. (2016k). Scalability of surrogate-assisted multi-objective optimization of antenna structures exploiting variable-fidelity EM-simulation models, *Engineering Opt.* **48**, 1778–1792.

Koziel, S., Leifsson, L. (2012). Knowledge-based airfoil shape optimization using space mapping, *AIAA Applied Aerodynamics Conf.*

Koziel, S., Leifsson L. (eds.). (2013). *Surrogate-Based Modeling and Optimization. Applications in Engineering*, Springer, Berlin, Germany.

Koziel, S., Michalewicz, Z. (1998). A decoder-based evolutionary algorithm for constrained parameter optimization problems, in (Bäck, T., Eiben, A.E., Schoenauer, M., Schwefel, H.-P. eds.), *Proc. Parallel Problem Solving from Nature, Lecture Notes in Computer Sci.*, **1498**, 231–240, Springer-Verlag, Berlin Heidelberg, Germany.

Koziel, S., Ogurtsov, S. (2011a). Simulation-driven design in microwave engineering: application case studies, in (Yang X.S., Koziel, S. eds.) *Computational*

Optimization and Applications in Engineering and Industry, Series: Studies in Computational Intelligence, 57–98, Springer-Verlag, Berlin Heidelberg, Germany.

Koziel, S., Ogurtsov, S. (2011b). Rapid design optimization of antennas using space mapping and response surface approximation models, *Int. J. RF & Microwave CAE*, **21**, 611–621.

Koziel, S., Ogurtsov, S. (2012a). Model management for cost-efficient surrogate-based optimization of antennas using variable-fidelity electromagnetic simulations, *IET Microwaves Ant. Prop.*, **6**, 1643–1650.

Koziel S., Ogurtsov, S. (2012b). Fast simulation-driven design of microwave structures using improved variable-fidelity optimization technique, *Eng. Opt.*, **44**, 1007–1019.

Koziel, S., Ogurtsov, S. (2012c). Reduced-cost design optimization of antenna structures using adjoint sensitivity, *Microwave Opt. Technol. Lett.*, **54**, 2594–2597.

Koziel, S., Ogurtsov, S. (2013a). Multi-objective design of antennas using variable-fidelity simulations and surrogate models, *IEEE Trans. Ant. Prop.*, **61**, 5931–5939.

Koziel, S., Ogurtsov, S. (2013b) Computational-budget-driven automated microwave design optimization using variable-fidelity electromagnetic simulations, *Int. J. RF Microwave Comp. Aid. Eng.*, **23**, 349–356.

Koziel, S., Ogurtsov, S. (2013c). Rapid optimization of omnidirectional antennas using adaptively adjusted design specifications and kriging surrogates, *IET Microwaves, Ant. Prop.*, **7**, 1194–1200.

Koziel, S., Ogurtsov, S. (2014a). *Antenna design by simulation-driven optimization*. Springer, Berlin, Germany.

Koziel, S., Ogurtsov, S. (2014b). Design optimization of antennas using electromagnetic simulations and adaptive response correction technique, *IET Microwaves, Ant. Prop.*, **8**, 180–185.

Koziel, S., Ogurtsov, S. (2015). Fast simulation-driven optimization of planar microstrip antenna arrays using surrogate superposition models, *Int. J. RF Microwave Comp. Aid. Eng.*, **25**, 371–381.

Koziel, S., Bandler, J.W., Madsen, K. (2005). Towards a rigorous formulation of the space mapping technique for engineering design, *Proc. Int. Symp. Circuits, Syst.*, **1**, 5605–5608.

Koziel, S., Bandler, J.W., Madsen, K. (2006). A space mapping framework for engineering optimization: theory and implementation, *IEEE Trans. Microwave Theory Tech.*, **54**. 3721–3730.

Koziel, S., Cheng, Q.S., Bandler, J.W. (2008a). Space mapping, *IEEE Microwave Magazine*, **9**, 105–122.

Koziel, S., Bandler, J.W., Madsen, K. (2008b). Quality assessment of coarse models and surrogates for space mapping optimization, *Opt. Eng.*, **9**, 375–391.

Koziel, S., Bandler, J.W., Madsen, K. (2009). Space mapping with adaptive response correction for microwave design optimization, *IEEE Trans. Microwave Theory Tech.*, **57**, 478–486.

Koziel, S., Bandler, J.W., Cheng, Q.S. (2010a). Robust trust-region space-mapping algorithms for microwave design optimization, *IEEE Trans. Microwave Theory Tech.*, **58**, 2166–2174.

Koziel, S., Cheng, Q.S., Bandler, J.W. (2010b). Implicit space mapping with adaptive selection of preassigned parameters, *IET Microwaves, Ant. Prop.*, **4**, 361–373.

Koziel, S., Bandler, J.W., Cheng, Q.S. (2010c). Adaptively constrained parameter extraction for robust space mapping optimization of microwave circuits, *IEEE MTT-S Int. Microwave Symp. Dig.*, 205–208.

Koziel, S., Echeverría Ciaurri, D., Leifsson, L. (2011a). Surrogate-Based Methods, in (Koziel, S., Yang, X.-S. eds.) *Computational Optimization, Methods and Algorithms*, 33–59, Springer-Verlag, Berlin Heidelberg, Germany.

Koziel, S., Bandler, J.W., Cheng, Q.S. (2011b). Tuning space mapping design framework exploiting reduced EM models, *IET Microwaves, Ant. Prop.*, **5**, 1219–1226.

Koziel, S., Leifsson, L., Ogurtsov, S. (2013a). Reliable EM-driven microwave design optimization using manifold mapping and adjoint sensitivity, *Microwave Opt. Technol. Lett.*, **55**, 809–813.

Koziel, S., Leifsson, L., Couckuyt, I., Dhaene, T. (2013b). Reliable reduced cost modeling and design optimization of microwave filters using co-kriging, *Int. J. Numerical Modelling: Electronic Devices and Fields*, **26**, 493–505.

Koziel, S., Ogurtsov, S., Bandler, J.W., Cheng, Q.S. (2013c). Reliable space-mapping optimization integrated with EM-based adjoint sensitivities, *IEEE Trans. Microwave Theory Techn.*, **61**, 3493–3502.

Koziel, S., Ogurtsov, S., Couckuyt, I., Dhaene, T. (2013d) Variable-fidelity electromagnetic simulations and co-kriging for accurate modeling of antennas, *IEEE Trans. Ant. Prop.*, **61**, 1301–1308.

Koziel, S., Yang, X.S., Zhang, Q.J., (eds.) (2013e). *Simulation-driven design optimization and modeling for microwave engineering.* London: Imperial College Press.

Koziel, S., Ogurtsov, S., Zieniutycz, W., Sorokosz, L. (2014a). Simulation-driven design of microstrip antenna subarrays, *IEEE Trans. Ant. Prop.*, **62**, 3584–3591.

Koziel, S., Bekasiewicz, A., Couckuyt, I., Dhaene, T. (2014b). Efficient multi-objective simulation-driven antenna design using co-kriging, *IEEE Trans. Ant. Prop.*, **62**, 5900–5905.

Koziel, S., Bekasiewicz, A., Zieniutycz, W. (2014c). Expedited EM-driven multi-objective antenna design in highly-dimensional parameter spaces, *IEEE Ant. Wireless Prop. Lett.*, **13**, 631–634.

Koziel, S., Ogurtsov, S., Cheng, Q.S., Bandler, J.W. (2014d). Rapid electromagnetic-based microwave design optimisation exploiting shape-preserving response prediction and adjoint sensitivities, *IET Microwaves, Ant. Prop.*, **8**, 775–781.

Koziel, S., Bekasiewicz, A., Kurgan, P. (2014e). Rapid EM-driven design of compact RF circuits by means of nested space mapping, *IEEE Microwave Wireless Comp. Lett.*, **24**, 364–366.

Koziel, S., Ogurtsov, S., Zieniutycz, W., Bekasiewicz, A. (2015a). Design of a planar UWB dipole antenna with an integrated balun using surrogate-based optimization, *IEEE Ant. Wireless Prop. Lett.*, **14**, 366–369.

Koziel, S., Bekasiewicz, A., Kurgan, P. (2015b). Rapid multi-objective simulation-driven design of compact microwave circuits, *IEEE Microwave Wireless Comp. Lett.*, **25**, 277–279.

Koziel, S., Bekasiewicz, A., Leifsson, L. (2015c). Multi-objective design optimization of planar Yagi-Uda antenna using physics-based surrogates and rotational design space reduction, *Int. Conf. Comp. Sci.*, in *Procedia Computer Sci.*, **51**, 825–833.

Koziel, S., Bekasiewicz, A., Kurgan, P. (2016a). Rapid design and size reduction of microwave couplers using variable-fidelity EM-driven optimization, *Int. J. RF Microwave Comp. Aid. Eng.*, **26**, 27–35.

Koziel, S., Bekasiewicz, A., Leifsson, L. (2016b). Rapid EM-driven antenna dimension scaling through inverse modeling, *IEEE Ant. Wireless Prop. Lett.*

Koziel, S., Bekasiewicz, A., Kurgan, P., Bandler, J.W. (2016c). Rapid multi-objective design optimization of compact microwave couplers by means of physics-based surrogates, *IET Microwaves, Ant. Prop.*

Koziel, S., Tesfahunegn, Y.A., Leifsson, L. (2016d). Variable-fidelity CFD models and co-kriging for expedited multi-objective aerodynamic design optimization, *Eng. Comp.*

Kraus, J.D., Marhefka, R.J. (2002). *Antennas for All Applications.* 3rd ed., McGraw-Hill, New York, US.

Kroese, D.P., Taimre, T., Botev, Z.I. (2011). *Handbook of Monte Carlo Methods.* John Wiley & Sons, Hoboken, New Jersey, US.

Kuhn, H.W., Tucker, A.W. (1951). Nonlinear programming, in (J. Neyman, Ed.) *Proc. Berkeley Symp. Mathematical Statistics Probability*, University of California Press, Berkeley, pp. 481–492.

Kukkonen, S., Lampinen, J. (2004). An extension of generalized differential evolution for multi-objective optimization with constraints, *Parallel Problem Solving from Nature*, 752–761.

Kuwahara, Y. (2005). Multiobjective optimization design of Yagi–Uda antenna, *IEEE Trans. Ant. Prop.*, **53**, 1984–1992.

Kurgan, P., Bekasiewicz, A. (2014). A robust design of a numerically demanding compact rat-race coupler, *Microwave Opt. Technol. Lett.*, **56**, 1259–1263.

Kurgan, P., Kitlinski, M. (2009). Novel microstrip low-pass filters with fractal defected ground structures, *Microwave Opt. Technol. Lett.*, **51**, 2473–2477.

Kurgan, P., Filipcewicz, J., Kitlinski, M. (2012). Development of a compact microstrip resonant cell aimed at efficient microwave component size reduction, *IET Microwaves, Ant. Prop.*, **6**, 1291–1298.

Lai, M.-I., Jeng, S.-K. (2006). Compact microstrip dual-band bandpass filters design using genetic-algorithm techniques, *IEEE Trans. Microwave Theory Tech.*, **54**, 160–168.

Laurenceau, J. Sagaut, P. (2008). Building efficient response surfaces of aerodynamic functions with kriging and cokriging, *AIAA J.*, **46**, 498–507.

Leary, S., Bhaskar, A., Keane, A. (2003). Optimal orthogonal-array-based latin hypercubes, *J. Applied Statistics*, **30**, 585–598.

Leifsson, L., Koziel, S. (2015a). *Simulation-driven aerodynamic design using variable-fidelity models*, Imperial College Press, London, UK.

Leifsson, L., Koziel, S. (2015b). Variable-resolution shape optimization: low-fidelity model selection and scalability, *Int. J. Mathematical Modeling Numerical Opt.*, **6**, 1–21.

Leoviriyakit, K., Kim, S., and Jameson, A. (2003). Viscous aerodynamic shape optimization of wings including planform variables, *App. Aerodynamics Conf.*

Leung, K.W., Ng, H.K. (2005). The slot-coupled hemispherical dielectric resonator antenna with a parasitic patch: applications to the circularly polarized antenna and wide-band antenna, *IEEE Trans. Ant. Prop.*, **53**, 1762–1769.

Leung, T.M., Zingg, D.W. (2012). Aerodynamic shape optimization of wings using a parallel Newton-Krylov approach, *AIAA J.*, **50**, 540–550.

Levin, D. (1998). The approximation power of moving least-squares, *Math. Computation.* **67**, 1517–1531.

Li, T., Zhai, H., Li, G., Li, L., Liang, C. (2012). Compact UWB band-notched antenna design using interdigital capacitance loading loop resonator, *IEEE Ant. Wireless Prop. Lett.*, **11**, 724–727.

Li, W.T., Shi, X.W., Hei, Y.Q., Liu, S.F., Zhu, J. (2010). A hybrid optimization algorithm and its application for conformal array pattern synthesis, *IEEE Trans. Ant. Prop.*, **58**, 3401–3406.

Lim, W.H., Isa, N.A.M. (2014). An adaptive two-layer particle swarm optimization with elitist learning strategy, *Information Sci.*, **273**, 49–72.

Liu, B., Aliakbarian, H., Zhongkun M., Vandenbosch, G.A.E., Gielen, G., Excell, P. (2014a). An efficient method for antenna design optimization based on evolutionary computation and machine learning techniques, *IEEE Trans. Ant. Prop.*, **62**, 7–18.

Liu, B., Zhang, Q., Gielen, G. (2014b) A Gaussian process surrogate model assisted evolutionary algorithm for medium scale expensive black box optimization problems, *IEEE Trans. Evol. Comp.*, **18**, pp. 180–192.

Liu, J., Han, Z., Song, W. (2012). Comparison of infill sampling criteria in kriging-based aerodynamic optimization, *Int. Congress Aeronautical Sci.*

Liu, L., Cheung, S.W., Yuk, T.I. (2013). Compact MIMO antenna for portable devices in UWB applications, *IEEE Trans. Ant. Prop.*, **61**, 4257–4264.

Lizzi, L., Viani, F., Azaro, R., Massa, A. (2007). Optimization of a spline-shaped UWB antenna by PSO, *IEEE Ant. Wireless Prop. Lett.*, **6**, 182–185.

Lizzi, L., Viani, F., Azaro, R., Massa, A. (2008). A PSO-driven spline-based shaping approach for ultra-wideband (UWB) antenna synthesis, *IEEE Trans. Ant. Prop.*, **56**, 2613–2621.

Lizzi, L., Viani, F., Massa, A. (2009). Dual-band spline-shaped PCB antenna for Wi-Fi applications, *IEEE Ant. Wireless Prop. Lett.*, **8**, 616–619.

Lollini, P., Bondavalli, A., Di Giandomenico, F. (2009). A decomposition-based modeling framework for complex systems, *IEEE Trans. Reliability*, **58**, 20–33.

Lophaven, S.N., Nielsen, H.B., Søndergaard, J. (2002). DACE: a Matlab kriging toolbox. Technical University of Denmark.

Luenberger, D.G. (2003). *Linear and Nonlinear Programming.* 2nd ed., Kluwer Academic Publishers, Boston, US.

Madavan, N.K. (2002). Multiobjective optimization using a pareto differential evolution approach, *Congress Evol. Comp.*, **2**, 1145–1150, IEEE Service Center, New Jersey, US.

Mader, C.A., Martins, J.R.R.A. (2012). Derivatives for time-spectral computational fluid dynamics using an automatic differentiation adjoint, *AIAA J.*, **50**, 2809–2819.

Makarov, S. (2002). *Antenna EM Modeling with Matlab.* Wiley-Interscience, New-York, US.

Manica, L., Rocca, P., Poli, L., Massa, A. (2009). Almost time-independent performance in time-modulated linear arrays, *IEEE Ant. Wireless Prop. Lett.*, **8**, 843–846.

Marheineke, N., Pinnau, R., Reséndiz, E. (2012). Space mapping-focused control techniques for particle dispersions in fluids, *Opt. Eng.*, **13**, 101–120.

Matheron, G. (1963). Principles of geostatistics, *Economic Geology*, **58**, 1246–1266.

MATLAB, (2012). MathWorks, Inc., 3 Apple Hill Drive, Natick, MA, US.

McKay M, Conover W, Beckman R. (1979). A comparison of three methods for selecting values of input variables in the analysis of output from a computer code, *Technometrics.* **21**, 239–245.

Meng, J., Xia, L. (2007). Support-vector regression model for millimeter wave transition, *Int. J. Infrared and Milimeter Waves*, **28**, 413–421.

Mezura-Montes, E., Velázquez-Reyes, J., Coello Coello, C.A. (2006a). A comparative study of differential evolution variants for global optimization, *Conf. Genetic Evolutionary Comp.*

Mezura-Montes, E. Reyes-Sierra, M. Coello Coello, C.A. (2006b). Multi-objective Optimization Using Differential Evolution: A Survey of the State-of-the-Art, in (Chakraborty, U.K. ed.) *Advances in Differential Evolution*, 173–196.

Mezura-Montes, E., Miranda-Varela, M.E., Gómez-Ramón, R.C. (2010). Differential evolution in constrained numerical optimization: An empirical study, *Inform. Sciences*, **180**, 4223–4262.

Michalewicz, Z. (1996). *Genetic Algorithms + Data Structures = Evolution Programs*, Springer-Verlag, Berlin Heidelberg, Germany.

Miettinen, K. (1999). *Nonlinear Multiobjective Optimization.* Boston: Kluwer.

Minsky, M.I., Papert, S.A. (1969). *Perceptrons: An Introduction to Computational Geometry.* The MIT Press, Cambridge.

Namiki, T. (1999) A new FDTD algorithm based on alternating-direction implicit method, *IEEE Microwave Theory Tech.*, **47**, 2003–2007.

Nelder, J.A., Mead, R. (1965). A simplex method for function minimization, *Computer J.*, **7**, 308–313.

Nguyen, D.T., Lee, D.H., Park, H.C. (2012). Very compact printed triple band-notched UWB antenna with quarter-wavelength slots, *IEEE Ant. Wireless Prop. Lett.*, **11**, 411–414.

Nguyen, N.T., Boriskin, A., Rolland, A., Le Coq, L., Sauleau, R. (2013). Shaped lens-like dome for UWB antennas with a Gaussian-like radiation pattern, *IEEE Trans. Ant. Prop.*, **61**, 1658–1664.

Nickabadi, A., Ebadzadeh, M.M., Safabakhsh, R. (2011). A novel particle swarm optimization algorithm with adaptive inertia weight, *Applied Soft Computing*, **11**, 3658–3670.

Nie, W., Luo, S., Guo, Y.-X., Fan, Y. (2014). Miniaturized rat-race coupler with harmonic suppression, *IEEE Microwave Wireless Comp. Lett.*, **24**, 754–756.

Nielsen, H.B., (1999). Damping parameter in Marquardt's method. IMM DTU. Report IMM-REP-1999-05.

Nocedal, J., Wright, S. (2006). *Numerical Optimization*, 2nd ed., Springer, New York, US.

Noel, M.M. (2012). A new gradient based particle swarm optimization algorithm for accurate computation of global minimum, *Applied Soft Computing*, **12**, 353–359.

Noman, N., Iba, H. (2005). Enhancing differential evolution performance with local search for high dimensional function optimization, in (Beyer, H.-G. ed.) *Proceedings of the 7th annual conference on Genetic and evolutionary computation*, 967–974, ACM, New York, US.

Ogurtsov, S., Koziel, S. (2011). Design of microstrip to substrate integrated waveguide transitions with enhanced bandwidth using protruding vias and EM-driven optimization, *Int. Rev. Prog. Applied Comp. EM*, 91–96.

O'Hagan, A. (1978). Curve fitting and optimal design for predictions, *J. Royal Statistical Society B*, **40**, 1–42.

OptiFDTD (2014), Optiwave Systems Inc., Capella Court 7, Ottawa, ON X2E 7X1.

Palmer, K., Tsui, K.-L. (2001). A minimum bias latin hypercube design, *IIE Trans.*, **33**, 793–808.

Pan, Y.-M., Leung, K.W., Lu, K. (2014). Compact quasi-isotropic dielectric resonator antenna with small ground plane, *IEEE Ant. Wireless Prop. Lett.*, **62**, 577–585.

Parsopoulos, K.E., Tasoulis, D.K., Vrahatis. M.N. (2004a). Multiobjective optimization using parallel vector evaluated particle swarm optimization, *Proc. Int. Conf. Artificial Intelligence App.*, **2**, 823–828, ACTA Press.

Parsopoulos, K.E., Tasoulis, D.K., Pavlidis, N.G., Plagianakos, V.P. Vrahatis, M.N. (2004b). Vector evaluated differential evolution for multiobjective optimization, *Congress Evol. Comp.*, **1**, 204–211.

Pattipati, B., Sankavaram, C., Pattipati, K.R. (2011). System identification and estimation framework for pivotal automotive battery management system characteristics, *IEEE Trans. Systems, Man, and Cybernetics, Part C*, **41**, 869–884.

Pavlidis, N.G., Parsopoulos, K.E., Vrahatis, M.N. (2005). Computing Nash equilibria through computational intelligence methods, *J. Comp. Applied Math.*, **175**, 113–136.

Pérez, V.M., Renaud, J.E., Watson, L.T. (2002). Interior point sequential approximate optimization methodology. *Proc. Symp. Multidisciplinary Analysis Opt.*, Atlanta, GA, AIAA-2002-5505.

Peterson, A.F., Ray S.L., Mittra, R. (1998). *Computational Methods for Electromagnetics.* Oxford University Press and IEEE Press, Oxford & New York, UK, US.

Petko, J.S., Werner, D.H. (2007). An autopolyploidy-based genetic algorithm for enhanced evolution of linear polyfractal arrays, *IEEE Trans. Ant. Prop.*, **55**, 583–593.

Petosa, A. (2007). *Dielectric Resonator Antenna Handbook.* Artech House, Norwood.

Polak, E., Ribiére, G. (1969). Note sur la convergence de méthodes de directions eonjuguées, *Revue Francaise d'lnformatique et de Recherche Opirationnelle*, **16**, 35–43.

Poli, L., Rocca, P., Oliveri, G., Massa, A. (2014). Failure correction in time-modulated linear arrays, *IET Radar, Sonar Nav.*, **8**, 195–201.

Pozar, D.M. (2012). *Microwave Engineering*, 4th ed., Hoboken: John Wiley & Sons.

Price, K., Storn, R.M., Lampinen, J.A. (2005). *Differential Evolution: A Practical Approach to Global Optimization.* Springer-Verlag, Berlin Heidelberg, Germany.

Priess, M., Koziel, S., Slawig, T. (2011). Surrogate-based optimization of climate model parameters using response correction, *J. Comp. Sci.*, **2**, 335–344.

Queipo, N.V., Haftka, R.T., Shyy, W., Goel, T., Vaidynathan, R., Tucker, P.K. (2005). Surrogate based analysis and optimization, *Prog. Aerospace Sci.*, **41**, 1–28.

Qian, Y., Deal, W.R., Kaneda, N., Itoh, T. (1998). Microstrip-fed quasi-Yagi antenna with broadband characteristics, *Electron. Lett.*, **34**, 2194–2196.

Qing, X., Chen, Z.N. (2009). Compact coplanar waveguide-fed ultra-wideband monopole-like slot antenna, *IET Microwaves Ant. Prop.*, **3**, 889–898.

Rajaguru, P., Stoyanov, S., Lu, H., Bailey, C. (2013). Application of kriging and radial basis function for reliability optimization in power modules, *ASME. J. Electron. Packag.*, **135**, 021009-1–021009-13.

Ramos, R., Saldanha, R., Takahashi, R., Moreira, F. (2003). The real-biased multiobjective genetic algorithm and its application to the design of wire antennas, *IEEE T. Magn.*, **39**, 1329–1332.

Rashedi, E., Nezamabadi-pour, H., Saryazdi, S. (2009). GSA: A gravitational search algorithm, *Inf. Sci.*, **179**, 2232–2248.

Rasmussen, C.E., Williams, C.K.I. (2006). *Gaussian Processes for Machine Learning.* MIT Pressm, Cambridge, MA, US.

Rayas-Sanchez, J.E. (2004). EM-based optimization of microwave circuits using artificial neural networks: the state-of-the-art, *IEEE Trans. Microwave Theory Tech.* **52**, 420–435.

Rechenberg, I. (1965). Cybernetic solution path of an experimental problem. Royal Aircraft Establishment, Library Translation Number 1122, UK, Farnborough.

Redhe, M., Nilsson, L. (2004). Optimization of the new Saab 9-3 exposed to impact load using a space mapping technique, *Structural Multidisciplinary Opt.*, **27**, 411–420.

Reyes-Sierra, M., Coello Coello, C.A. (2006). Multi-objective particle swarm optimizers: a survey of the state-of-the-art, *Int. J. Comp. Intelligence Res.*, **2**, 287–308.

Robič, T., Filipič, B. (2005). DEMO: Differential evolution for multiobjective optimization, in (Coello Coello, C.A., Aguirre, A.H., Zitzler, E. eds.) *Evolutionary Multi-Criterion Optimization*, 520–533.

Robinson, T.D., Eldred, M.S., Willcox, K.E., Haimes, R. (2008). Surrogate-based optimization using multifidelity models with variable parameterization and corrected space mapping, *AIAA J.*, **46**, 2316–2326.

Rocca, P., Oliveri, G., Massa, A. (2011a). Differential Evolution as applied to electromagnetics, *IEEE Ant. Prop. Mag.*, **53**, 38–49.

Rocca, P., Poli, L., Oliveri, G., Massa, A. (2011b). A multi-stage approach for the synthesis of sub-arrayed time modulated linear arrays, *IEEE Trans. Ant. Prop.*, **59**, 3246–3254.

Rocca, P., Poli, L., Oliveri, G., Massa, A. (2012). Adaptive nulling in time-varying scenarios through time-modulated linear arrays, *IEEE Ant. Wireless Prop. Lett.*, **11**, 101–104.

Rosenbrock, H.H. (1960), An automatic method for finding the greatest or least value of a function, *Computer J.*, **3**, 175–184.

Roux, W.J., Stander, N., Haftka, R.T. (1998). Response surface approximations for structural optimization, *Int. J. Numerical Methods Eng.*, **42**, 517–534.

Roy, G.G., Das, S., Chakraborty, P., Suganthan, P.N. (2011). Design of non-uniform circular antenna arrays using a modified invasive weed optimization algorithm, *IEEE Trans. Ant. Prop.*, **59**, 110–118.

Rojo-Alvarez, J.L., Camps-Valls, G., Martinez-Ramon, M., Soria-Olivas, E., Navia-Vazquez, A., Figueiras-Vidal, A.R. (2005). Support vector machines framework for linear signal processing, *Signal Processing*, **85**, 2316–2326.

Sans, M., Selga, J., Rodriguez, A., Bonache, J., Boria, V.E., Martin, F. (2014). Design of planar wideband bandpass filters from specifications using a two-step aggressive space mapping (ASM) optimization algorithm, *IEEE Trans. Microwave Theory Tech.*, **62**, 3341–3350.

Sarker, R., Abbass, H., Newton, C. (2002). Solving two multi-objective optimization problems using evolutionary algorithm, in (Mohammadian, M., Sarker, R., Yao, X., eds.), *Computational Intelligence in Control.* IdeaGroup Publishing.

Schwefel, H.-P. (1968). Experimentelle optimierung einer zweiphasendüse tell 1. AEG Research Institute Project MHD-Staustrahlrohr 11034/68, Technical report 35.

Schwefel, H.-P. (1988). Collective Intelligence in Evolving Systems, in (Wolff, W., Soeder, C.-J., Drepper, F.R. eds.) *Ecodynamics*, pp. 95–100, Springer-Verlag, Berlin Heidelberg.

Schwefel, H.-P., Rudolph, G. (1995). Contemporary evolution strategies, in (Moran, F., Moreno, A., Merelo, J.J., Chacon, P. eds.) *Advances in Artificial Life*, 891–907, Springer-Verlag, Berlin Heidelberg, Germany.

Selleri, S., Mussetta, M. Pirinoli, P., Zich, R.E., Matekovits, L. (2008). Differentiated meta-PSO methods for array optimization, *IEEE Trans. Ant. Prop.*, **56**, 67–75.

Shaker, G.S.A., Bakr, M.H., Sangary, N., Safavi-Naeini, S. (2009). Accelerated antenna design methodology exploiting parameterized Cauchy models, *J. Progress EM Res. B*, **18**, 279–309.

Shao, J., Fang, G., Kai Tan, Y.J., Yin, H. (2013). A novel compact tapered-slot antenna for GPR applications, *IEEE Ant. Wireless Prop. Lett.*, **12**, 972–975.

Shelokar, P.S., Siarry, P., Jayaraman, V.K., Kulkarni, B.D. (2007). Particle swarm and ant colony algorithms hybridized for improved continuous optimization, *App. Math. Comp.*, **188**, 129–142.

Simpson, T.W., Pelplinski, J.D., Koch, P.N., Allen, J.K. (2001). Metamodels for computer-based engineering design: survey and recommendations, *Eng. with Computers*, **17**, 129–150.

Sivasubramani, S., Swarup, K.S. (2011). Multi-objective harmony search algorithm for optimal power flow problem, *Int. J. Electrical Power Energy Syst.*, **33**, 745–752.

Smola, A.J. Schölkopf, B. (2004). A tutorial on support vector regression, *Statistics Comp.*, **14**, pp. 199−222.

Smolders, A.B., Johannsen, U. (2011). Axial ratio enhancement for circularly-polarized millimeter-wave phased-arrays using a sequential rotation technique, *IEEE Trans. Ant. Prop.*, **59**, 3465–3469.

Sobieszczanski-Sobieski, J., Haftka, R.T. (1997). Multidisciplinary aerospace design optimization: survey of recent developments, *Struct. Opt.*, **14**, 1–23.

Søndergaard, J. (2003) *Optimization using surrogate models — by the space mapping technique.* Ph.D. Thesis, Informatics and Mathematical Modelling, Technical University of Denmark, Lyngby.

Sonnet ***em*** (2013). Sonnet Software, Elwood Davis Road 100, North Syracuse, NY 13212.

Spalart, P.R., Allmaras, S.R. (1992). A one-equation turbulence model for aerodynamic flows, *Aerospace Sci. Meeting Exhibit*, AIAA Paper 92-0439.

Spence, T.G., Werner, D.H. (2006). A novel miniature broadband/multiband antenna based on an end-loaded planar open-sleeve dipole, *IEEE Trans. Ant. Prop.*, **54**, 3614–3620.

Srigiriraju, K.C. (2000). *Noninferior Surface Tracing Evolutionary Algorithm (NSTEA) for Multi Objective Optimization.* Master's thesis, North Carolina State University, Raleigh, North Carolina.

Srinivasan, D., Seow, T.H. (2003). Particle swarm inspired evolutionary algorithm (PS-EA) for multiobjective optimization problem, *Congress Evol. Comp.*, **3**, 2292–2297, IEEE Press.

Storn R., Price, K. (1997). Differential evolution — a simple and efficient heuristic for global optimization over continuous spaces, *J. Global Opt.*, **11**, 341–359.

Suzuki, J. (1995). Finite Markov chain analysis of genetic algorithms, *IEEE Trans. Syst. Man. Cybernetics.*, **25**, 1995.

Taflove, A., Hagness, S.C. (2005). Computational Electrodynamics: The Finite-Difference Time-Domain Method, 3rd ed., Artech House, Boston, MA, US.

Talbi, E.-G. (2009). *Metaheuristics: From Design to Implementation*, John Wiley & Sons, Hoboken, New Jersey, US.

Tan, K., Khor, E., Lee, T. (2005). *Multiobjective Evolutionary Algorithms and Applications*. Springer-Verlag, London, UK.

Tapia, R.A. (1978). Quasi-Newton methods for equality constrained optimization: Equivalence of existing methods and a new implementation, in (O. Mangasarian, R. Meyer, S. Robinson, eds.) *Nonlinear Programming*, **3**, pp. 125–164, Academic Press, New York.

Toal, D.J.J., Keane, A.J. (2011). Efficient Multipoint Aerodynamic Design Optimization via Cokriging, *J. Aircraft*, **48**, 1685–1695.

Toivanen, J.I., Makinen, R.A.E., Jarvenpaa, S., Yla-Oijala, P., Rahola, J. (2009). Electromagnetic sensitivity analysis and shape optimization using method of moments and automatic differentiation, *IEEE Trans. Ant. Prop.*, **57**, 168–175.

Tonkin, M.J., Larson, S.P. (2002). Kriging water levels with a regional-linear and point-logarithmic drift, *Ground Water*, **40**, 185–193.

Toropov, V.V., Filatov, A.A., Polynkin, A.A. (1993). Multiparameter structural optimization using FEM and multipoint explicit approximations, *Struct. Optim.*, **6**, 7–14.

Tseng, C.-H., Chen, H.-J. (2008). Compact rat-race coupler using shunt-stub-based artificial transmission lines, *IEEE Microwave Wireless Comp. Lett.*, **18**, 734–736.

Tu, S., Cheng, Q.S., Zhang, Y., Bandler, J.W., Nikolova, N.K. (2013). Space mapping optimization of handset antennas exploiting thin-wire models, *IEEE Trans. Ant. Prop.*, **61**, 3797–3807.

Valderas, D., Sancho, J.I., Puente, D., Ling, C., Chen, X. (2011). *Ultrawideband Antennas: Design and Applications*, Imperial College Press, London, UK.

van Coevorden, C., Garcia, S., Pantoja, M., Bretones, A., Martin, R. (2005). Microstrip-patch array design using a multiobjective GA, *IEEE Ant. Wireless Prop. Lett.*, **4**, 100–103.

van Laarhoven, P.J., Aarts, E.H. (1987). *Simulated Annealing: Theory and Applications.* Springer, Netherlands.

Viani, F., Lizzi, L., Azaro, R., Massa, A. (2008). A miniaturized UWB antenna for wireless dongle devices, *IEEE Ant. Wireless Prop. Lett.*, **7**, 714–717.

Wang, L., Xu, L., Chen, X., Yang, R., Han, L., Zhang, W. (2014). A compact ultrawideband diversity antenna with high isolation, *IEEE Ant. Wireless Prop. Lett.*, **13**, 35–38.

Weigand, S., Huff, G.H., Pan, K.H., Bernhard, J.T. (2003). Analysis and design of broad-band single-layer rectangular U-slot microstrip patch antennas, *IEEE Trans. Ant. Prop.*, **51**, 457–468.

Wild, S.M., Regis, R.G., Shoemaker, C.A. (2008). ORBIT: Optimization by radial basis function interpolation in trust-regions, *SIAM J. Sci. Comp.*, **30**, 3197–3219.

Wu, J., Zhao, Z., Nie, Z., Liu, Q.-H. (2014). Bandwidth enhancement of a planar printed quasi-Yagi antenna with size reduction, *IEEE Trans. Ant. Prop.*, **62**, 463–467.

XFDTD (2014). Remcom, Inc., South Allen 315, Suite 416, State College, PA 16801.

Xia, L., Xu, R.M., Yan, B. (2007). LTCC interconnect modeling by support vector regression, *Prog. EM Res.*, **69**, 67–75.

Yang, S., Li, M., Liu, X., Zheng, J. (2013). A grid-based evolutionary algorithm for many-objective optimization, *IEEE Trans. Evol. Comp.*, **17**, 721–736.

Yang, X.S., (2005). Engineering optimization via nature-inspired virtual bee algorithms, *Lecture Notes in Computer Science*, 317–323.

Yang, X.S. (2008). *Nature-inspired metaheuristic algorithms.* Luniver Press.

Yang, X.-S. (2010a). *Engineering optimization: an introduction with metaheuristic applications*, John Wiley & Sons, Hoboken, New Jersey, US.

Yang, X.-S. (2010b). A New Metaheuristic Bat-Inspired Algorithm, in (González, J.R., Pelta, D.A., Cruz, C., Terrazas, G., Krasnogor, N. eds.) *Nature Inspired Cooperative Strategies for Optimization.* 65–74, Springer, Berlin Heidelberg, Germnay.

Yang, X.S. (2013). Multiobjective firefly algorithm for continuous optimization, *Eng. with Computers*, **29**, 175–184.

Yang, X.-S. (ed.). (2014). *Cuckoo Search and Firefly Algorithm: Theory and Applications.* Springer International Publishing.

Yang, X.-S., Deb, S. (2013). Multiobjective cuckoo search for design optimization, *Comp. Operations Res.*, **40**, 1616–1624.

Yang, X.S., Ng, K.T., Yeung, S.H., Man, K.F. (2008). Jumping genes multiobjective optimization scheme for planar monopole ultrawideband antenna, *IEEE Trans. Ant. Prop.*, **56**, 3659–3666.

Ye, K.Q. (1998). Orthogonal column latin hypercubes and their application in computer experiments, *J. American Statistical Assoc.*, **93**, 1430–1439.

Yelten, M.B., Zhu, T., Koziel, S., Franzon, P.D., Steer, M.B. (2012). Demystifying surrogate modeling for circuits and systems, *IEEE Circuits Syst. Mag.*, **12**, 45–63.

Yeung, S.H., Man, K.F., Luk, K.M., Chan, C.H. (2008). A trapeizform U-slot folded patch feed antenna design optimized with jumping genes evolutionary algorithm, *IEEE Trans. Ant. Prop.*, **56**, 571–577.

Zhang, Y., Nikolova, N.K., Meshram, M.K. (2012). Design optimization of planar structures using self-adjoint sensitivity analysis, *IEEE Trans. Ant. Prop.*, **60**, 3060–3066.

Zitzler, E., Thiele, L. (1999). Multiobjective evolutionary algorithms: A comparative case study and the strength Pareto approach, *IEEE Trans. Evol. Comp.*, **3**, 257–271.

Zhang, K., Han, Z., (2013). Support vector regression-based multidisciplinary design optimization in aircraft conceptual design. *AIAA Aerospace Sciences Meeting.* AIAA paper 2013–1160.

Zhigljavsky, A., Žilinskas, A. (2008). *Stochastic Global Optimization*, Springer US.

Index

D

E

F

G

H

I

Q

U

V

W

X

Y

Z